Short Version

Fifth Edition

Microbiological Applications

A Laboratory Manual in General Microbiology

Short Version *Fifth Edition*

Microbiological Applications

A Laboratory Manual in General Microbiology

Harold J. Benson
Pasadena City College

Wm. C. Brown Publishers

Book Team
Editor *Edward G. Jaffe*
Developmental Editor *Janette S. Stecki*
Production Editor *Kay J. Brimeyer*
Designer *K. Wayne Harms*
Art Editor *Janice M. Roerig*
Visuals Processor *Joyce E. Watters*

WCB

Wm. C. Brown Publishers
President *G. Franklin Lewis*
Vice President, Publisher *George Wm. Bergquist*
Vice President, Publisher *Thomas E. Doran*
Vice President, Operations and Production *Beverly Kolz*
National Sales Manager *Virginia S. Moffat*
Advertising Manager *Ann M. Knepper*
Marketing Manager *Craig S. Marty*
Editor in Chief *Edward G. Jaffe*
Production Editorial Manager *Colleen A. Yonda*
Production Editorial Manager *Julie A. Kennedy*
Publishing Services Manager *Karen J. Slaght*
Manager of Visuals and Design *Faye M. Schilling*

Some of the laboratory experiments included in this text may be hazardous if materials are handled improperly or if procedures are conducted incorrectly. Safety precautions are necessary when you are working with chemicals, glass test tubes, hot water baths, sharp instruments, and the like, or for any procedures that generally require caution. Your school may have set regulations regarding safety procedures that your instructor will explain to you. Should you have any problems with materials or procedures, please ask your instructor for help.

Cover illustration:
Electron micrograph utilizing false color transmission of *Pneumococcus pneumoniae*. Provided by Science Photo Library/Visuals Unlimited.

Photo credits for Figure 2.1:
left—Tom E. Adams/Visuals Unlimited;
right—M. Abbey/Visuals Unlimited

ISBN 0-697-05762-3

Printed in the United States of America by Wm. C. Brown Publishers, 2460 Kerper Boulevard, Dubuque, IA 52001

10 9 8 7 6 5

Contents

PART 9

MEDICAL MICROBIOLOGY AND IMMUNOLOGY

Preface

An attempt has been made in this fifth edition of *Microbiological Applications* to do the following: (1) update the classification and nomenclature of organisms to conform with *Bergey's Manual of Systematic Bacteriology,* Volumes 1 and 2; (2) clarify the procedures for some of the more difficult experiments; (3) upgrade illustrations where necessary; (4) incorporate a few new experiments; and (5) eliminate any experiments that ceased to be of value. Since most of the experiments in the previous edition have served us well, it was hoped that these changes could be accomplished without disturbing the unaltered material.

Due to the fact that the last edition of this book was published within a short time after the publication of *Bergey's Manual,* there was insufficient time to incorporate much of the new information available to us from the first two volumes. In this edition we have been able to correct this shortcoming. In addition to upgrading the classification throughout this book, a new exercise, "Data Applications to Systematics," has replaced Exercise 41 of the previous edition, which was entitled "Use of Bergey's Manual." The new exercise exploits the new *Bergey's Manual* and also takes advantage of the computer in identifying unknown bacteria.

A problem that has been troublesome to some users of this book is the length and complexity of the exercises that pertain to the applications of physiological tests in the identification of unknown bacteria (Exercises 38, 39, and 40). Admittedly, a considerable amount of information is covered in these three exercises. To make these exercises more understandable and less cumbersome to both students and instructors alike, we have incorporated several procedural flow diagrams for guidance.

The most significant new exercises in this edition are Exercises 30, 32, and 50. I would like to express my thanks here to Kathy Talaro for Exercises 30 and 50, and to Dorothy Maron and Bruce Ames for Exercise 32.

Only one experiment was dropped from the previous edition: The Caries Susceptibility Test. Although it was requested that some others also be dropped, they were retained for the time being; that which seems useless to some instructors is often indispensable to others.

Most of the experiments in this edition are still being performed as in the past; however, minor changes have been included where applicable. Requests have been made that greater substitution of organisms be provided. This has been done. We have also had requests that more color be incorporated, especially with respect to photomicrographs of stained slides that students can use to evaluate their own slides. This, too, has been accomplished. And finally, most of the illustrations have been upgraded to improve clarity.

The author is indebted to all who have so graciously offered their suggestions for this edition. Ahmad Kamal of Olive-Harvey College, John H. Boulet of Triton College, and Judith Kandel of California State University, Fullerton, were the principal critical readers for this edition. Although most of their suggestions have been incorporated, some had to be postponed to the next edition.

I would also like to express my gratitude to Mary Timmer, our laboratory technician, for her help with cultures and stained slides that made certain color photos and photomicrographs possible.

Introduction

These laboratory exercises have been developed to guide you in your daily experiences in microbiology so that you will understand fully the principles involved. Since you will be working with unseen living forms, it will be necessary that you develop a set of techniques that are new to you. Not only will these techniques determine the success or failure of your scientific probing, but they will also be essential to protect you and others around you against potentially harmful forms.

Before you come to class, read the exercise that is to be performed that day so that you understand what is to be done. The laboratory session will begin with a short discussion to brief you on the availability of materials and procedures. Feel free to ask questions when you do not understand the instructions or the principles that are involved.

Always plan your work to avoid serious time-consuming mistakes. Keep an accurate record of what you do. An orderly notebook will pay dividends in the long run. *Before you start to record data on the Laboratory Report sheets, which are at the back of the manual, remove them from the binding.* Trying to shift from the front of the book to the back is inconvenient and time-consuming. Before handing in the Laboratory Reports, trim the perforations from the binding edge with a pair of scissors. These ragged edges make handling of the sheets very difficult.

Assume that all the organisms you work with are disease-producing forms (pathogens). Many of them will be harmless, but some are not. The following safeguards should be observed at all times:

1. A laboratory coat or apron must be worn at all times in the laboratory. Not only will it protect your clothing from accidental contamination, but it will also protect expensive blouses, sweaters, etc., against stains used daily in the lab. When leaving the laboratory, remove the coat or apron.
2. Long hair must be secured in a ponytail to prevent injury from Bunsen burners and contamination of culture material.
3. Lunches, coats, and books that are not required for this course must be stored in lockers provided for that purpose.
4. Before you start your work at the beginning of the period, scrub down the top of your laboratory table with a disinfectant. This will reduce the danger of contaminating your bacterial cultures with dustborne microorganisms.

   To protect the next students who will use your table, repeat this scrub-down procedure at the end of the period to remove any organisms that might have been unknowingly spilled on the tabletop.
5. Keep only those materials on your desktop that you need to perform your experiment. Workspace is minimal at best; cluttering it with unneeded items such as lunches, books, and purses will only hamper your laboratory efficiency.
6. No eating or smoking is permitted in the laboratory. Make it a habit to keep your hands away from your face. Gummed labels should never be moistened with your tongue; use tap water instead.
7. To avoid burns, beware of hot Bunsen burners and tripods. Report all accidents and injuries as soon as possible.
8. Place old cultures in receptacles that are to be autoclaved. Do not allow your locker to become filled with cultures that have ceased to be of value in your work.
9. It is the responsibility of everyone at each table to see that all materials are put away at the end of the periods. Beakers, Bunsen burners, graduates, etc., must always be returned to lockers.
10. Whenever bacterial cultures are accidentally spilled on the floor, notify the instructor so that proper disinfection procedures can be assured.
11. Do not remove cultures, reagents, or other materials from the laboratory at any time unless specific permission has been granted.
12. Before leaving the laboratory at the end of the period, wash your hands with soap and water.

Part 1 Microscopy

Microscopes of various types are available to the microbiologist. Each type has its specific applications, and limitations. No instrument functions best in all applications. The principal types are the brightfield, phase-contrast, fluorescence, and electron microscopes.

Since the brightfield and phase-contrast microscopes are the most widely used instruments in basic microbiology courses, two exercises (1 and 2) are devoted to these instruments. Understanding the basic principles and applications of these two types of microscopes is a fundamental goal in this course. The manner in which they are used will play an important role in your success at microscopic observation.

Exercise 3 pertains to the use of an ocular micrometer in determining the sizes of microscopic organisms. The technique outlined here applies to phase-contrast as well as to brightfield microscopes.

Before either type of microscope is used in the laboratory, the questions of the Laboratory Reports for Exercises 1 and 2 should be answered. If each exercise is read over very carefully before using either instrument, a clear understanding of proper microscopic techniques will result.

Brightfield Microscopy 1

A microscope that allows light rays to pass directly through to the eye without being deflected by an intervening opaque plate in the condenser is called a **brightfield microscope.** This is the conventional type of microscope encountered by students in beginning courses in biology; it is also the first type to be used in this laboratory.

All brightfield microscopes have certain things in common, yet they differ somewhat in mechanical operation. An attempt will be made in this exercise to point out the similarities and differences of various makes so that you will know how to use the instrument that is available to you. Before attending the first laboratory session in which the microscope will be used, read this exercise and answer all the questions on the Laboratory Report. Your instructor may require that the Laboratory Report be handed in prior to doing any laboratory work.

Care of the Instrument

Microscopes represent considerable expense and can be damaged rather easily if certain precautions are not observed. The following suggestions cover most hazards.

Transport When carrying your microscope from one part of the room to another, use both hands when holding the instrument, as illustrated in figure 1.1. If it is carried with only one hand and allowed to dangle at your side, there is always the danger of collision with furniture or some other object. And, incidentally, under no circumstances should one attempt to carry two microscopes at one time.

Clutter Keep your workstation uncluttered while doing microscopic work. Keep unnecessary books, lunches, and other unneeded objects away from your work area. A clear work area promotes efficiency and results in fewer accidents.

Electric Cord Microscopes have been known to tumble off of tabletops when students have entangled a foot in a dangling electric cord. Don't let the light cord on your microscope dangle in such a way as to hazard entanglement.

Lens Care At the beginning of each laboratory period check the lenses to make sure they are clean. At the end of each lab session be sure to wipe any immersion oil off the oil immersion lens if it has been used. More specifics about lens care are itemized on pages 5 and 6.

Dust Protection In most laboratories dustcovers are used to protect the instruments during storage. If one is available, place it over the microscope at the end of the period.

Components

Before we discuss the procedures for using a microscope, let's identify the principal parts of the microscope as illustrated in figure 1.2.

Framework All microscopes have a basic frame structure which includes the **arm** and **base.** To this framework all other parts are attached. On many of the older microscopes the base is not rigidly attached to the arm as is the case in figure 1.2; instead, a pivot point is present which enables one to tilt the arm backward to adjust the eyepoint level.

Stage The horizontal platform that supports the microscope slide is called the *stage.* Note that it

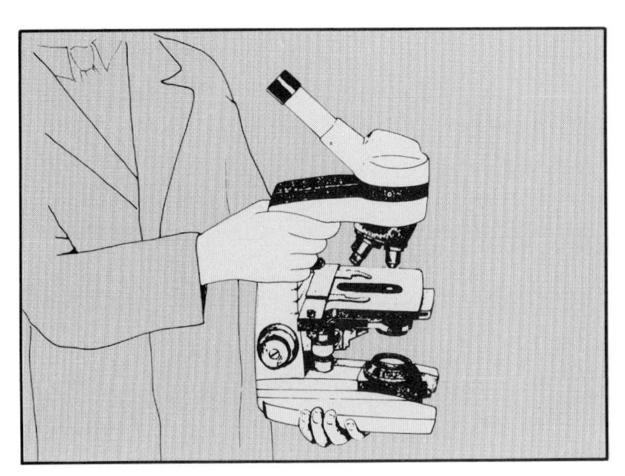

Figure 1.1 The microscope should be held firmly with both hands while carrying it.

has a clamping device, the **mechanical stage,** which is used for holding and moving the slide around on the stage. Note, also, the location of the **mechanical stage control** in figure 1.2.

Light Source In the base of most microscopes is positioned some kind of light source. Ideally, the lamp should have a voltage control to vary the intensity of light. The microscope in figure 1.2 has the type of lamp that is conrolled by a detached transformer unit (not shown). The light source on the microscope base in figure 1.4 has an electronic voltage control that is built into the lamp housing.

Most microscopes utilize a **neutral density filter** in addition to a voltage control. It is usually needed when observing objects at lower magnifications. Note that a lever is located on the lamp housing of the microscope in figure 1.2 that allows one to move this filter in and out of position.

Lens Systems All microscopes have three lens systems: the oculars, the objectives, and the con-

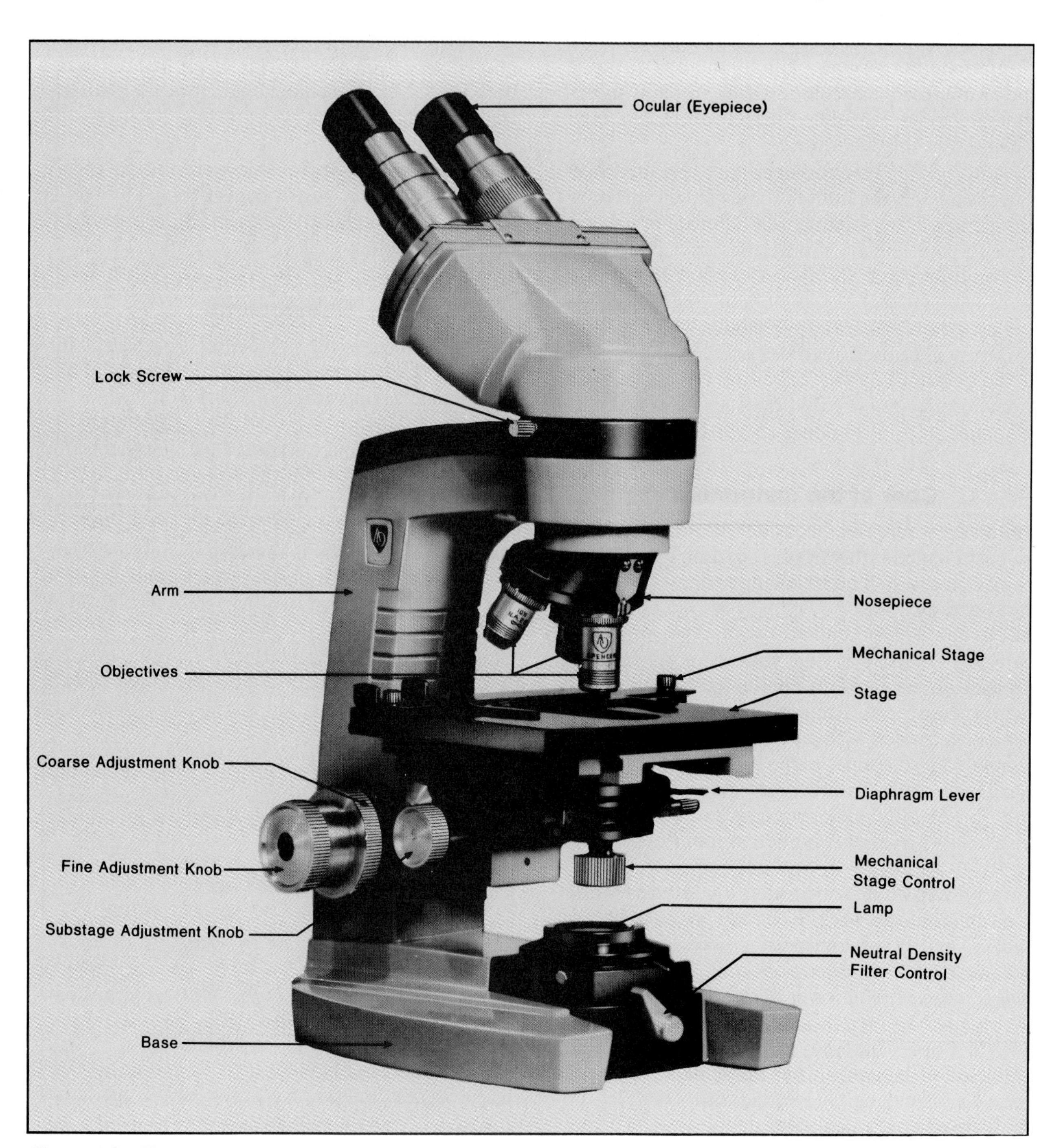

Figure 1.2 The compound microscope.

denser. Figure 1.3 illustrates the light path through these three systems.

The **ocular,** or eyepiece, which is at the top of the instrument, consists of two or more internal lenses and usually has a magnification of 10×. Although the microscope in figure 1.2 has two oculars (binocular), a microscope often has only one.

Three or more **objectives** are usually present. Note that they are attached to a rotatable **nosepiece** which makes it possible to move them into position over a slide. Objectives on most laboratory microscopes have magnifications of 10×, 45×, and 100×, designated as **low power, high-dry,** and **oil immersion,** respectively.

The third lens system is the **condenser,** which is located under the stage. Its position is best shown in figure 1.3. It collects and directs the light from the lamp to the slide being studied. The **substage adjustment knob** shown in figure 1.2 enables one to move the condenser up and down. A **diaphragm lever** is also provided to control the amount of light exiting from the condenser under the stage.

Focusing Knobs The concentrically arranged **coarse adjustment** and **fine adjustment knobs** on the side of the microscope are used for bringing objects into focus when studying an object on a slide.

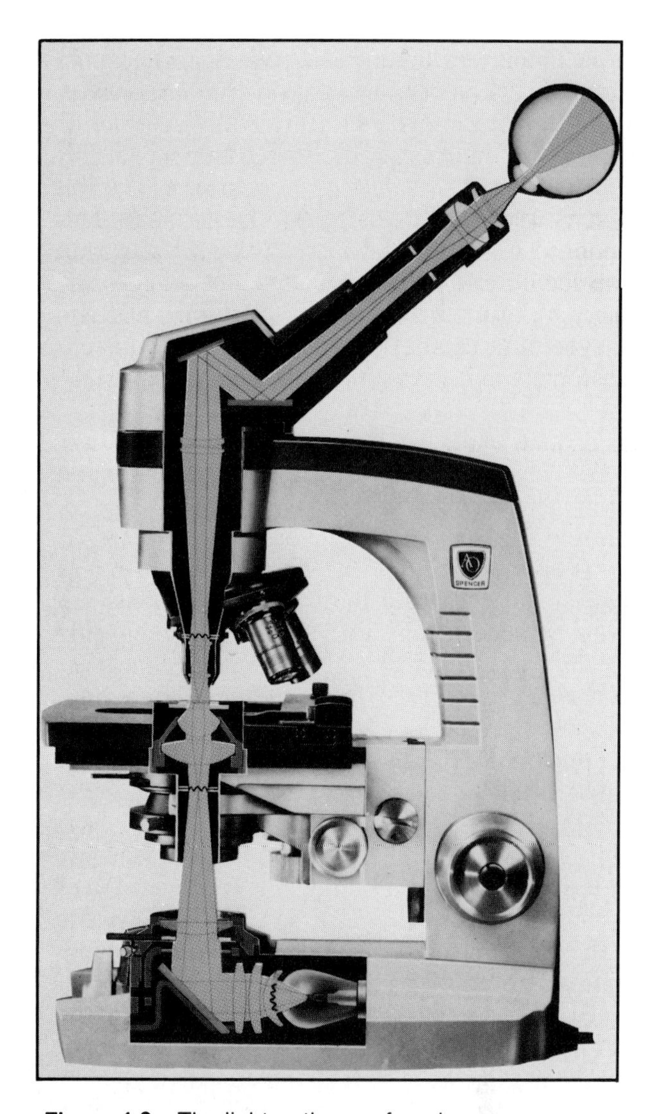

Figure 1.3 The light pathway of a microscope.

Resolution

The resolution limit, or **resolving power,** of a microscope lens is a function of its numerical aperture, the wavelength of light, and the design of the condenser. The maximum resolution of the best microscopes is around 0.2 μm. This means that two small objects that are 0.2 μm apart will be seen as separate entities; objects closer than that will be seen as a single object.

To get the maximum amount of resolution from a lens system, the following factors must be taken into consideration:

- A **blue filter** should be in place over the light source because the short wavelength of blue light provides maximum resolution.
- The **condenser** should be kept at its highest position where it enables a maximum amount of light to enter the objective.
- The **diaphragm** should not be stopped down too much. Although stopping down improves contrast it reduces the numerical aperture.
- **Immersion oil** should be used between the slide and the 100× objective.

Lens Care

Keeping the lenses of your microscope clean is a constant concern. Unless all lenses are kept free of dust, oil, and other contaminants, they are unable to achieve the degree of resolution that is intended. Consider the following suggestions for cleaning the various lens components.

Cleaning Tissues Only lint-free, optically safe tissues should be used to clean lenses. Tissues free of abrasive grit fall in this category. Booklets of lens tissue are most widely used for this purpose. Although several types of boxed tissues are also safe, *use only the type of tissue that is recommended by your instructor.*

Solvents Various liquids can be used for cleaning microscope lenses. Green soap with warm water works very well. Xylene is universally acceptable. Alcohol and acetone are also recommended, but often with some reservations. Acetone is a powerful solvent that could possibly dissolve the lens mounting cement in some objective lenses if it were used too liberally. When it is used it should be used sparingly. Your instructor will inform you as to what solvents can be used on the lenses of your microscope.

Oculars The best way to determine if your eyepiece is clean is to rotate it between the thumb and forefinger as you look through the microscope. A rotating pattern will be evidence of dirt.

If cleaning the top lens of the ocular with lens tissue fails to remove the debris, one should try cleaning the lower lens with lens tissue and blowing off any excess lint with an air syringe. *Whenever the ocular is removed from the microscope, it is imperative that a piece of lens tissue be placed over the open end of the microscope as illustrated in figure 1.5.*

Objectives Objective lenses often become soiled by materials from slides or fingers. A piece of lens tissue moistened with green soap and water or one of the other solvents mentioned above will usually remove whatever is on the lens. Sometimes a cotton swab with a solvent will work better than lens tissue. At any time that the image on a slide is blurred or cloudy, assume at once that the objective you are using is soiled.

Condenser Dust often accumulates on the top surface of the condenser; thus, wiping it off occasionally with lens tissue is desirable.

Procedures

If your microscope has three objectives you have three magnification options: (1) low-power or 100× magnification, (2) high-dry magnification which is 450× with a 45× objective, and (3) 1000× magnification with the oil immersion objective. Note that the magnification seen through an objective is calculated by multiplying the power of the ocular by the power of the objective.

Whether you use the low-power objective or the oil immersion objective will depend on how much magnification is necessary. Generally speaking, however, it is best to start with the low-power objective and progress to the higher magnifications as your study progresses. Consider the following suggestions for setting up your microscope and making microscopic observations.

Viewing Setup If your microscope has a rotatable head, such as the one being used by the two students in figure 1.6, there are two ways that you can use the instrument. Note that the student on the left has the arm of the microscope near him, and the other student has the arm away from her. With this type of microscope, the student on the right has the advantage in that the stage is easier to observe. Note, also, that when focusing the instrument she is able to rest her arm on the table. The manufacturers of this type of microscope intended that the instrument be used in the way demonstrated by the young lady. If the head is not rotatable, it will be necessary to use the other position.

Low-Power Examination The main reason for starting with the low-power objective is to enable you to explore the slide to look for the object you are planning to study. Once you have found what

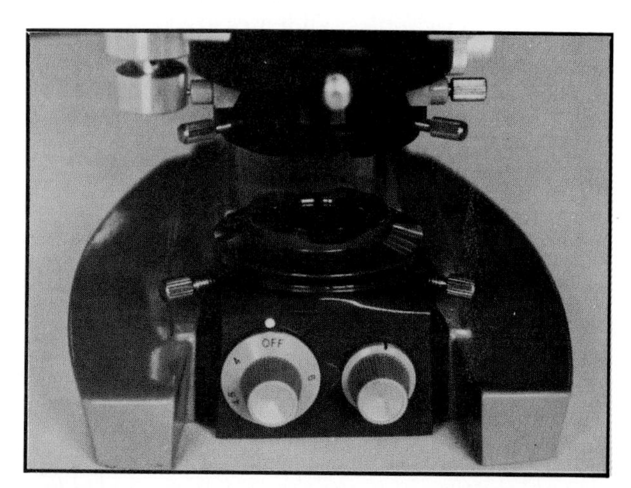

Figure 1.4 The left knob controls voltage; the other knob controls the neutral density filter.

Figure 1.5 After cleaning the lenses, a blast of air from an air syringe removes residual lint.

you are looking for, you can proceed to higher magnifications. Use the following steps when exploring a slide with the low-power objective:

1. Position the slide on the stage with the material to be studied on the *upper* surface of the slide. Figure 1.7 illustrates how the slide must be held in place by the mechanical stage retainer lever.
2. Turn on the light source, using a *minimum* amount of voltage. If necessary, reposition the slide so that the stained material on the slide is in the *exact center* of the light source.
3. Check the condenser to see that it has been raised to its highest point.
4. If the low-power objective is not directly over the center of the stage, rotate it into position. Be sure that as you rotate the objective into position it clicks into its locked position.
5. Turn the coarse adjustment knob to lower the objective *until it stops.* A built-in stop will prevent the objective from touching the slide.
6. While looking down through the ocular or oculars, bring the object into focus by turning the fine adjustment focusing knob. Don't readjust the coarse adjustment knob. If you are using a binocular microscope it will also be necessary to adjust the distance between the two eyepieces to match your eyes.
7. Manipulate the diaphragm lever to reduce or increase the light intensity to produce the clearest, sharpest image. Note that as you close down the diaphragm to reduce the light intensity, the contrast improves and the depth of field increases.
8. Once an image is visible, move the slide about to search out what you are looking for. The slide is moved by turning the knobs that move the mechanical stage.
9. Check the cleanliness of the ocular, using the procedure outlined earlier.
10. Once you have identified the structures to be studied and wish to increase the magnification, you may proceed to either high-dry or oil immersion magnification.

High-Dry Examination To proceed from low-power to high-dry magnification, all that is necessary is to rotate the high-dry objective into position and open up the diaphragm somewhat. It may be necessary to make a minor adjustment with the fine adjustment knob to sharpen up the image, but *the coarse adjustment knob should not be touched.*

If a microscope is of good quality, only minor focusing adjustments are needed when changing from low power to high-dry because all the objectives will be **parfocalized.** Nonparfocalized microscopes do require considerable focusing adjustments when changing objectives.

High-dry objectives should only be used on slides that have cover glasses; without them, images are often unclear. When increasing the lighting, be sure to open up the diaphragm first instead of increasing the voltage on your lamp; reason: lamp life is greatly extended when used at low voltage. If the field is not bright enough after opening the diaphragm, feel free to increase the voltage. A final point: keep the condenser at its highest point.

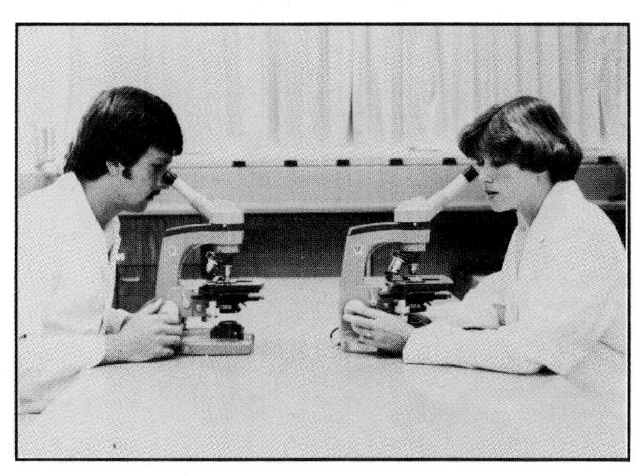

Figure 1.6 The microscope position on the right has the advantage of stage accessibility.

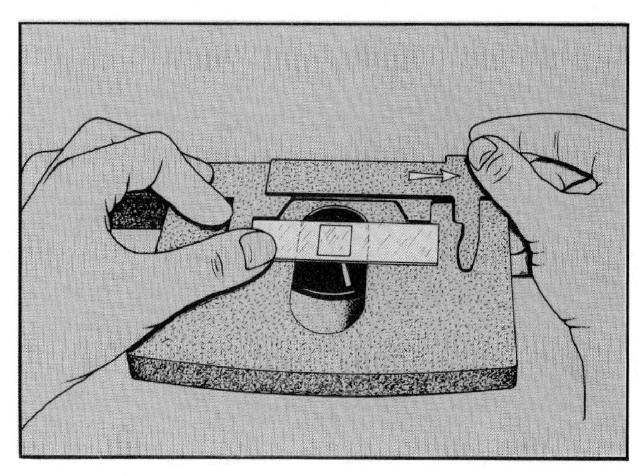

Figure 1.7 The slide must be properly positioned as the retainer lever is moved to the right.

Oil Immersion Techniques The oil immersion lens derives its name from the fact that a special mineral oil is interposed between the lens and the microscope slide. The oil is used because it has the same refractive index as glass, which prevents the loss of light due to the bending of light rays as they pass through air. The use of oil in this way enhances the resolving power of the microscope. Figure 1.8 reveals this phenomenon.

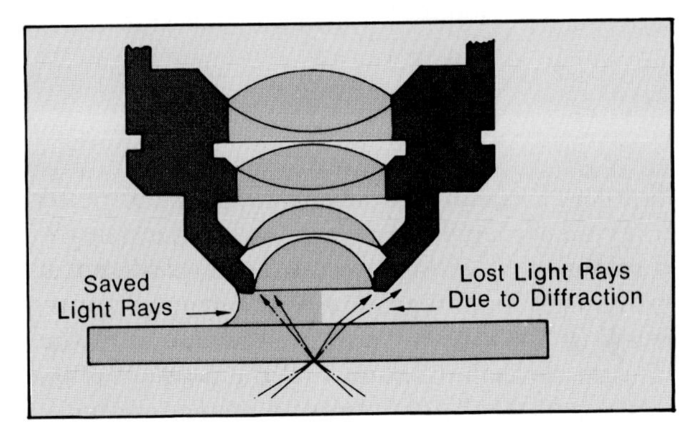

Figure 1.8 Immersion oil, having the same refractive index as glass, prevents light loss due to diffraction.

With parfocalized objectives one can go to oil immersion from either low power or high-dry. Once the microscope has been brought into focus at one magnification, the oil immersion lens can be rotated into position without fear of striking the slide.

Before rotating the oil immersion lens into position, however, a drop of immersion oil is placed on the slide. If the oil appears cloudy it should be discarded.

When using the oil immersion lens it is best to open the diaphragm as much as possible. Stopping down the diaphragm tends to limit the resolving power of the optics. In addition, the condenser might be at its highest point. If different colored filters are available for the lamp housing, it is best to use blue or greenish filters to enhance the resolving power.

Since the oil immersion lens will be used extensively in all bacteriological studies, it is of paramount importance that you learn how to use this lens properly. Using this lens takes a little practice due to the difficulties usually encountered in manipulating the lighting. A final comment of importance: at the end of the laboratory period remove all immersion oil from the lens tip with lens tissue.

Returning Microscope to Cabinet

When you take a microscope from the cabinet at the beginning of the period, you expect it to be clean and in proper working condition. The next person to use the instrument after you have used it will expect the same consideration. A few moments of care at the end of the period will insure these conditions. Check over this list of items at the end of the period before you return the microscope to the cabinet:

1. Remove the slide from the stage.
2. If immersion oil has been used, wipe it off the lens and stage with lens tissue.
3. Rotate the low-power objective into position.
4. If the microscope has been inclined, return it to an erect position.
5. If the microscope has a built-in movable lamp, raise the lamp to its highest position.
6. If the microscope has a long attached electric cord, wrap it around the base.
7. Adjust the mechanical stage so that it does not project too far on either side.
8. Replace the dustcover.
9. If the microscope has a separate transformer, return it to its designated place.
10. Return the microscope to its correct place in the cabinet.

Laboratory Report

Before the microscope is to be used in the laboratory, answer all the questions on the Laboratory Report. Preparation on your part prior to going to the laboratory will greatly facilitate your understanding. Your instructor may wish to collect this report at the *beginning of the period* on the first day that the microscope is to be used in class.

Phase-Contrast Microscopy 2

The difficulty that one encounters in trying to examine cellular organelles is that most protoplasmic material is completely transparent and defies differentiation. It is for this reason that stained slides are usually used in brightfield cytological studies. Since the staining of slides results in cellular death, it is obvious that when we study a stained slide we are observing artifacts, not actual living structures.

A microscope that is able to differentiate transparent protoplasmic structures without staining and killing them is the **phase-contrast microscope.** The first phase-contrast microscope was developed in 1933 by Frederick Zernike, and was originally referred to as the *Zernike microscope.* It is the instrument of choice for studying living protozoans and other types of transparent cells. Figure 2.1 illustrates the differences between brightfield and phase-contrast images. Note the greater degree of differentiation that can be seen inside of cells when they are observed with phase-contrast optics. In this exercise we will study the principles that function in this type of microscope; we will also see how different manufacturers have met the design challenges of these principles.

Image Contrast

Objects of a microscope field may be categorized as being either amplitude or phase objects. **Amplitude objects** (illustration 1, figure 2.2) show up as dark objects under the microscope because the amplitude (intensity) of light rays is reduced as the rays pass through the objects. **Phase objects** (illustration 2, figure 2.2), on the other hand, are completely transparent since light rays pass through them unchanged with respect to amplitude. As some of the light rays pass through phase objects, however, they are retarded by ¼ wavelength. This retardation, known as *phase shift,* occurs with no amplitude diminution; thus, the objects are transparent. Since most biological specimens are phase objects, lacking in contrast, it becomes necessary to apply dyes of various kinds to cells that are to be studied with a brightfield microscope. To un-

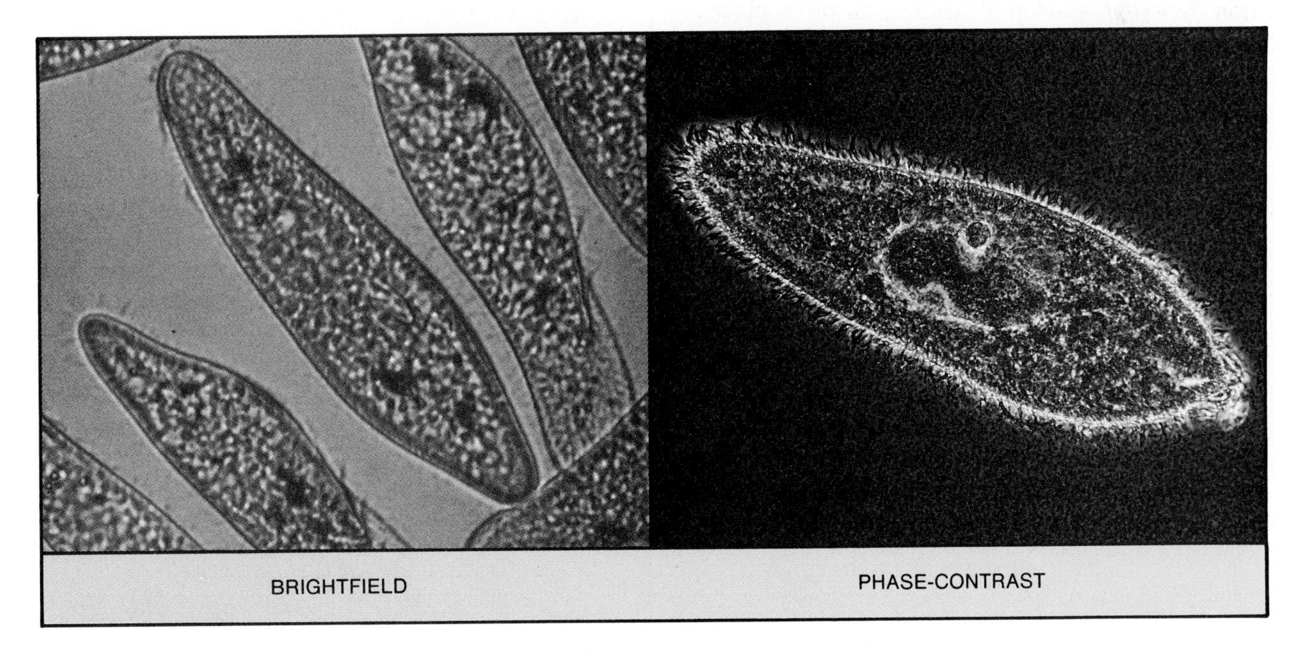

Figure 2.1 Comparison of brightfield and phase-contrast images.

derstand how Zernike took advantage of the ¼ wavelength phase shift in developing his microscope we must understand the difference between direct and diffracted light rays.

Direct and Diffracted Light Rays

Light rays passing through a transparent object emerge as either direct or diffracted rays. Those rays that pass straight through unaffected by the medium are called **direct rays.** They are unaltered in amplitude and phase. The balance of the rays that are bent by their slowing through the medium (due to density differences) emerge from the object as **diffracted rays.** It is these rays that are retarded ¼ wavelength. Illustration 3, figure 2.2, illustrates these two types of light rays.

An important characteristic of these light rays is that if the direct and diffracted rays of an object can be brought into exact phase, or *coincidence,* with each other, the resultant amplitude of the converged rays is the sum of the two waves. This increase in amplitude will produce increased brightness of the object in the field. On the other hand, if two rays of equal amplitude are in reverse phase (½ wavelength off), their amplitudes cancel each other to produce a dark object. This phenomenon is called *interference.* Illustration 4, figure 2.2, shows these two conditions.

The Zernike Microscope

In constructing his first phase-contrast microscope, Zernike experimented with various configurations of diaphragms and various materials that could be used to retard or advance the direct light rays. Figure 2.3 illustrates the optical system of a typical modern phase-contrast microscope. It differs from a conventional brightfield microscope by having (1) a different type of diaphragm and (2) a phase plate.

The diaphragm consists of an **annular stop** that allows only a hollow cone of light rays to pass up through the condenser to the object on the slide. The **phase plate** is a special optical disk located at the rear focal plane of the objective. It has a *phase ring* on it that advances or retards the direct light rays ¼ wavelength.

Note in figure 2.3 that the direct rays converge on the phase ring to be advanced or retarded ¼ wavelength. These rays emerge as solid lines from the object on the slide. This ring on the phase plate

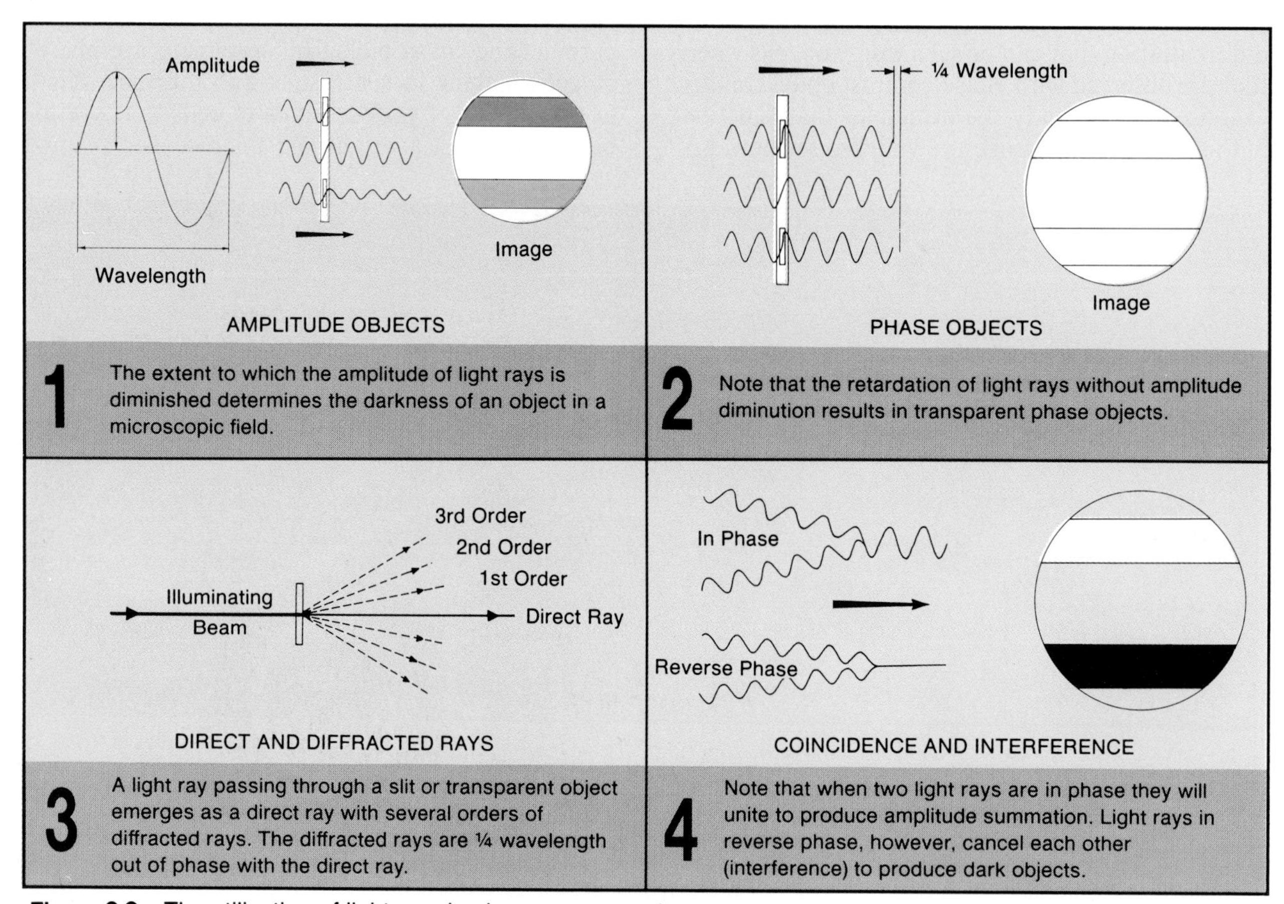

Figure 2.2 The utilization of light rays in phase-contrast microscopy.

is coated with a material that will produce the desired phase shift. The diffracted rays, on the other hand, which have already been retarded ¼ wavelength by the phase object on the slide, completely miss the phase ring and are not affected by the phase plate. It should be clear, then, that depending on the type of phase-contrast microscope, the convergence of diffracted and direct rays on the image plane will result in either a brighter image (amplitude summation) or a darker image (amplitude interference or reverse phase). The former is referred to as *bright phase* microscopy; the latter

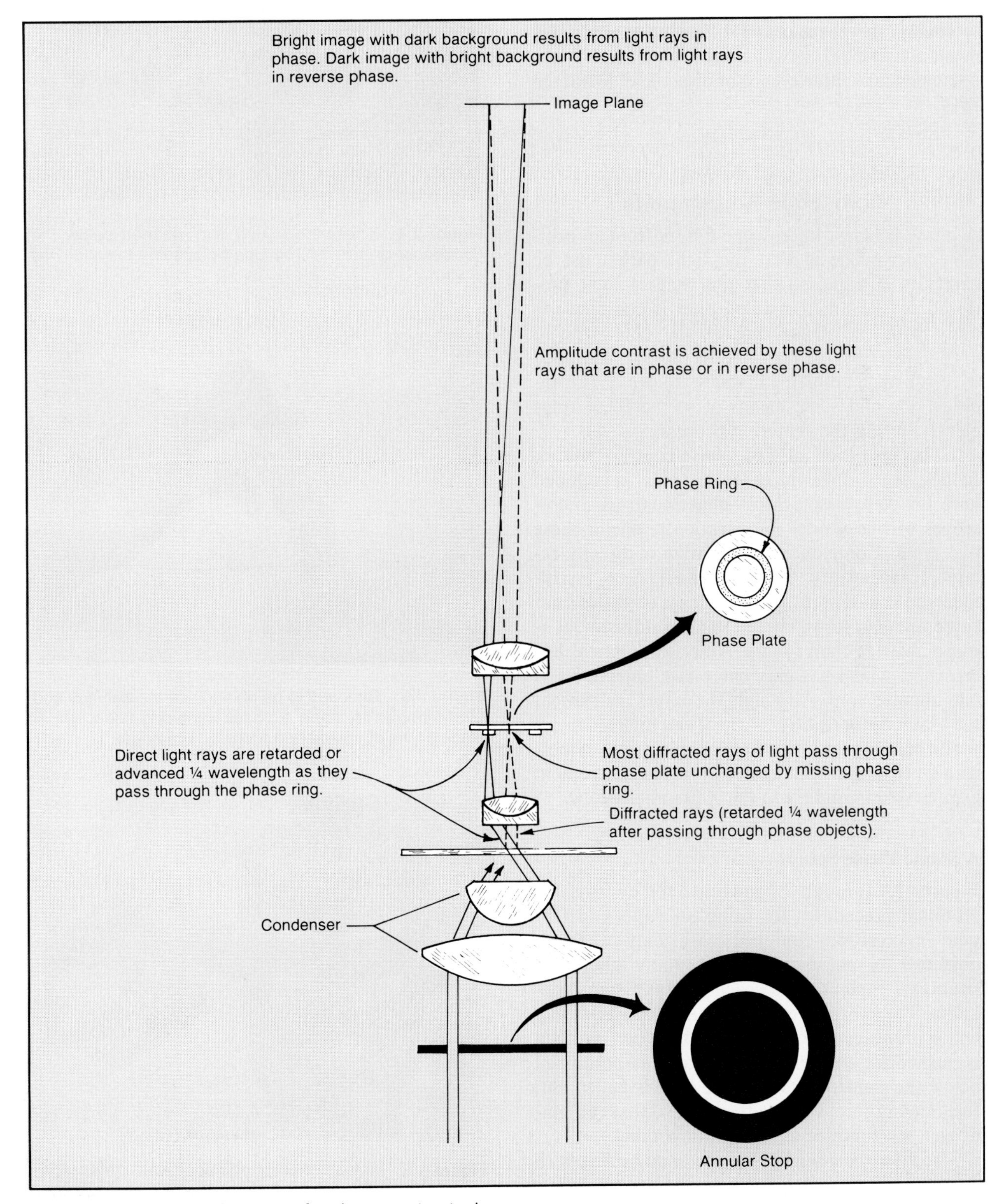

Figure 2.3 The optical system of a phase-contrast microscope.

as *dark phase* microscopy. The apparent brightness or darkness, incidentally, is proportional to the square of the amplitude; thus, the image will be four times as bright or dark as seen through a brightfield microscope.

It should be added here, parenthetically, that the phase plates of some microscopes have coatings to change the phase of the diffracted rays. In any event the end result will be the same: to achieve coincidence or interference of direct and diffracted rays.

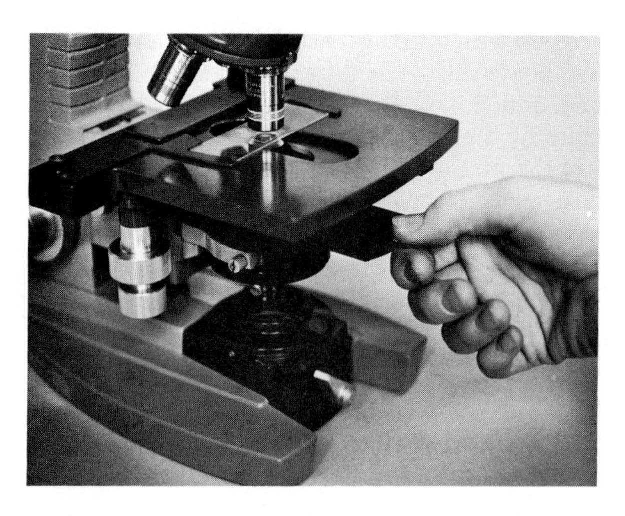

Figure 2.4 The annular ring is positioned below the condenser on this microscope by pushing the slideway inward.

Microscope Adjustments

A phase-contrast microscope differs from an ordinary microscope in that the light path must be carefully adjusted so that the cone of light produced by the annular diaphragm is centered exactly with the phase ring of the objective. To accomplish this, the instruments must be equipped with substage adjustment screws to center the annular stop and some means of seeing these rings clearly during the centering process.

The operation of two phase-contrast microscopes, one simple, the other complex, is included here for comparison. Most phase-contrast microscopes will be similar in operation to one of these two types. Complexity of operation is directly related to versatility. Since the Karl Zeiss instrument, described here, has three phase objectives and three annular stops, the method of adjustment is more involved than for the American Optical microscope, which has only one phase objective and one annular stop. Although the latter instrument may lack the versatility of the Zeiss microscope, it performs well, optically, in its single range. American Optical also has a more versatile arrangement comparable in quality to the Zeiss microscope.

Figure 2.5 One way to be able to see the annulus and phase ring is to insert a phase centering telescope in place of the eyepiece and focus on the rings.

A Single Phase Setup

Figures 2.4 through 2.7 illustrate some of the operational procedures for using an American Optical microscope equipped with only a single objective for phase-contrast. In many laboratory situations such a microscope is completely adequate. The substage condenser unit has a slideway which moves in and out (figure 2.4). When the slide is pushed in, an annular diaphragm is positioned below the condenser. When the slide is pulled out, the annular diaphragm is removed so that the condenser can function as a brightfield unit.

To view the alignment of the annular ring with the phase ring, two different units are available: the phase centering telescope and an aperture

Figure 2.6 Instead of being equipped with a telescope, some microscopes are equipped with an aperture viewing unit to focus on the two rings.

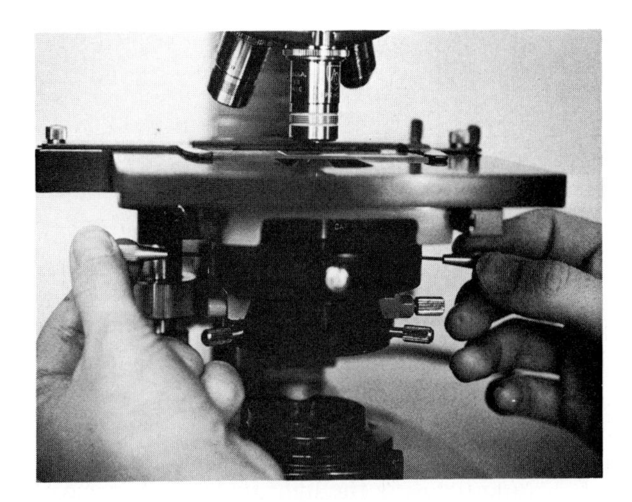

Figure 2.7 Alignment of the annulus diaphragm and phase ring is accomplished with a pair of small Allen-type screwdrivers on this American Optical microscope.

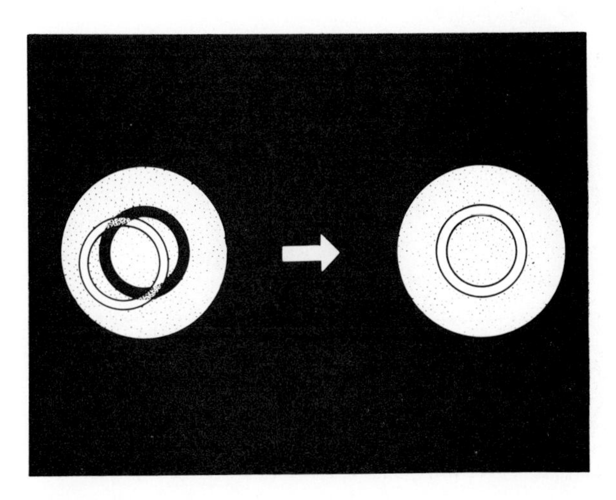

Figure 2.8 The image on the right illustrates the appearance of the rings when perfect alignment of phase ring and annulus diaphragm has been achieved.

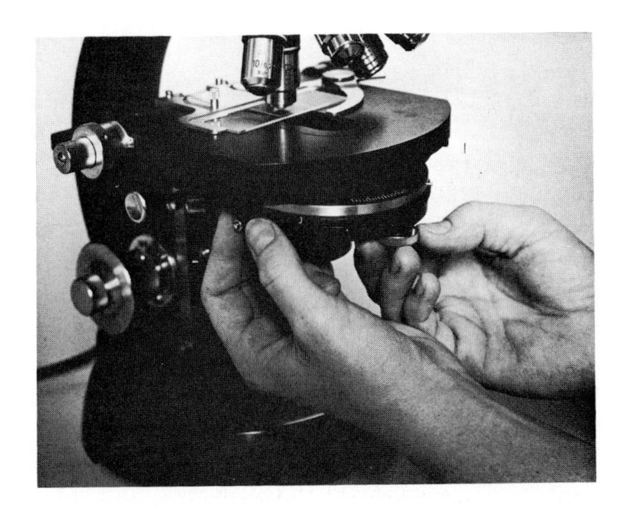

Figure 2.9 Alignment of the annulus and phase ring on this Zeiss microscope is achieved by adjusting the two knobs as shown.

viewing unit. Figure 2.5 illustrates the **phase centering telescope.** This is a special eyepiece that replaces the regular eyepiece during the centering procedure. After the rings are in alignment, as shown in figure 2.8, the telescope is replaced by the eyepiece. Figure 2.6 illustrates use of the **aperture viewing unit.** This component allows one to see the rings without replacing the regular eyepiece. The advantage of the latter unit over the telescope is that one can instantly check ring concentricity without eyepiece manipulation.

The actual alignment of the rings is accomplished with a pair of Allen-type screwdrivers, as shown in figure 2.7. In checking the alignment of the annular diaphragm and the phase ring, the following specifics should be observed:

1. If a phase centering telescope is used for focusing on the rings, it is necessary to slide the upper part in and out to bring the rings into focus. Although the annular diaphragm and phase ring appear much smaller than with the aperture viewing unit, the accuracy of alignment is just as precise.
2. To bring the rings into view with the aperture viewing unit, it is necessary to move the lever into position. If the rings are unclear, the little round knob within the lever is pushed in and out until they are in sharp focus.
3. To align the rings while viewing them, it is necessary to use both Allen-type screwdrivers simultaneously as shown in figure 2.7. Recessed Allen-head adjustment screws force the annular diaphragm into correct position. Once this adjustment is complete it should remain stable.

With this type of microscope, the condenser slideway is kept in the outward position when it is used as a conventional brightfield microscope. When it is desirable to observe an object under phase-contrast, the slideway is pushed inward. It must be remembered, however, that the annular diaphragm will work only with the phase objective for which it is matched.

A Multiple Phase Setup

Figure 2.10 illustrates a Zeiss phase-contrast microscope that is equipped with three phase-contrast objectives (plus two brightfield objectives) and three matching annular diaphragms. The substage unit consists of a turret which enables one to swing the correct annular diaphragm into position for each objective. Between the binocular eyepiece unit and the revolving nosepiece is another turret-type aperture viewing unit, which Zeiss calls the **Optovar.**

The Optovar has two knurled dials (wheels 1 and 2) that are used for observing the annular diaphragm and phase ring as well as for changing the magnification of the microscopic field.

Due to the fact that there are so many more adjustments on this type of a microscope than the simpler American Optical instrument described earlier, it is desirable to check the alignment of the components of the light path if the microscope has not been used for some time. Where an instrument is to be shared by many students it is essential that they be cautioned against accidentally altering the adjustments.

There are essentially two major concerns in preparing this microscope for use: proper adjustment of the light source and alignment of the annular diaphragm and phase ring. Proceed as follows:

Light Source Adjustments Evenness of illumination and proper centering of the light source are paramount considerations.

1. To produce an evenly illuminated light source in the base of the microscope it is necessary to do as follows:
   a. If the light source is a 15-watt bulb (type shown in figure 2.10), first loosen the bulb housing by turning the knurled ring. Then, while moving the bulb housing in and out and focusing the light image on a piece of

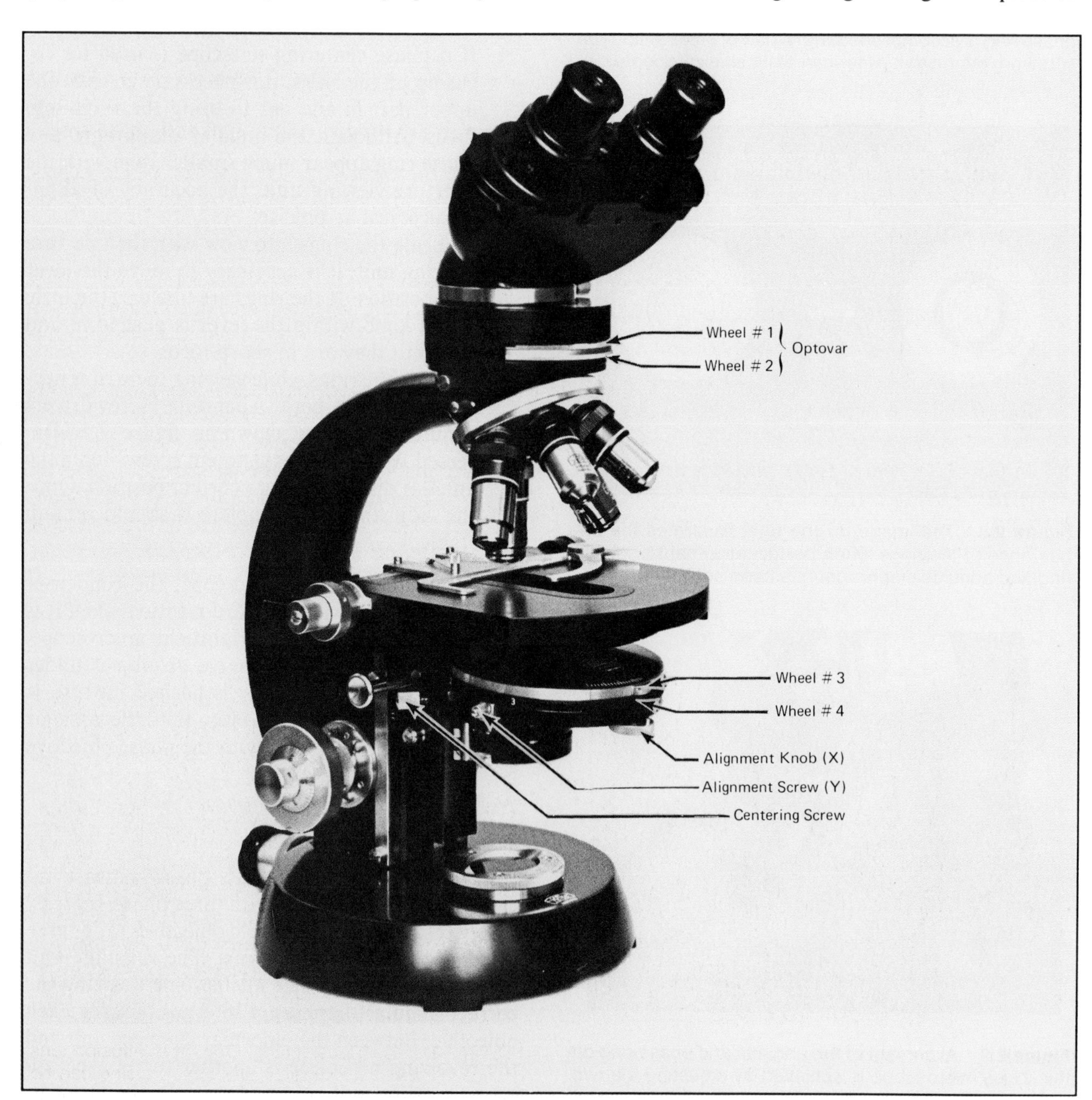

Figure 2.10 Adjustment components of a Zeiss phase-contrast microscope.

translucent white paper over the light source of the base, refasten the bulb in place when the projected light is uniform.

b. If the illuminator contains a 60-watt bulb, as illustrated in figure 2.11, it will be necessary to remove the lamp housing from the microscope to adjust the bulb position. Once the bulb is removed, turn the knob (see figure 2.11) that removes the diffusion lens from in front of the bulb. Then loosen another screw (see figure 2.12) which allows you to move the bulb fore and aft within the lamp housing. Now, project the light on a wall ten feet away and slide the outer housing until the filament grid comes into sharp focus. Once the grid is in focus, lock the screw that fixes the two parts together and return the diffusion lens to front of bulb position. Return the lamp housing to the microscope; place a piece of translucent white paper over the light source and adjust the base diaphragm to produce an evenly illuminated spot of light.

2. Position a blue filter in the light path. If the microscope does not have a filter holder built into the substage unit, the filter may be placed over the light source on the base.
3. Place a slide preparation on the stage.
4. Rotate wheel 4 to the brightfield position (J).
5. Rotate wheel 1 of the Optovar to position 1× (no magnification).
6. Move the lowest powered phase objective (phase 1 or 2) into position.
7. Bring the image of the specimen on the slide into focus with the coarse and fine adjustment knobs.
8. Adjust the binocular component for your eyes. This is accomplished by setting each ocular on the same number that is read on the scale between the oculars when the eyepieces are adjusted to match the distance between your eyes.
9. Adjust the iris diaphragm of the condenser (wheel 3) for optimum lighting.
10. Turn the base diaphragm, reducing the light to the least amount of light.
11. While looking into the eyepieces, move the condenser up and down until you see the outline of the diaphragm distinctly.
12. Center the diaphragm by turning the two knurled centering screws. Refer to figure 2.10 to locate these screws.

Annulus and Phase Ring Alignment Now that the light source is properly adjusted, proceed as follows to center the annular diaphragm and phase ring.

1. Rotate wheel 4 to the number which corresponds to the phase objective being used (1 or 2). This positions the proper annular diaphragm for the objective.
2. Rotate wheel 1 of the Optovar to PH. This brings a lens into position, which allows one to see the two rings while looking down into the microscope.
3. Raise the condenser unit by turning the substage adjustment knob until the two rings come into focus. If the rings are indistinct after raising the condenser, bring them into focus by rotating wheel 2 of the Optovar. (Instruments that lack the Optovar unit will be focused with a centering telescope as described previously for the American Optical microscope.)

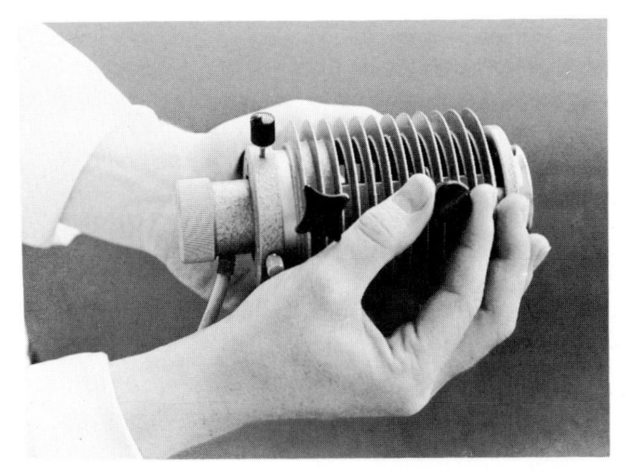

Figure 2.11 Lens adjustment. It is necessary to remove this type of lamp housing from the Zeiss microscope for focusing. The diffusion lens of the housing is moved out of position by turning the knob as illustrated.

Figure 2.12 Bulb positioning. Once the diffusion lens has been moved to one side, the bulb housing can be moved fore and aft for focusing after the lock screw is loosened as shown.

4. Center the annular diaphragm and phase ring by manipulating knobs X and Y as shown in figure 2.9. The left-hand knob (Y) must be unscrewed slightly before it will move. Once the rings are centered as shown in figure 2.8, this knob should be locked tightly in position.

Examining the Slide

Having centered the light source and aligned the phase elements, you are now ready to examine slide preparations. Keep in mind that from now on most of the adjustments described earlier should not be altered. It may be necessary to go through many of the steps to rectify the misalignment. Before attempting to study material under phase-contrast keep these points in mind:

- Use only optically perfect slides and cover glasses (no bubbles or striae in the glass).
- Be sure that slides and cover glasses are perfectly clean (no grease or chemicals on them).
- Use wet mount slides instead of hanging drop preparations. The latter leave much to be desired. Culture broths containing bacteria and blood cell suspensions are suitable for wet mounts.
- Avoid slides in which Canada balsam or other similar mounting media have been used. Poor image contrast will result with certain types of mounting media. Special preparations are available for phase work.
- In general, limit observations to living cells. In most instances stained slides are less satisfactory.

To bring the object into view under phase-contrast with the Zeiss microscope, once all the adjustments are completed, follow this sequence:

1. Rotate wheel 1 of the Optovar to 1.0 (no magnification) and wheel 4 (condenser turret) to same number as phase objective being used (1 or 2).
2. Bring the object into focus, using the coarse adjustment knob first.
3. Adjust the light to a desirable level with the rheostat (if present) or the base diaphragm.
4. If greater magnification is desired, rotate wheel 1 of the Optovar to 1.25, 1.6, or 2. It will be necessary to increase the light supply as the magnification is increased.
5. If still greater magnification is desired, rotate PH 2 or PH 3 objectives into position. Wheel 4 must also be rotated to corresponding numbers 2 or 3. When PH 3 objective is used, immersion oil must be placed on top of the condenser as well as on top of the cover glass.
6. If you wish to compare a phase image with a brightfield image, rotate wheel 4 to the J position.

Laboratory Report

This exercise may be used in conjunction with Part 2 in studying various types of organisms. Organelles in protozoans and algae will show up more distinctly than with ordinary brightfield instruments. After reading this exercise and doing any special assignments made by your instructor, answer the questions on combined Laboratory Report 2, 3 that pertain to this exercise.

3 Microscopic Measurements

If an ocular micrometer is properly installed into the eyepiece of your microscope, it is a simple procedure to measure the size of microorganisms that are seen on the microscope slide. An **ocular micrometer** consists of a circular disk of glass which has graduations engraved on one surface. These graduations appear as shown in illustration B, figure 3.4. In some microscopes, the ocular has to be disassembled so that the disk can be placed on a shelf in the ocular tube between the two lenses; however, in most of the better microscopes, such as the one in figure 3.4, the ocular micrometer is simply inserted into the bottom of the ocular as shown in figure 3.1. Before one can use the ocular micrometer it is necessary to calibrate it for each of the objectives by using a stage micrometer. The principal purpose of this exercise is to show you how to calibrate an ocular micrometer for the various objectives on your microscope.

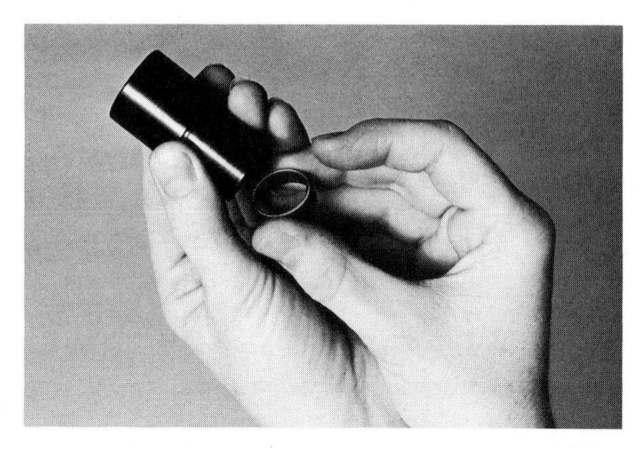

Figure 3.1 Ocular micrometer with retaining ring is inserted into base of eyepiece.

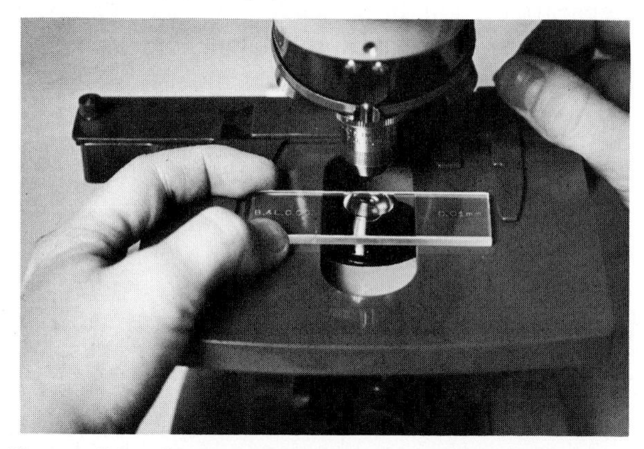

Figure 3.2 Stage micrometer is positioned by centering small glass disk over light source.

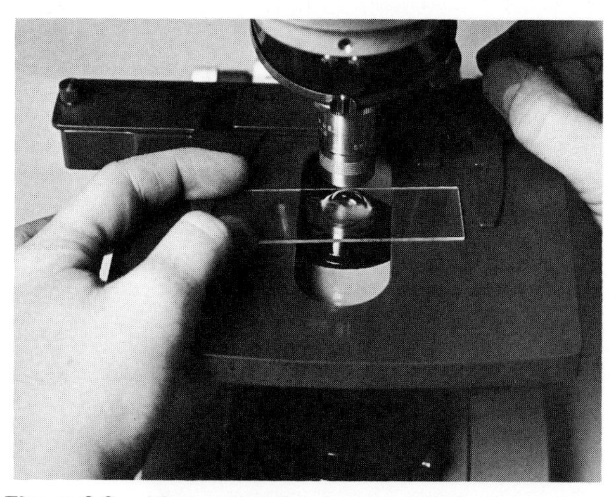

Figure 3.3 After calibration is completed, stage micrometer is replaced by slide with organisms to be measured.

Calibration

The distance between the lines of an ocular micrometer is an arbitrary measurement that only has meaning if the ocular micrometer is calibrated for the objective that is being used. A **stage micrometer,** also known as an *objective micrometer,* has scribed lines on it that are exactly 0.01 mm (10 micrometers) apart. Illustration C, figure 3.4, reveals the appearance of these graduations.

To calibrate the ocular micrometer for a given objective, it is necessary to superimpose the two scales and determine how many of the ocular graduations coincide with one graduation on the stage micrometer scale. Illustration A in figure 3.4 shows how the two scales appear when they are properly aligned in the microscope. In this case, seven ocular divisions match up with one stage micrometer division of 0.01 mm to give an ocular value of 0.01/7 or 0.00143 mm. Since there are 1000 micrometers in 1 millimeter, these divisions are 1.43 μm apart. With this information known, the stage micrometer is replaced with a slide of organisms to be measured. Illustration D shows how the field might appear with the ocular micrometer in the eyepiece. To determine the size of an organism, then, it is a

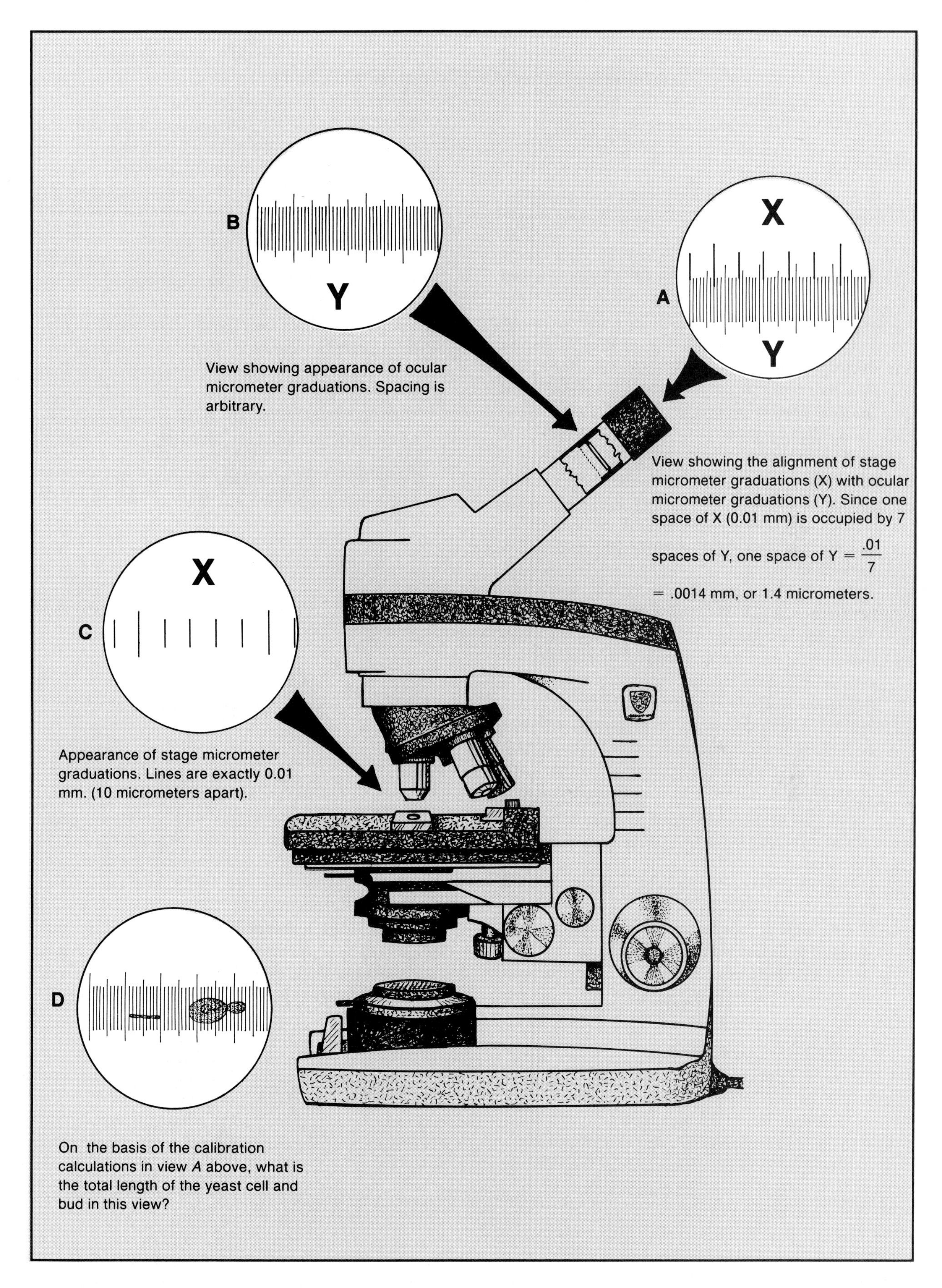

Figure 3.4 Calibration of ocular micrometer.

simple matter to count the graduations and multiply this number by the known distance between the graduations. When calibrating a microscope for a specific magnification, proceed as follows:

Materials:

ocular micrometer or eyepiece that contains an ocular micrometer
stage micrometer

1. If eyepieces are available that contain ocular micrometers, replace the eyepiece in your microscope with one of them.

   If it is necessary to insert an ocular micrometer in your eyepiece, find out from your instructor whether it is to be inserted below the bottom lens or placed between the two lenses within the eyepiece. In either case, great care must be taken to avoid dropping the eyepiece or reassembling the lenses incorrectly. *Only with your instructor's prior approval shall eyepieces be disassembled.* Be sure that the graduations are on the upper surface of the glass disk.
2. Place the stage micrometer on the stage and center it exactly over the light source.
3. With the low-power (10×) objective in position, bring the graduations of the stage micrometer into focus, *using the coarse adjustment knob.* Reduce lighting.
   Note: If the microscope has an automatic stop, do not use it as you normally would for regular microscope slides. The stage micrometer slide is too thick to allow it to function properly.
4. Rotate the eyepiece until the graduations of the ocular micrometer lie parallel to the lines of the stage micrometer.
5. If the **low-power objective** is the objective to be calibrated proceed to step 8.
6. If the **high-dry objective** is to be calibrated, swing it into position and proceed to step 8.
7. If the **oil immersion lens** is to be calibrated, place a drop of immersion oil on the stage micrometer, swing the oil immersion lens into position, and bring the lines into focus; then, proceed to the next step.
8. Move the stage micrometer laterally until the lines at one end coincide. Then look for another line on the ocular micrometer that coincides with one on the stage micrometer. Occasionally one stage micrometer division will include an even number of ocular divisions, as shown in illustration A. In most instances, however, several stage graduations will be involved. In this case, divide the number of stage micrometer divisions by the number of ocular divisions that coincide. The figure you get will be that part of a stage micrometer division that is seen in an ocular division. This value must then be multiplied by 0.01 mm to get the amount of each ocular division.

   Example: 3 divisions of the stage micrometer line up with 20 divisions of the ocular micrometer.

   Each ocular division =

   $$\frac{3}{20} \times 0.01 = 0.0015 \text{ mm, or } 1.5\ \mu\text{m}.$$
9. Replace the stage micrometer with slides of organisms to be measured.

Measuring Assignments

Slides of various organisms that are studied in the next two units may be referred to this exercise. It would be well for the student to measure representatives of protozoans, algae, fungi, and bacteria to note their differences.

There is **no Laboratory Report** for this exercise.

Important: Remove the ocular micrometer from your microscope at the end of the laboratory period.

Part 2 Survey of Microorganisms

The four exercises of this unit pertain to a study of the overall population of the microbial world. Too often, in our serious concern with the direct applications of microbiology to human welfare, we neglect the large number of interesting free-living microorganisms that abound in the water, soil, and air. In this unit we will study representatives in three kingdoms (Monera, Protista, and Myceteae) that include the majority of microorganisms in fresh water and air. Note in the following five-kingdom list that the subdivisions that are in **bold type** are the groups that will be encountered in this study. Although there are no phyla listed under Kingdom Animalia, certain species of invertebrates are usually encountered when studying pond water populations. It is for this reason that Exercise 5—"Microscopic Invertebrates"—has been included: to satisfy student curiosity pertaining to the identity of some of these behemoths of the microbial world.

Kingdom MONERA
- Subkingdom I: **Cyanobacteria**
- Subkingdom II: **Bacteria**

Kingdom PROTISTA
- Subkingdom I: **Protozoa**
- Subkingdom II: **Algae**

Kingdom MYCETEAE
- Division 1: Gymnomycota
- Division 2: Mastigomycota
- Division 3: **Amastigomycota**

Kingdom PLANTAE

Kingdom ANIMALIA

The principal source of protozoa, algae, and cyanobacteria in Exercise 4 will be pond waters. Water and bottom debris from various ponds will be available in specimen bottles on the demonstration table. Wet mount slides from these samples will be studied with brightfield or phase-contrast microscopes to identify the various genera. It is necessary to study these microorganisms *simultaneously* because the organisms will be identified as they are encountered, not in any particular sequence.

The fungi (Amastigomycota) for Exercise 6 will be collected from the air. Plates of media that have been exposed to the air will be the principal source of these organisms. Microscope slides will be made from these plate cultures so that detailed structure of molds and yeasts can be observed. For our study of bacteria in Exercise 7 we will expose nutrient agar plates and tubes of nutrient broth to the air, the hands, and objects in the environment. After the plates and broth tubes have been incubated, microscope slides will be made from them for microscopic examination.

4 Protozoa, Algae, and Cyanobacteria

In this exercise a study will be made of protozoans, algae, and cyanobacteria that are found in pond water. Bottles that contain water and bottom debris from various ponds will be available for study. Illustrations and text provided in this exercise will be used to assist you in your attempt to identify the various types that are encountered. Unpigmented, moving microorganisms will probably be protozoans; greenish or golden brown organisms are usually algae; and blue-green organisms will be cyanobacteria. Supplementary books on the laboratory bookshelf will also be available for assistance in identifying the organisms that are not described in the text of this exercise. If you encounter invertebrates and are curious as to their identification, refer to Exercise 5; however, keep in mind that our prime concern here is only with protozoans, algae, and cyanobacteria.

Materials:

bottles of pond water samples
microscope slides and cover glasses
rubber-bulbed pipettes and forceps
china marking pencil
reference books

To study the microorganisms of pond water it will be necessary to make wet mount slides. The procedure for making such slides is relatively simple. All that is necessary is to place a drop of suspended organisms on a microscope slide and cover it with a cover glass. If several different cultures are available, the number of the bottle should be recorded on the slide with a china marking pencil. As you prepare and study your slides, observe the following guidelines.

• Clean the slide and cover glass with soap and water, rinse thoroughly, and dry. Do not attempt to study a slide that lacks a cover glass.

• When using a pipette, insert it into the bottom of the bottle to get a maximum number of organisms. Very few organisms will be found swimming around in mid-depth of the bottle.

• Use forceps to remove filamentous algae from specimen bottles. Avoid putting too much material on the slides.

• Explore the slide with the low-power objective. Reduce the lighting with the iris diaphragm. Keep the condenser high. When you find an organism of interest, swing the high-dry objective into position and adjust lighting. If your microscope has phase-contrast elements, use them.

• Refer to figures 4.1 through 4.6 to identify the various organisms that you encounter. Record your observations on the Laboratory Report.

Protozoa

The Subkingdom **Protozoa** includes all the animallike microorganisms of the Kingdom Protista. All of the representatives in this subkingdom are single-celled organisms; however, some of them do form colonial aggregates.

Externally, the cells are covered with a cell membrane, or pellicle; cell walls are absent and distinct nuclei with nuclear membranes are present. Specialized organelles such as contractile vacuoles, cytostomes, mitochondria, ribosomes, flagella, and cilia may also be present.

All protozoa reproduce asexually by cell division. Some exhibit various degrees of sexual reproduction. Their ability to form *cysts,* which are resistant dormant stages, enables them to survive drought, heat, and freezing.

The Subkingdom Protozoa is divided into three phyla: Sarcomastigophora, Ciliophora, and Sporozoa. Type of locomotion plays an important role in classification here.

Phylum Sarcomastigophora

Members within this phylum have been subdivided into two subphyla: Sarcodina and Mastigophora.

Sarcodina (*Amoebae*) Members of this subphylum move about by the formation of flowing protoplasmic projections called *pseudopodia.* The formation of pseudopodia is commonly referred to as *amoeboid movement.* Illustrations 5, 6, 7, and 8 in figure 4.1 are representative amoebae.

Mastigophora (*Zooflagellates*) These protozoans possess whiplike structures called *flagella.* There

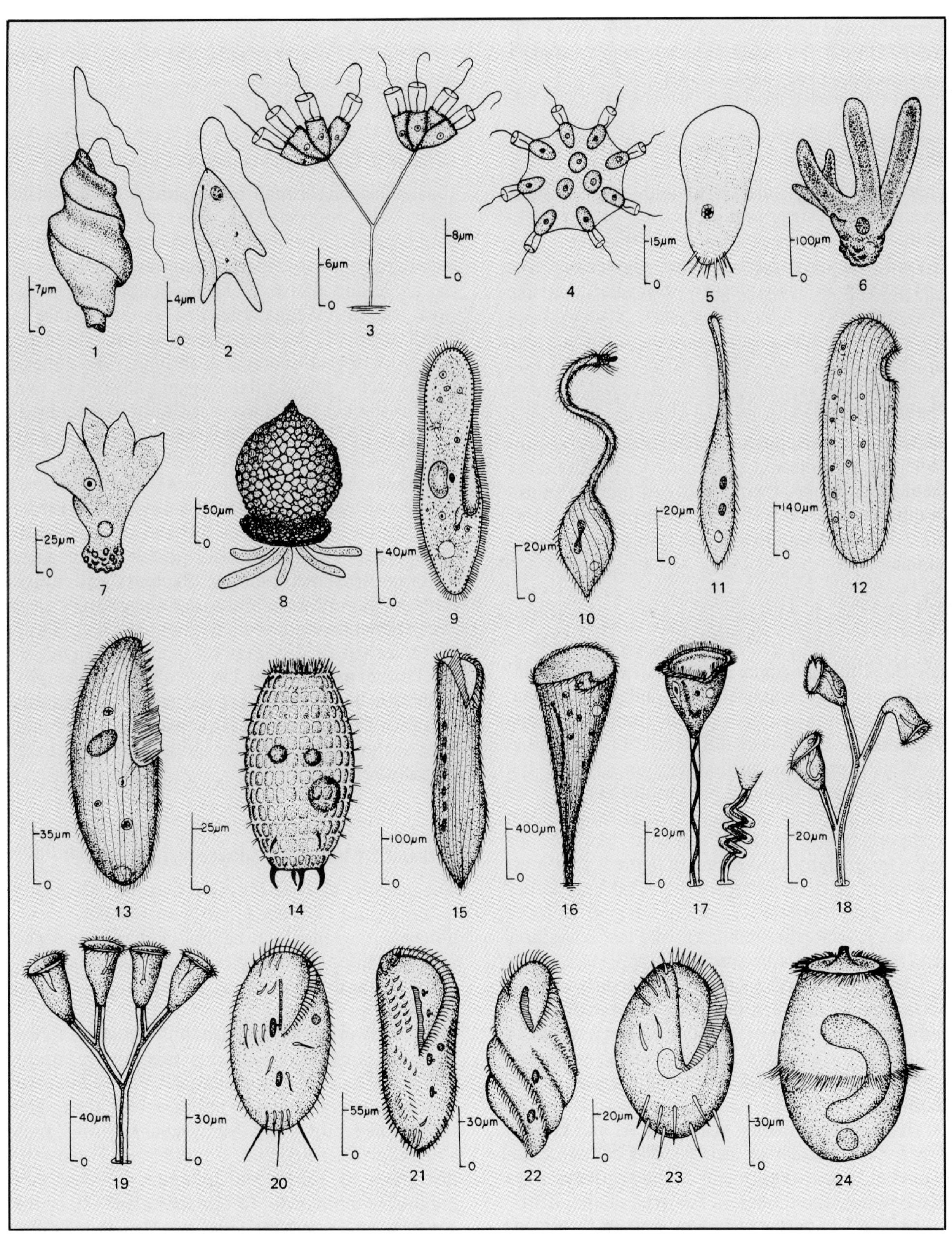

1. *Heteronema*
2. *Cercomonas*
3. *Codosiga*
4. *Protospongia*
5. *Trichamoeba*
6. *Amoeba*
7. *Mayorella*
8. *Diffugia*
9. *Paramecium*
10. *Lacrymaria*
11. *Lionotus*
12. *Loxodes*
13. *Blepharisma*
14. *Coleps*
15. *Condylostoma*
16. *Stentor*
17. *Vorticella*
18. *Carchesium*
19. *Zoothamnium*
20. *Stylonychia*
21. *Onychodromos*
22. *Hypotrichidium*
23. *Euplotes*
24. *Didinium*

Figure 4.1 Protozoans.

is considerable diversity among the members of this group. Only a few representatives (illustrations 1 through 4) are seen in figure 4.1.

Phylum Ciliophora

These microorganisms are undoubtedly the most advanced and structurally complex of all protozoans. Evidence seems to indicate that they have evolved from the zooflagellates. Movement and food-getting is accomplished with short hairlike structures called *cilia.* Illustrations 9 through 24 are typical ciliates.

Phylum Sporozoa

Members of this phylum lack locomotor organelles. All are internal parasites. As indicated by their group name, their life cycles include spore-forming stages. *Plasmodium,* the malarial parasite, is the most significant pathogenic sporozoan of humans.

Algae

The Subkingdom **Algae** includes all the photosynthetic eukaryotic organisms in Kindgom **Protista.** Being true protists, they differ from the plants (*Plantae*) in that tissue differentiation is lacking. In Whittaker's five-kingdom system some of the algae have been included with protozoans.

The algae may be unicellular, as those shown in the top row of figure 4.2; colonial, like the four in the lower right-hand corner of figure 4.2; or multicellular, as those in figure 4.3. The undifferentiated algal structure is often referred to as a *thallus.* It lacks the stem, root, and leaf structures that result from tissue specialization.

These microorganisms are universally present where ample moisture, favorable temperature, and sufficient sunlight exist. Although a great majority of them live submerged in water, some grow on soil; others grow on the bark of trees or the surfaces of rocks.

Algae have distinct, visible nuclei and chloroplasts. **Chloroplasts** are packets that contain chlorophyll a and other pigments. Photosynthesis takes place within these bodies. The size, shape, distribution, and numbers of chloroplasts vary considerably from species to species; in some instances a single chloroplast may occupy most of the cell space.

Although there are seven divisions of algae, only five will be listed here. Since two groups, the cryptomonads and red algae, are not usually encountered in freshwater ponds, they have not been included in this exercise.

Division 1 Euglenophycophyta (Euglenoids)

Illustrations 1 through 6 in figure 4.2 are typical euglenoids, representing four different genera within this relatively small group. All of them are flagellated and appear to be intermediate between the algae and protozoa. Protozoanlike characteristics seen in the euglenoids are (1) the absence of a cell wall, (2) the presence of a gullet, (3) the ability to ingest food (not through the gullet), (4) the ability to assimilate organic substances, and (5) the absence of chloroplasts in some species. In view of these facts, it becomes readily apparent why many zoologists often group the euglenoids with the zooflagellates.

The absence of a cell wall makes these protists very flexible in movement. Instead of a cell wall they possess a semirigid outer **pellicle,** which gives the organism a definite form. Photosynthetic types contain chlorophylls a and b, and they always have a red **stigma** (eyespot) which is light sensitive. Their characteristic food storage compound is a lipopolysaccharide, **paramylum.** The photosynthetic euglenoids can be bleached experimentally by various means in the laboratory. The colorless forms that develop, however, cannot be induced to revert back to phototrophy.

Division 2 Chlorophycophyta (Green Algae)

The majority of algae observed in ponds will belong to this group. They are grass-green in color, resembling the euglenoids in having chlorophylls a and b. They differ from euglenoids in that they synthesize **starch** instead of paramylum for food storage.

The diversity of this group is too great to explore its subdivisions in this preliminary study; however, the small, flagellated *Chlamydomonas* (illustration 8, figure 4.2) appears to be the archetype of the entire group. Many colonial forms, such as *Pandorina, Eudorina, Gonium,* and *Volvox* (illustrations 14, 15, 19, and 20, figure 4.2) consist of organisms similar to *Chlamydomonas.* It is the general consensus that from this flagellated form all the filamentous algae have evolved.

Except for *Vaucheria* and *Tribonema* in figure 4.3, all of these filamentous forms are Chlorophycophyta. All of the nonfilamentous, nonflagellated algae in figure 4.4 also are green algae.

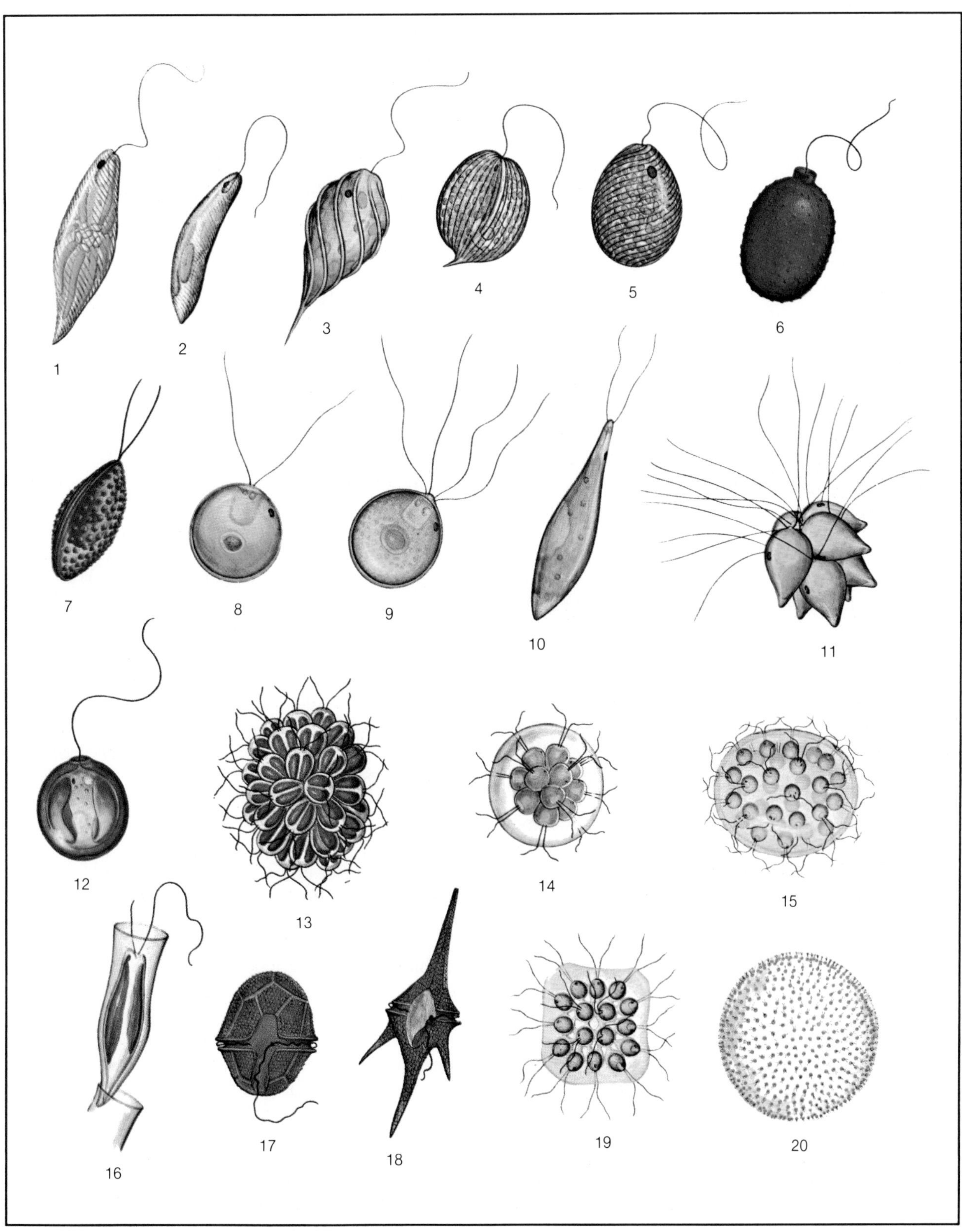

Courtesy of the U.S. Environmental Protection Agency, Office of Research & Development, Cincinnati, Ohio 45268.

1. *Euglena* (700X)
2. *Euglena* (700X)
3. *Phacus* (1000X)
4. *Phacus* (350X)
5. *Lepocinclis* (350X)
6. *Trachelomonas* (1000X)
7. *Phacotus* (1500X)
8. *Chlamydomonas* (1000X)
9. *Carteria* (1500X)
10. *Chlorogonium* (1000X)
11. *Pyrobotrys* (1000X)
12. *Chrysococcus* (3000X)
13. *Synura* (350X)
14. *Pandorina* (350X)
15. *Eudorina* (175X)
16. *Dinobyron* (1000X)
17. *Peridinium* (350X)
18. *Ceratium* (175X)
19. *Gonium* (350X)
20. *Volvox* (100X)

Figure 4.2 Flagellated algae.

A unique group of green algae are the **desmids** (illustrations 16 through 20, figure 4.4). With the exception of a few species, the cells of desmids consist of two similar halves, or semicells. The two halves usually are separated by a constriction, the *isthmus*.

Division 3 Chrysophycophyta
(Golden Brown Algae)

This large diversified division consists of over 6000 species. They differ from the euglenoids and green algae in that (1) food storage is in the form of **oils** and **leucosin,** a polysaccharide, (2) **chlorophylls a** and **c** are present, and (3) **fucoxanthin,** a brownish pigment, is present. It is the combination of fucoxanthin, other yellow pigments, and the chlorophylls that causes most of these algae to appear golden brown.

Representatives of this division are seen in figures 4.2, 4.3, and 4.5. In figure 4.2, *Chrysococcus, Synura,* and *Dinobyron* are typical flagellated chrysophycophytes. *Vaucheria* and *Tribonema* are the only filamentous chrysophycophytes shown in figure 4.3.

All of the organisms in figure 4.5 are chrysophycophytes and fall into a special category called the **diatoms.** The diatoms are unique in that they have hard cell walls of pectin, cellulose, or silicon that are constructed in two halves. The two halves fit together like lid and box. It is postulated by some that our petroleum reserves were formulated by the accumulation of oil from dead diatoms over millions of years.

Division 4 Phaeophycophyta (Brown Algae)

With the exception of three freshwater species, all algal protists of this division exist in salt water (*marine*); thus, it is unlikely that you will encounter any phaeophycophytes in this laboratory experience. These algae have essentially the same pigments seen in the chrysophycophytes, but they appear brown because of the masking effect of the greater amount of fucoxanthin. Food storage in the brown algae is in the form of **laminarin,** a polysaccharide, and **mannitol,** a sugar alcohol. All species of brown algae are multicellular and sessile. Most seaweeds are brown algae.

Division 5 Pyrrophycophyta (Fire Algae)

The principal members of this division are the **dinoflagellates.** Since the majority of these protists are marine, only two freshwater forms are shown in figure 4.2: *Peridinium* and *Ceratium* (illustrations 17 and 18). Most of these protists possess cellulose walls of interlocking armor plates, as in *Ceratium.* Two flagella are present: one is directed backward in swimming and the other moves within a transverse groove. Many marine dinoflagellates are bioluminescent. Some species of marine *Gymnodinium* produce the red tides when present in large numbers.

These algae have **chlorophylls a** and **c** and several xanthophylls. Foods are variously stored in the form of **starch, fats,** and **oils.**

CYANOBACTERIA

The blue-green bacteria, or *Cyanobacteria,* comprise Division I of the Kingdom Monera. Although these microorgamisms were formerly referred to as algae, their prokaryotic-type nucleus definitely sets them apart from the eukaryotic algae.

Although some bacteria are phototrophic, the difference between phototrophic bacteria and cyanobacteria is that the cyanobacteria have chlorophyll a and the phototrophic bacteria do not. Bacteriochlorophyll is the photosynthetic pigment in the phototrophic bacteria.

Over 1000 species of cyanobacteria have been reported. They are present in almost all moist environments from the tropics to the poles, including both freshwater and marine. Figure 4.6 illustrates only a random few that are most commonly seen.

The designation of these bacteria as "blue-green" is somewhat misleading in that many cyanobacteria are actually black, purple, red, and various shades of green instead of blue-green. These different colors are produced by the varying proportions of the numerous pigments present. These pigments are **chlorophyll a, carotene, xanthophylls,** blue **c-phycocyanin,** and red **c-phycoerythrin.** The last two pigments are unique to the cyanobacteria.

Cellular structure is considerably different from the eukaryotic algae. As stated earlier, nuclear membranes in cyanobacteria are absent. The nuclear material consists of DNA granules in a more or less colorless area in the center of the cell.

Unlike the algae, the pigments of the cyanobacteria are not contained in chloroplasts; instead, they are located in granules (**phycobilisomes**) that are attached to membranes (**thylakoids**) that permeate the cytoplasm.

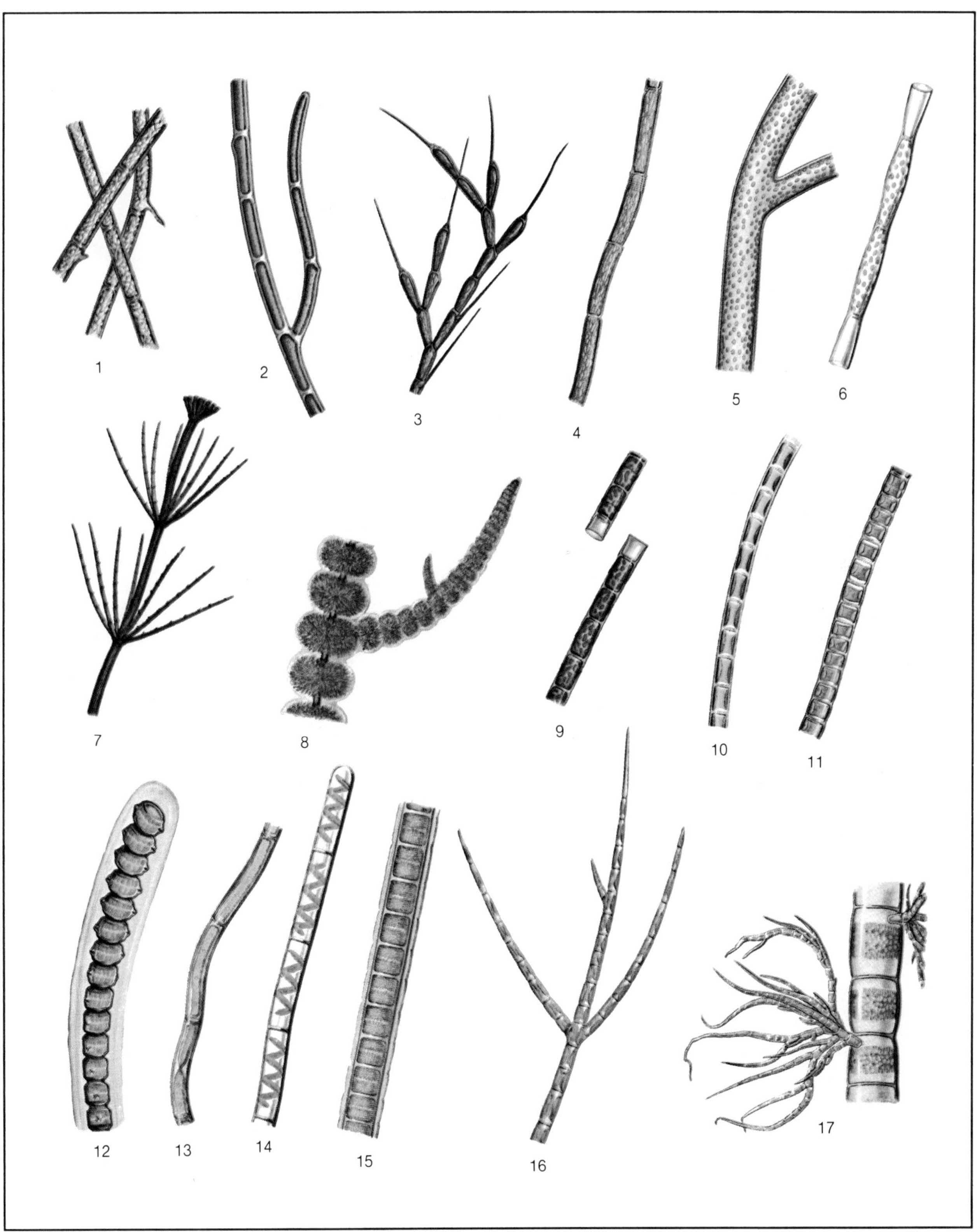

Courtesy of the U.S. Environmental Protection Agency, Office of Research & Development, Cincinnati, Ohio 45268.

1. *Rhizoclonium* (175X)
2. *Cladophora* (100X)
3. *Bulbochaete* (100X)
4. *Oedogonium* (350X)
5. *Vaucheria* (100X)
6. *Tribonema* (300X)
7. *Chara* (3 X)
8. *Batrachospermum* (2 X)
9. *Microspora* (175X)
10. *Ulothrix* (175X)
11. *Ulothrix* (175X)
12. *Desmidium* (175X)
13. *Mougeotia* (175X)
14. *Spirogyra* (175X)
15. *Zygnema* (175X)
16. *Stigeoclonium* (300X)
17. *Draparnaldia* (100X)

Figure 4.3 Filamentous algae.

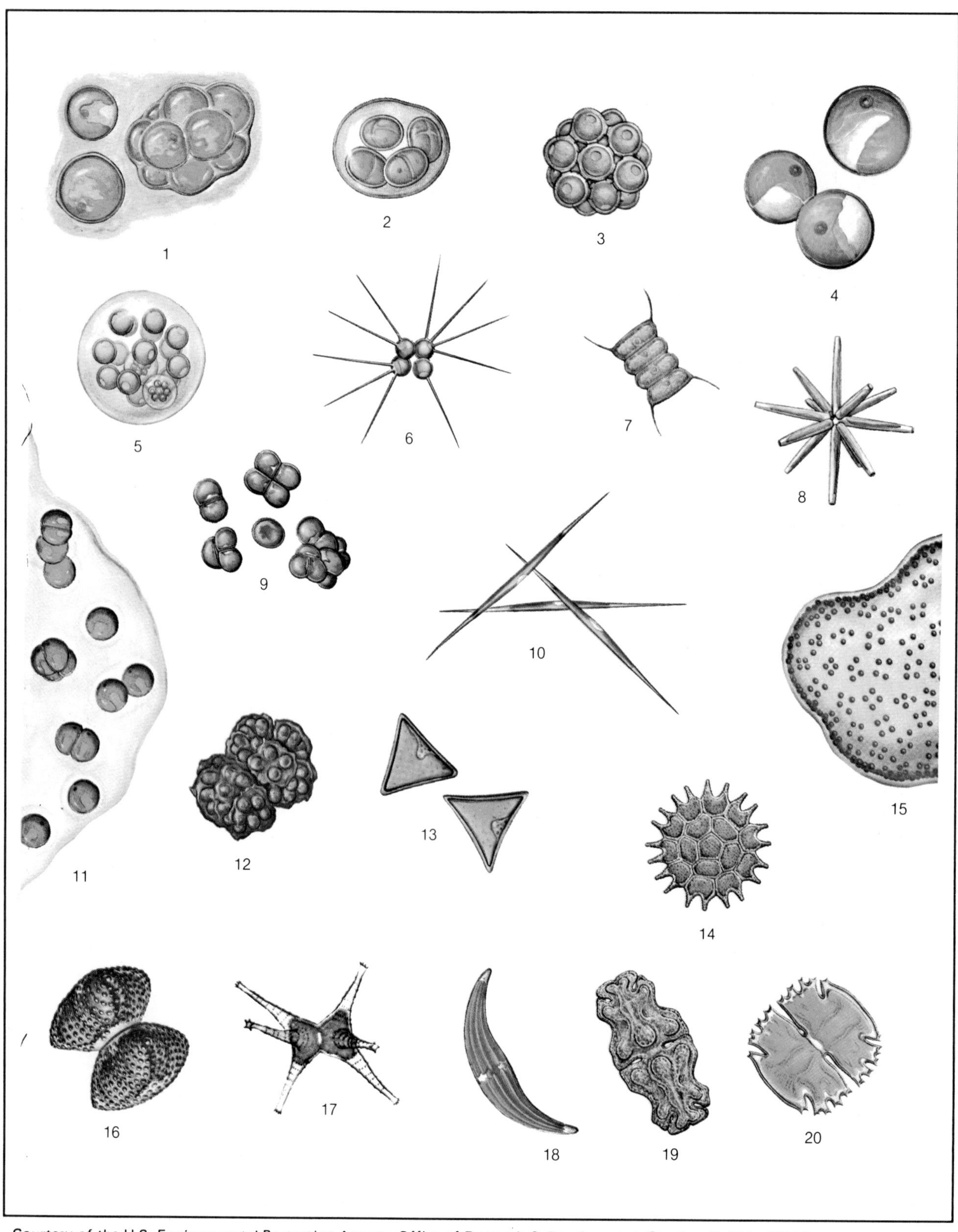

Courtesy of the U.S. Environmental Protection Agency, Office of Research & Development, Cincinnati, Ohio 45268.

1. *Chlorococcum* (700X)
2. *Oocystis* (700X)
3. *Coelastrum* (350X)
4. *Chlorella* (350X)
5. *Sphaerocystis* (350X)
6. *Micractinium* (700X)
7. *Scendesmus* (700X)
8. *Actinastrum* (700X)
9. *Phytoconis* (700X)
10. *Ankistrodesmus* (700X)
11. *Pamella* (700X)
12. *Botryococcus* (700X)
13. *Tetraedron* (1000X)
14. *Pediastrum* (100X)
15. *Tetraspora* (100X)
16. *Staurastrum* (700X)
17. *Staurastrum* (350X)
18. *Closterium* (175X)
19. *Euastrum* (350X)
20. *Micrasterias* (175X)

Figure 4.4 Nonfilamentous and nonflagellated algae.

1. *Diatoma* (1000X)
2. *Gomphonema* (175X)
3. *Cymbella* (175X)
4. *Cymbella* (1000X)
5. *Gomphonema* (2000X)
6. *Cocconeis* (750X)
7. *Nitschia* (1500X)
8. *Pinnularia* (175X)
9. *Cyclotella* (1000X)
10. *Tabellaria* (175X)
11. *Tabellaria* (1000X)
12. *Synedra* (350X)
13. *Synedra* (175X)
14. *Melosira* (750X)
15. *Surirella* (350X)
16. *Stauroneis* (350X)
17. *Fragillaria* (750X)
18. *Fragillaria* (750X)
19. *Asterionella* (175X)
20. *Asterionella* (750X)
21. *Navicula* (750X)
22. *Stephanodiscus* (750X)
23. *Meridion* (750X)

Figure 4.5 Diatoms.

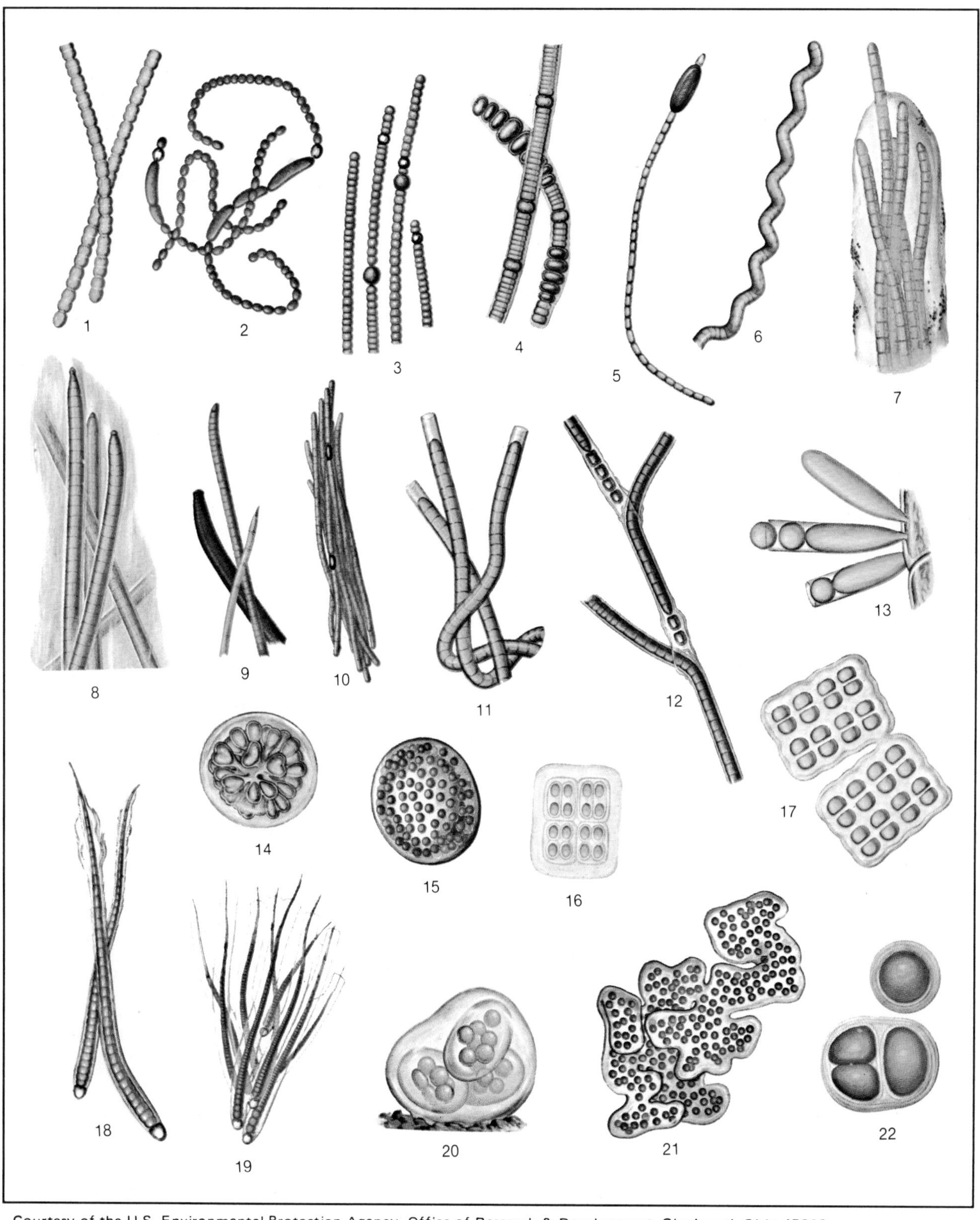

1. *Anabaena* (350X)
2. *Anabaena* (350X)
3. *Anabaena* (175X)
4. *Nodularia* (350X)
5. *Cylindrospermum* (175X)
6. *Arthrospira* (700X)
7. *Microcoleus* (350X)
8. *Phormidium* (350X)
9. *Oscillatoria* (175X)
10. *Aphanizomenon* (175X)
11. *Lyngbya* (700X)
12. *Tolypothrix* (350X)
13. *Entophysalis* (1000X)
14. *Gomphosphaeria* (1000X)
15. *Gomphosphaeria* (350X)
16. *Agmenellum* (700X)
17. *Agmenellum* (175X)
18. *Calothrix* (350X)
19. *Rivularia* (175X)
20. *Anacystis* (700X)
21. *Anacystis* (175X)
22. *Anacystis* (700X)

Figure 4.6 Cyanobacteria.

Microscopic Invertebrates 5

While looking for protozoa, algae, and cyanobacteria in pond water, one invariably encounters large, transparent, complex microorganisms that, to the inexperienced, appear to be protozoans. In most instances these "monsters" are rotifers (illustrations 13 through 17, figure 5.1); in some cases they are copepods, daphnia, or any one of the other forms illustrated in figure 5.1.

All of the organisms illustrated in figure 5.1 are multicellular with organ systems. If organ systems are present, then the organisms cannot be protists, since organs indicate the presence of tissue differentiation. Collectively, these microscopic forms are designated as "invertebrates." It is to prevent you from misinterpreting some of these invertebrates as protozoans that these organisms are described here.

In using figure 5.1 to identify what you consider might be an invertebrate, keep in mind that there are considerable size differences. A few invertebrates, such as *Dugesia* and *Hydra,* are macroscopic in adult form, but microscopic when immature. Be sure to judge size differences by reading the scale beside each organism. The following phyla are listed according to degree of complexity, the simplest first.

Phylum Coelenterata

(Illust. 1)

Members of this phylum are almost exclusively marine. The only common freshwater form shown in figure 5.1 is *Hydra.* In addition, there are a few less-common freshwater genera similar to the marine hydroids.

The hydras are quite common in ponds and lakes. They are usually attached to rocks, twigs, or other substrata. Around the mouth at the free end are five tentacles of variable length, depending on the species. Smaller organisms, such as *Daphnia,* are grasped by the tentacles and conveyed to the mouth. These animals have a digestive cavity that makes up the bulk of the interior. Since no anus is present, undigested remains of food are expelled through the mouth.

Phylum Platyhelminthes

(Illust. 2, 3, 4, 5)

The invertebrates of this phylum are commonly referred to as **flatworms.** The phylum contains two parasitic classes and one class of free-living organisms, the *Turbellaria.* It is the organisms of this class that are encountered in fresh water. The four genera of this class shown in figure 5.1 are *Dugesia, Planaria, Macrostomum,* and *Provortex.* The characteristics common to all these organisms are dorsoventral flatness, ciliated epidermis, ventral mouth, and eyespots on the dorsal surface near the anterior end. As in the coelenterates, undigested food must be ejected through the mouth since no anus is present. Reproduction may be asexual by fission or fragmentation; generally, however, reproduction is sexual, each organism having both male and female reproductive organs. Species identification of the turbellarians is exceedingly difficult and is based to a great extent on the details of the reproductive system.

Phylum Nematoda

(Illust. 6)

The members of this phylum are the **roundworms.** They are commonly referred to as *nemas* or *nematodes.* They are characteristically round in cross section, have an external cuticle without cilia, lack eyes, and have a tubular digestive system complete with mouth, intestine, and anus. The males are generally much smaller than the females and have a hooked posterior end. The number of named species is only a fraction of the total nematodes in existence. Species identification of these invertebrates requires very detailed study of many minute anatomical features, which requires complete knowledge of anatomy.

Phylum Aschelminthes

This phylum includes the classes *Gastrotricha* an*ᶜ* *Rotifera.* Most of the members of this phylum *ᵃ* microscopic. Their proximity to the nematod*ᶜ* classification is due to the type of body cavity (*pseudocoel*) which is present in both phyla.

The **gastrotrichs** (illust. 7, 8, 9, 10) range from 10 to 540 μm in size. They are very similar to the ciliated protozoans in size and habits. The typical gastrotrich is elongate, flexible, forked at the posterior end, and covered with bristles. The digestive system consists of an anterior mouth surrounded by bristles, a pharynx, intestine, and posterior anus. Species identification is based partially on the shape of the head, tail structure and size, and distribution of spines. Overall length is also an important identification characteristic. They feed primarily on unicellular algae.

The **rotifers** (illust. 13, 14, 15, 16, 17) are most easily differentiated by the wheellike arrangement of cilia at the anterior end and the presence of a chewing pharynx within the body. They are considerably diversified in food habits: some feed on algae and protozoa, others on juices of plant cells, and some are parasitic. They play an important role in keeping waters clean. They also serve as food for small worms and crustaceans, being an important link in the food chain of fresh waters.

Phylum Annelida
(Illust. 18)

This phylum includes three classes: *Oligochaeta, Polychaeta,* and *Hirudinea.* Since polychaetes are primarily marine and the leeches (Hirudinea) are mostly macroscopic and parasitic, only the oligochaete is represented in figure 5.1. Some oligochaetes are marine, but the majority are found in fresh water and soil. These worms are characterized by body segmentation, bristles (*setae*) on each segment, an anterior mouth, and a roundish protrusion—the *prostomium*—anterior to the mouth. Although most oligochaetes breathe through the skin, some aquatic forms possess gills at the posterior end or along the sides of the segments. Most oligochaetes feed on vegetation; some feed on the muck of the bottoms of polluted waters, aiding in purifying such places.

Phylum Tardigrada
(Illust. 11, 12)

These invertebrates are of uncertain taxonomic position. They appear to be closely related to both the Annelida and Arthropoda. They are commonly referred to as the **water bears.** They are generally no more than 1 mm long, with a head, four trunk segments, and four pairs of legs. The ends of the legs may have claws, fingers, or disklike structures. The anterior end has a retractable snout with teeth. Eyes are often present. Sexes are separate and females are oviparous. The animals are primarily herbivorous. Locomotion is by crawling, not swimming. During desiccation of their habitat they contract to form barrel-shaped *tuns,* and they are able to survive years of dryness, even in extremes of heat and cold. Widespread distribution is due to dispersal of the tuns by the wind.

Phylum Arthropoda
(Illust, 19, 20, 21, 22, 23)

This phylum contains most of the known Animalia, almost a million species. Representatives of three groups of the class *Crustacea* are shown in figure 5.1: Cladocera, Ostracoda, and Copepoda. The characteristics these three have in common are jointed appendages, an exoskeleton, and gills.

The **cladocera** are represented by *Daphnia* and *Latonopsis* in figure 5.1. They are commonly known as **water fleas.** All cladocera have a distinct head. The body is covered by a bivalvelike carapace. There is often a distinct cervical notch between the head and body. A compound eye is usually present which is movable. They have many appendages: antennules, antennae, mouth parts, and four to six pairs of legs.

The **ostracods** are bivalved crustaceans that are distinguished from minute clams by the absence of lines of growth on the shell. Their bodies are not distinctly segmented. They have seven pairs of appendages. The end of the body terminates with a pair of *caudal furca.*

The **copepods** represented here are *Cyclops* and *Canthocamptus.* They lack the shell-like covering of the ostracods and cladocera; instead they exhibit distinct body segmentation. They may have three simple eyes or a single median eye. Eggs are often seen attached to the abdomen of females.

Laboratory Report
There is no Laboratory Report for this exercise.

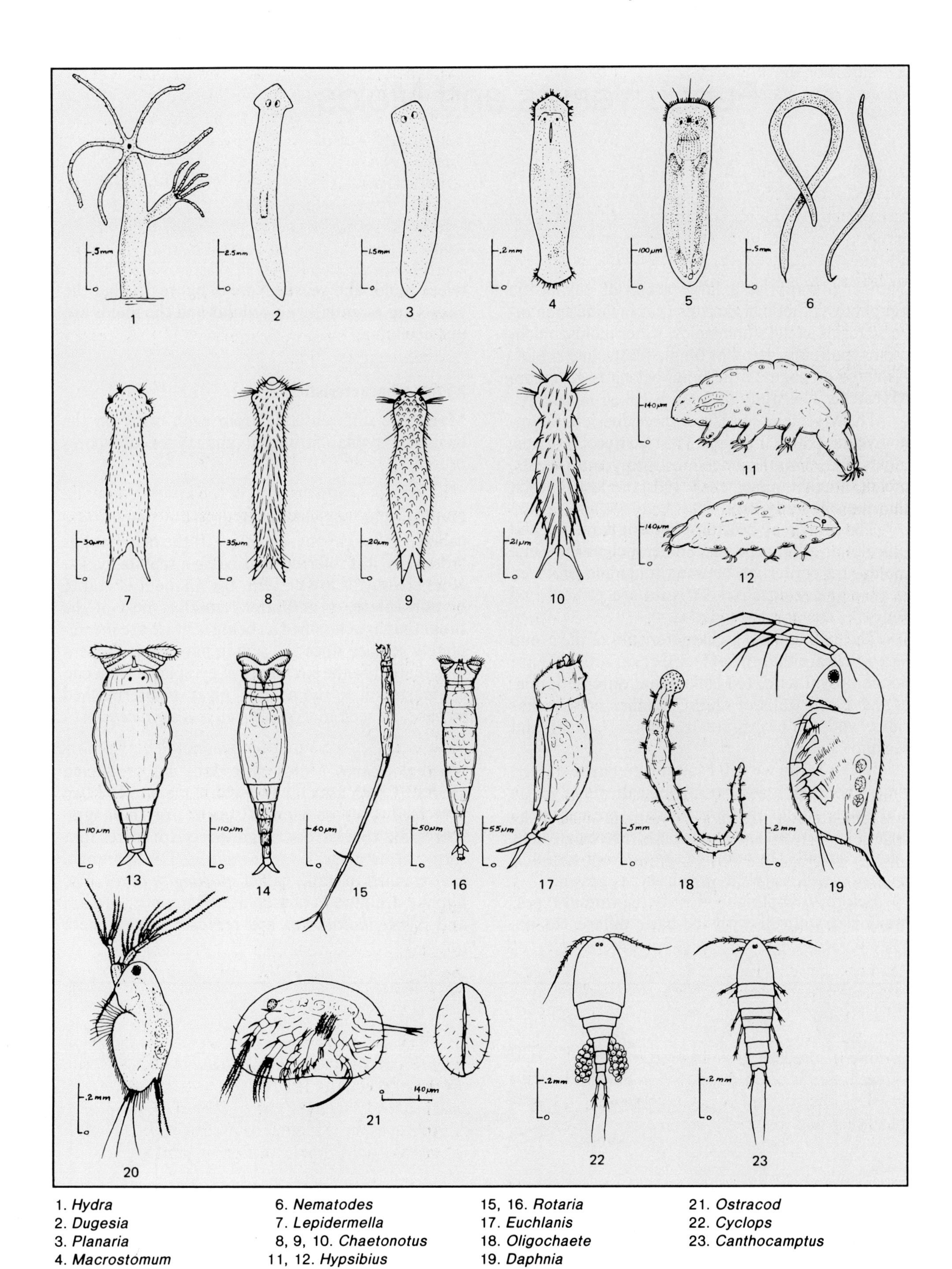

1. *Hydra*
2. *Dugesia*
3. *Planaria*
4. *Macrostomum*
5. *Provortex*
6. *Nematodes*
7. *Lepidermella*
8, 9, 10. *Chaetonotus*
11, 12. *Hypsibius*
13, 14. *Philodina*
15, 16. *Rotaria*
17. *Euchlanis*
18. *Oligochaete*
19. *Daphnia*
20. *Latonopis*
21. *Ostracod*
22. *Cyclops*
23. *Canthocamptus*

Figure 5.1 Microscopic invertebrates.

6 Fungi: Yeasts and Molds

The fungi comprise a large group of eukaryotic nonphotosynthetic organisms that include such diverse forms as the slime molds, water molds, mushrooms, puffballs, bracket fungi, yeasts, and molds. As noted on page 25 the fungi belong to Kingdom **Myceteae.** The study of fungi is called **mycology.**

The Myceteae consists of three divisions: Gymnomycota (slime molds), Mastigomycota (water molds and others), and Amastigomycota (yeasts, molds, bracket fungi, etc.). It is the last division that we will study here.

The fungi may be saprophytic or parasitic, and unicellular or filamentous. Some, such as the slime molds, are borderline between fungi and protozoa in that amoeboid characteristics are present and fungilike spores are produced.

The distinguishing characteristics of the group as a whole are that they (1) are eukaryotic, (2) are nonphotosynthetic, (3) lack tissue differentiation, (4) have cell walls of chitin or other polysaccharides, and (5) propagate by spores (sexual and asexual).

In this study we will examine prepared stained slides and slides made from living cultures of yeasts and molds. Molds that are normally present in the air will be cultured and studied macroscopically and microscopically. In addition, an attempt also will be made to identify the various types grown.

Before attempting to identify the various types, familarize yourself with the basic differences between molds and yeasts. Note in figure 6.1 that the yeasts are essentially unicellular and the molds are multicellular.

Molds Characteristics

Molds are differentiated from each other on the basis of hyphal structure and types of spores present.

Hyphae The individual filaments of molds are called *hyphae* (hypha, singular). If the filament has crosswalls, it is referred to as being a **septate hypha.** If no crosswalls are present, the filament is said to be **nonseptate,** or **aseptate.** Actually, most of the fungi that are classified as being septate are incompletely septate since the septae have central openings that allow the streaming of cytoplasm from one compartment to the next. A mass of intermeshed hyphae, as seen macroscopically, is a *mycelium.*

Asexual Spores Molds reproduce by producing spores by both asexual and sexual means. The two principal types of asexual spores are sporangiospores and conidia. **Sporangiospores** are spores that form within a sac, or *sporangium.* The sporangia are attached to stalks called *sporangiophores.* The hyphae are always nonseptate. *Rhizopus, Mucor,* and *Syncephalastrum* are representative genera

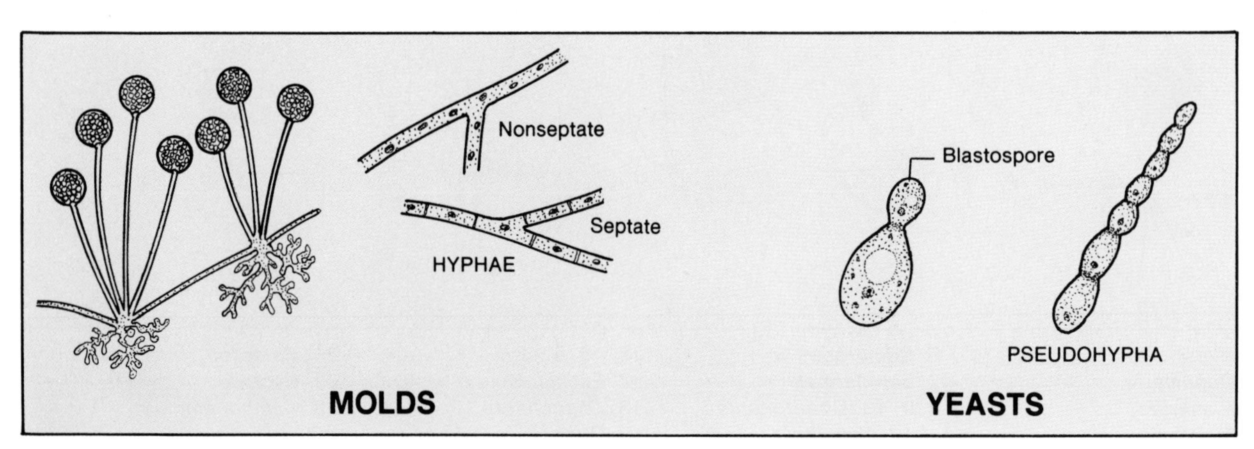

Figure 6.1 Comparison of mold and yeast structure.

that possess these spores (see illustration 1, figure 6.2).

Conidia are asexual spores that form on specialized hyphae called *conidiophores*. Small conidia are called *microconidia;* large multicellular conidia are known as *macroconidia.*. The following types of conidia are shown in figure 6.2:

- **Phialospores:** Conidia of this type are produced by vase-shaped cells called *phialides.* Note in figure 6.2 that *Penicillium* and *Gliocadium* produce this type of conidia.
- **Blastoconidia:** Conidia of this type are produced by budding from cells of preexisting conidia, as in *Cladosporium.*
- **Arthrospores:** This type of conidia forms by separation from preexisting hyphal cells. Example: *Oospora.*
- **Chlamydospores:** These spores are large, thick-walled, round or irregular structures formed within or terminally on a hypha. Common to most fungi, they generally form on old cultures. Example: *Candida albicans.*

Sexual Spores Three kinds of sexual spores are seen in the Amastigomycota: **zygospores, ascospores,** and **basidiospores.** All three are shown in figure 6.3. Classification of the Amastigomycota is based on these spores. Description of each type of sexual spore follows in the description of each of the subdivisions.

Yeast Characteristics

The yeasts reproduce by a process called budding instead of by fission. A **bud,** or **blastospore,** forms as an outpouching of the parent cell. It is easily detected, visually, by its small size as compared to the parent cell. It may separate from the original cell or remain attached. In some yeasts, buds may remain attached to each other to form a *pseudohypha.*

Yeasts and molds cannot be separated by classification. Many species of fungi have both yeast and mold phases. Such fungi are said to be *biphasic.*

Amastigomycota

The Amastigomycota consists of four subdivisions that are separated on the basis of the type of sexual reproductive spores that are present.

Subdivision Zygomycotina

These fungi have nonseptate hyphae and produce zygospores. **Zygospores** are formed by the union of

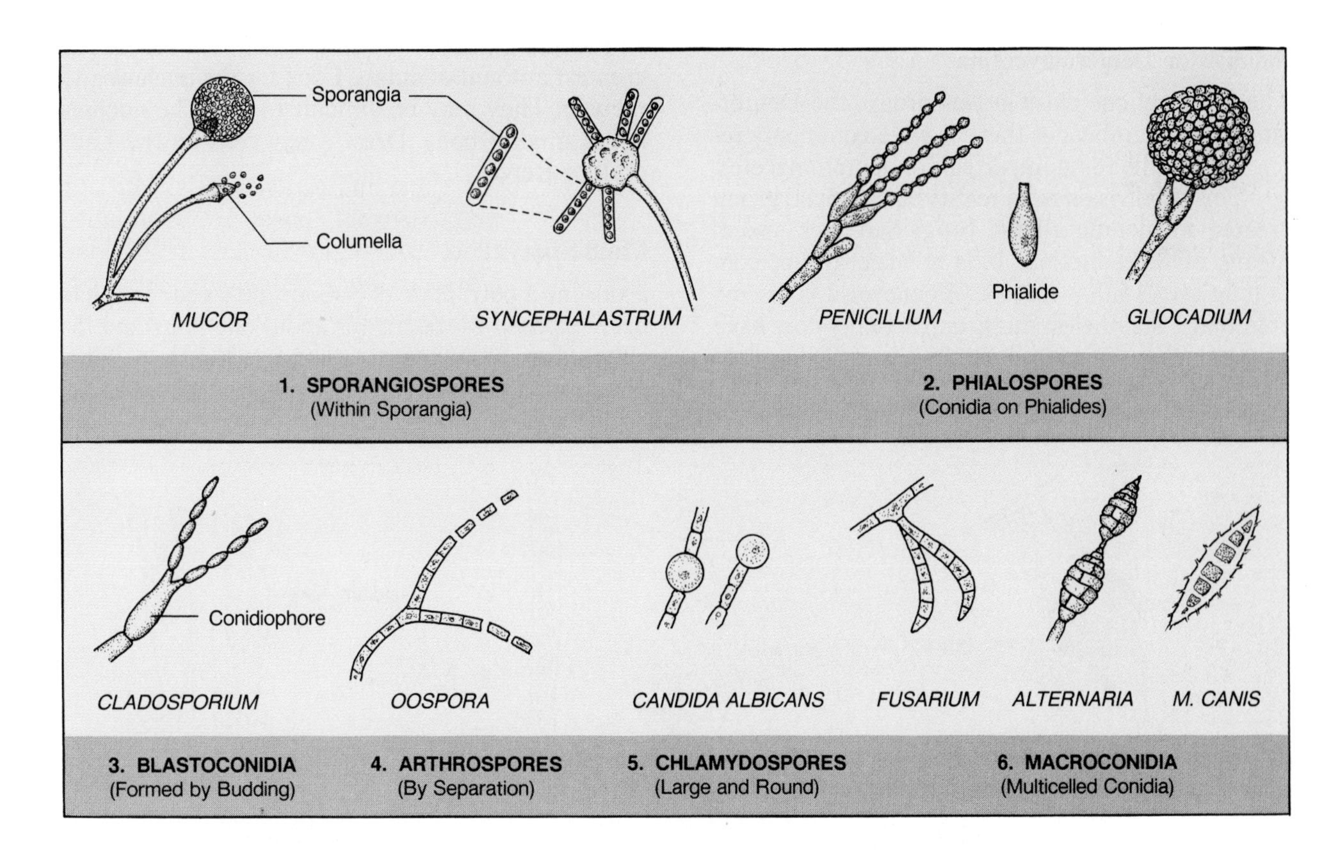

Figure 6.2 Types of asexual spores seen in fungi.

nuclear material from the hyphae of two different strains. These fungi also produce sporangiospores.

Subdivision Ascomycotina

All fungi in this subdivision are grouped in one class, the Ascomycetes; consequently, they are commonly referred to as the ascomycetes. They are also called *sac fungi.* All of them have septate hyphae; most have chitinous walls.

The characteristic sexual **ascospores** of this class are produced in oval sacs called *asci* (ascus, singular). Those fungi that produce a single ascus are the *ascomycetes yeasts.* Other ascomycetes that produce numerous asci in complex fruiting bodies include such organisms as *Penicillium* and powdery mildews.

Subdivision Basidiomycotina

All fungi in this subdivision belong to one class, the Basidiomycetes. Puffballs, mushrooms, smuts, rust, and shelf fungi on dead tree branches are also basidiomycetes. The sexual spores of this class are **basidiospores,** which are produced on club-shaped bodies called *basidia.* A basidium is considered by some to be a modified type of ascus.

Subdivision Deuteromycotina

There is only one class in this group, the Deuteromycetes. Members of this class are commonly referred to as the fungi imperfecti or deuteromycetes.

This subdivision is, in reality, an artificial group created to include all the fungi that lack sexual means of reproduction. It is a large group, containing over 15,000 species. Whenever it is discovered that a member in this group actually does have a sexual means of reproduction, it is moved up into one of the other subdivisions.

Laboratory Procedures

Several options are provided here for the study of molds and yeasts. The procedures to be followed will be outlined by your instructor.

Yeast Study

The organism *Saccharomyces cerevisiae,* which is used in bread making and alcohol fermentation, will be used for this study. Either prepared slides or living organisms may be used.

Materials:

prepared slides of *Saccharomyces cerevisiae*
broth cultures of *Saccharomyces cerevisiae*
microscope slides and cover glasses

Prepared Slides If prepared slides are used, they may be examined under high-dry or oil immersion. One should look for typical **blastospores** and **ascospores.** Space is provided on the Laboratory Report for drawing the organisms.

Living Material If broth cultures of *Saccharomyces cerevisiae* are available they should be examined on a wet mount slide under a phase-contrast or brightfield microscope. Two or three loopfuls of the organisms should be placed on the slide with an inoculating loop. Oil immersion will reveal the greatest amount of detail. Look for the **nucleus** and **vacuole.** They may be difficult to see. The nucleus is the smaller body. Draw a few cells on the Laboratory Report.

Mold Study

Examine a petri plate of Sabouraud's agar that has been exposed to the air for about one hour and incubated at room temperature for 3–5 days. This medium has a low pH, which makes it selective for

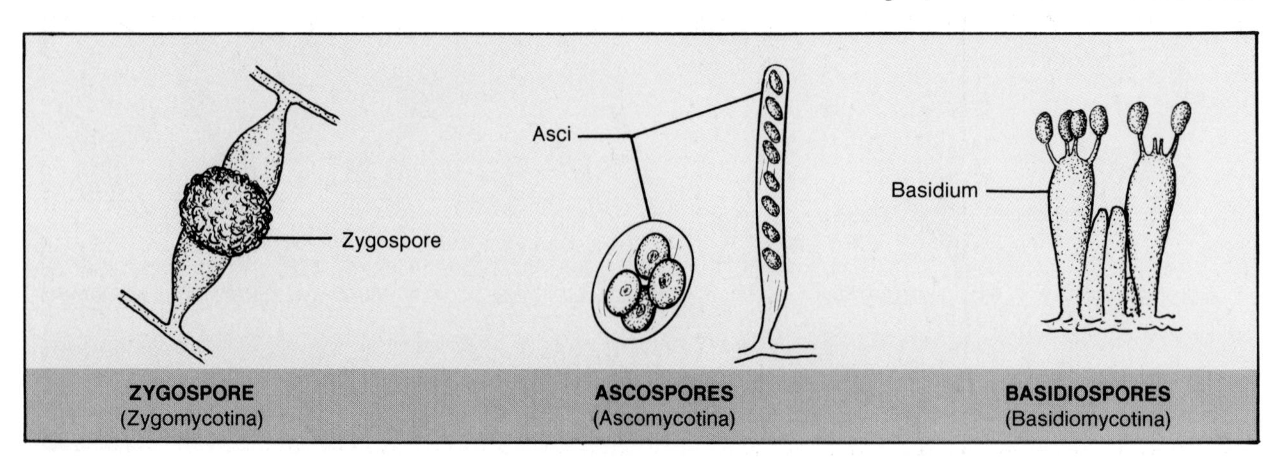

Figure 6.3 Types of sexual spores seen in the Amastigomycota.

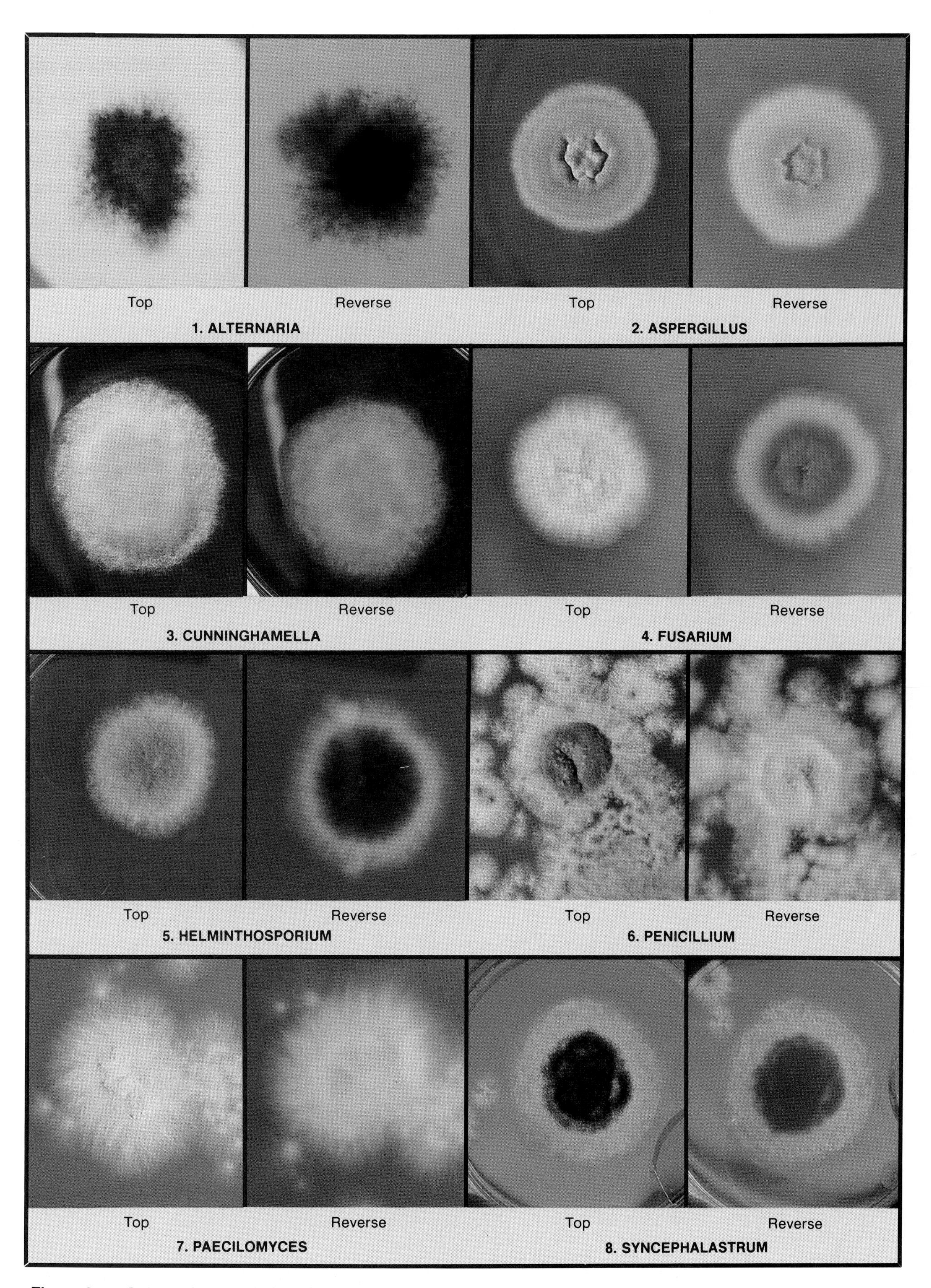

Figure 6.4 Colony characteristics of some of the more common molds.

molds. A good plate will have many different colored colonies. Note the characteristic "cottony" nature of the colonies. Also, look at the bottom of the plate and observe how the colonies differ in color here. The identification of molds is based on surface color, backside color, hyphal structure, and types of spores. Figure 6.4 reveals how some of the more common molds appear when grown on Sabouraud's agar. Keep in mind in using figure 6.4 that the appearance of a mold colony can change appreciably as it gets older. The photographs in figure 6.4 are of colonies that are 10 to 21 days old.

Conclusive identification cannot be made unless a microscope slide is made to determine the type of hyphae and spores that are present. Figure 6.5 reveals, diagrammatically, the microscopic differences that one looks for in identifying various genera.

Two Options In making slides from mold colonies one can make wet mounts directly from the colonies by the procedure outlined here, or make cultured slides as outlined in Exercise 20. The following steps should be used for making stained slides directly from the colonies. Your instructor will indicate the number of identifications that are to be made.

Materials:

mold cultures on Sabouraud's agar
microscopic slides and cover glasses
lactophenol cotton blue stain
sharp pointed scalpels or dissecting needles

1. Place the uncovered plate on the stage of your microscope and examine the edge of a colored colony with the low-power objective. Look for hyphal structure and spore arrangement. Ignore the white colonies since they generally lack spores and are difficult to identify.
2. Consult figures 6.4 and 6.5 to make a preliminary identification based on colony characteristics and low-power magnification of hyphae and spores.
3. Make a wet mount slide by transferring a small amount of the culture with a sharp scalpel or dissecting needle to a drop of lactophenol cotton blue stain on a slide. Cover with a cover glass and examine under low-power and high-dry objectives. Refer again to figure 6.5 to confirm any conclusions drawn from your previous examination of the edge of the colony.
4. Repeat the above procedure for each different colony.

Laboratory Report

After recording your results on the Laboratory Report, answer all the questions.

Figure 6.5 Legend

1. ***Penicillium***—bluish-green; "brush" arrangement of phialospores.
2. ***Aspergillus***—bluish-green with sulfur yellow areas on the surface.
3. ***Verticillium***—pinkish-brown, elliptical microconidia.
4. ***Trichoderma***—green, resemble *Penicillium* macroscopically.
5. ***Gliocladium***—dark green; conidia (phialospores) borne on phialides, similar to *Penicillium;* grows faster than *Penicillium.*
6. ***Cladosporium*** (*Hormodendrum*)—light green to grayish surface; gray to black back surface; blastoconidia.
7. ***Pleospora***—tan to green surface with brown to black back; ascospores shown.
8. ***Scopulariopsis***—light brown; rough walled microconidia.
9. ***Paecilomyces***—yellowish brown; elliptical microconidia.
10. ***Alternaria***—black surface with gray periphery; black on reverse side; chains of macroconidia.
11. ***Helminthosporium***—black surface with grayish periphery; macroconidia shown.
12. ***Pullularia***—black, shiny, leathery surface; thick-walled; budding spores.
13. ***Diplosporium***—buff-colored wooly surface, reverse side has red center surrounded by brown.
14. ***Oospora*** (*Geotrichum*)—buff-colored surface; hyphae break up into thin-walled rectangular arthrospores.
15. ***Fusarium***—variants of yellow, orange, red, and purple colonies; sickle-shaped macroconidia.
16. ***Trichothecium***—white to pink surface; two-celled conidia.
17. ***Mucor***—white to dark gray mycelium; nonseptate; sporangia with sporangiospores.
18. ***Rhizopus***—white to dark gray; nonseptate; rootlike rhizoids; sporangiospores.
19. ***Syncephalastrum***—white to dark gray surface; nonseptate; sporangiospores.
20. ***Nigrospora***—white to gray surface; reverse side is black.
21. ***Montospora***—dark gray center with light gray periphery; yellow-brown conidia.

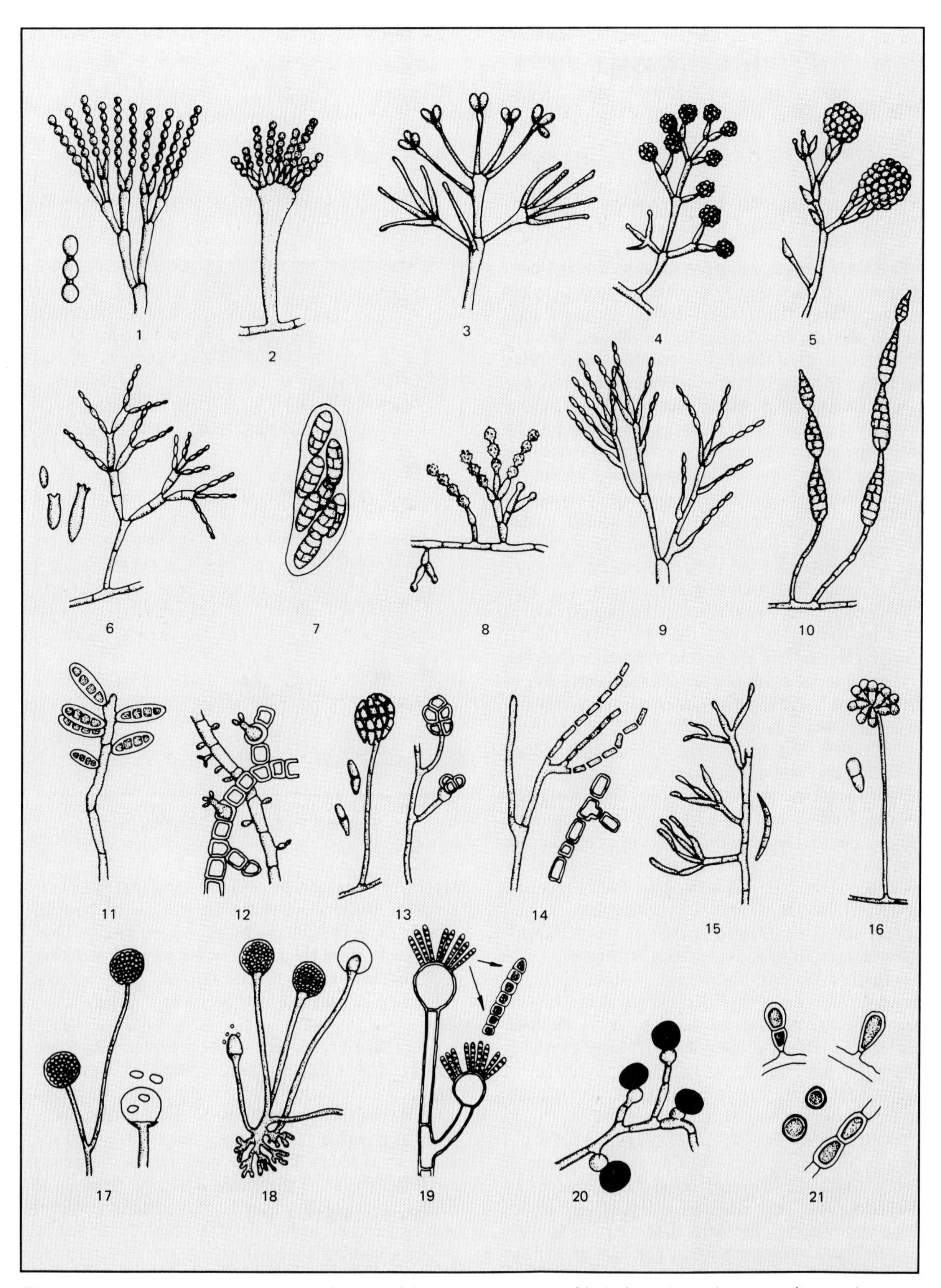

Figure 6.5 Microscopic appearance of some of the more common molds (refer to legend on opposite page).

7 Bacteria

Of all the microorganisms studied so far, the bacteria are the most widely distributed, the simplest in morphology, the smallest in size, the most difficult to classify, and the hardest organisms to identify. It is even difficult to provide a descriptive definition of what a bacterial organism is because of the considerable diversity in the group. About the only generalization that can be made for the entire group is that they are prokaryotic and are seldom photosynthetic. The few that are photosynthetic utilize a pigment which is chemically different from chlorophyll a. It is called bacteriochlorophyll. Probably the simplest definition that one can construct from these facts is *bacteria are prokaryons without chlorophyll a.*

Since they are prokaryons, the bacteria share the Kingdom Monera with the Cyanobacteria. Although the bacteria are generally smaller than the Cyanobacteria, some of the Cyanobacteria are in the bacteria's size range. Most bacteria are only 0.5 to 2.0 micrometers in diameter.

Figure 7.1 illustrates most of the shapes of bacteria that one would encounter. Note that they can be grouped into three types: rod, spherical, and curved. Rod-shaped bacteria may vary considerably in length; have square, round, or pointed ends; and may be motile or nonmotile. The spherical, or coccus-shaped, bacteria may occur singly, in pairs, in tetrads, in chains, and in irregular masses. The helical and curved bacteria exist as slender spirochaetes, spirillum, and bent rods (vibrios).

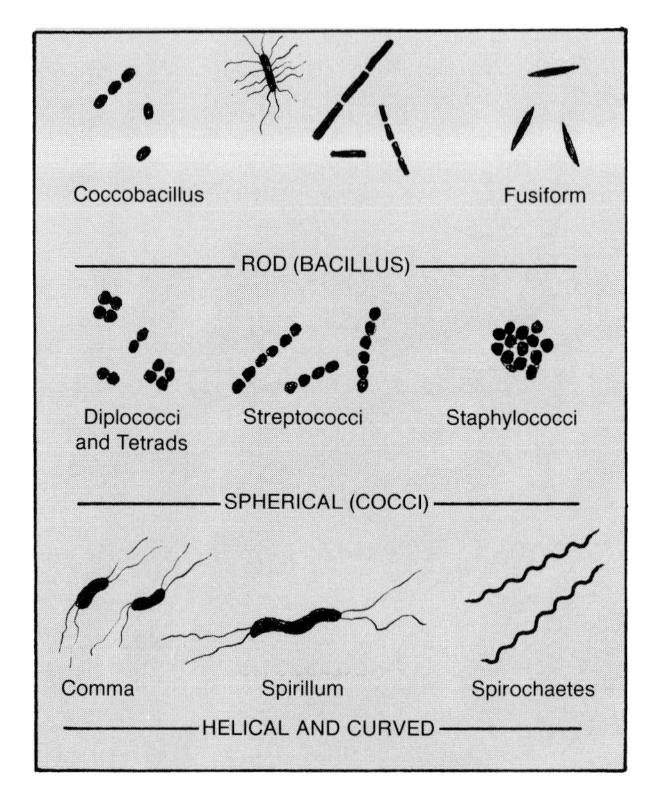

Figure 7.1 Bacterial morphology.

In this exercise an attempt will be made to demonstrate the ubiquitousness of these organisms. No attempt will be made to study detailed bacterial anatomy or physiology. Many exercises related to staining, microscopy, and physiology in subsequent laboratory periods will provide a clear understanding of these microorganisms.

Our concern here relates primarily to the widespread distribution of bacteria in our environment. Being thoroughly aware of their existence all around us is of prime importance if we are to develop those laboratory skills that we refer to, collectively, as *aseptic technique.* The awareness that bacteria are everywhere must be constantly in our minds when handling bacterial cultures. In the next laboratory period you will be handling tube cultures of bacteria, and unless you learn how to handle them in such a way as to keep foreign bacteria away from them, you will always be working with contaminated cultures. *Without pure cultures the study of bacteriology becomes a hopeless endeavor.*

In this exercise you will be provided with three kinds of sterile bacteriological media which will be exposed to the environment in various ways. To ensure that these exposures cover as wide a spectrum as possible, specific assignments will be made for each student. In some instances a moistened swab will be used to remove bacteria from some object; in other instances a petri plate of medium will be exposed to the air or a cough. You will be issued a number that will enable you to determine your specific assignment from the chart on the next page.

Materials:

Per student:
1 tube of nutrient broth
1 petri plate of trypticase soy agar (TSA)
1 sterile cotton swab
china marking pencil
Per two or more students:
1 petri plate of blood agar

1. Expose your TSA plate according to your assignment in the table below. *Label the bottom* of your plate with your initials, your assignment number, and the date.
2. Moisten a sterile swab by immersing it into a tube of nutrient broth and expressing most of the broth out of it by pressing the swab against the inside wall of the tube.
3. Rub the moistened swab over a part of your body such as a finger or ear, or some object such as a doorknob, telephone mouthpiece, etc., and return the swab to the tube of broth. It may be necessary to break off the stick end of the swab so that you can replace the cap on the tube.
4. Label the tube with your initials and the source of the bacteria.
5. Expose the blood agar plate by coughing onto it. Label the bottom of the plate with the initials of the individuals that cough onto it. Be sure to date the plate.
6. Incubate the plates and tube at 37° C for 48 hours.

Evaluation

After 48 hours incubation, examine the tube of nutrient broth and two plates. Shake the tube vigorously without wetting the cap. Is it cloudy or clear? Compare it with an uncontaminated tube of broth. What is the significance of a cloudy tube of broth? Do you see any colonies growing on the blood agar plate? Are the colonies all the same size and color? If not, what does this indicate? Group together a set of TSA plates representing all nine types of exposures. Record your results on the Laboratory Report.

Your instructor will indicate whether these tubes and plates are to be used for making slides in Exercise 10—"Simple Staining." If the plates and tubes are to be saved, containers will be provided for their storage in the refrigerator. Place the plates and tubes in the designated containers.

Laboratory Report

Complete the Laboratory Report for this exercise.

Exposure Method For TSA Plate	Student Number
1. To the air in laboratory for 30 minutes	1, 10, 19, 28
2. To the air in room other than laboratory for 30 minutes	2, 11, 20, 29
3. To the air outside of building for 30 minutes	3, 12, 21, 30
4. Blow dust onto exposed medium	4, 13, 22, 31
5. Moist lips pressed against medium	5, 14, 23, 32
6. Fingertips pressed lightly on medium	6, 15, 24, 33
7. Several coins pressed temporarily on medium	7, 16, 25, 34
8. Hair is combed over exposed medium (10 strokes)	8, 17, 26, 35
9. *Optional:* Any method not listed above	9, 18, 27, 36

Part 3 Microscopic Slide Techniques (Bacterial Morphology)

The eight exercises in this unit include the procedures for most of the slide techniques that one might employ in morphological studies of bacteria. A culture method in Exercise 15 also is included as a substitute for slide techniques when pathogens are encountered.

These exercises are intended to serve two equally important functions: (1) to help you to develop the necessary skills in making slides and (2) to introduce you to the morphology of bacteria. Although the title of each exercise pertains to a specific technique, the organisms chosen for each method have been carefully selected so that you can learn to recognize certain morphological features. For example, in the exercise on simple staining (Exercise 10) the organisms selected exhibit metachromatic granules, pleomorphism, and palisade arrangement of cells. In Exercise 12, "Gram Staining," you will observe the differences between cocci and bacilli, as well as learn how to execute the staining routine.

The importance of the mastery of these techniques cannot be overemphasized. Although one is seldom able to make species identification on the basis of morphological characteristics alone, it is a very significant starting point. This fact will become increasingly clear with subsequent experiments.

Although the steps in the various staining procedures may seem relatively simple, student success is often quite unpredictable. Unless your instructor suggests a variation in the procedure, try to follow the procedures exactly as stated, without improvisation. Photomicrographs in color have been provided for many of the techniques; use them as a guide to evaluate the slides you have prepared. Once you have mastered a specific technique feel free to experiment.

8 Negative Staining

The simplest way to make a slide of bacteria is to prepare a wet mount, much in the same manner that was used for studying protozoa and algae. Although this method will quickly produce a slide, finding the microorganisms under the microscope may be another matter, especially for a beginning microbiologist. The problem one encounters is that they are generally colorless and quite transparent. Unless the microscope diaphragm is carefully adjusted, the beginner usually has difficulty bringing these small organisms into focus.

A better way to observe bacteria for the first time is to prepare a slide by a process called *negative,* or *background, staining.* This method consists of mixing the microorganisms in a small amount of india ink or nigrosine and spreading the mixture over a clean slide. These two pigments are not really bacterial stains because they do not penetrate the microorganisms; instead, they obliterate the background, leaving the organisms transparent and visible in a darkened field. This technique can be useful for determining cell morphology and size. Since no heat is applied to the slide, there is less shrinkage of the cells and, consequently, more accurate cell size determinations result than with some other methods. This method is also useful for certain bacteria, such as spirochaetes, that are difficult to stain.

The source of organisms in this exercise will be the normal flora that exist between the teeth in the mouth. Figure 8.1 illustrates the procedure that will be followed. Success of this method depends on the following considerations: (1) the slide must be absolutely clean: the presence of any residual grease or dirt on the slide will produce an uneven smear; (2) the amount of india ink or nigrosine is critical; most students tend to use too much; and finally, (3) *the mixture must be dragged over the slide, not pushed.* If you keep these three facts in mind you should be able to produce a perfect slide the first time. Proceed as follows:

Materials:

microscope slides (with polished edges)
nigrosine solution or india ink
sterile toothpicks
dissecting needle
Bunsen burner
china marking pencil

1. Scrub two microscope slides clean with Bon Ami and add a very small drop (not over 1/8″ diameter) of nigrosine or india ink near the right end of one of the slides. See illustration 1, figure 8.1.
2. Remove a small amount of material from between your teeth with a toothpick and mix it into the drop on the slide. Complete emulsification of the organisms is important at this stage. If necessary, flame and cool a dissecting needle and use it to help break up the clumps of bacteria.
3. Place the edge of a spreader slide to the left of the suspension as shown in illustration 3, and move the spreader slide toward the suspension until contact is made as in illustration 4. Note that the suspension drop spreads along the back edge of the spreader slide.
4. Push the spreader slide away from the suspension as shown in illustration 5. Observe that the smear will be thick where it begins and feather out into a very thin smear at the end of the spreading stroke. Somewhere between the thickest end and the thinnest end will be an ideal thickness of suspension.
5. Allow the slide to dry at room temperature. Do not apply heat to the slide.
6. Examine under oil immersion. Select that part of the smear that shows the cells distinctly with a dark background.

Laboratory Report

Draw a few representative types of organisms on Laboratory Report 8–11. Look for yeasts and hyphae as well as bacteria. Spirochaetes may also be present.

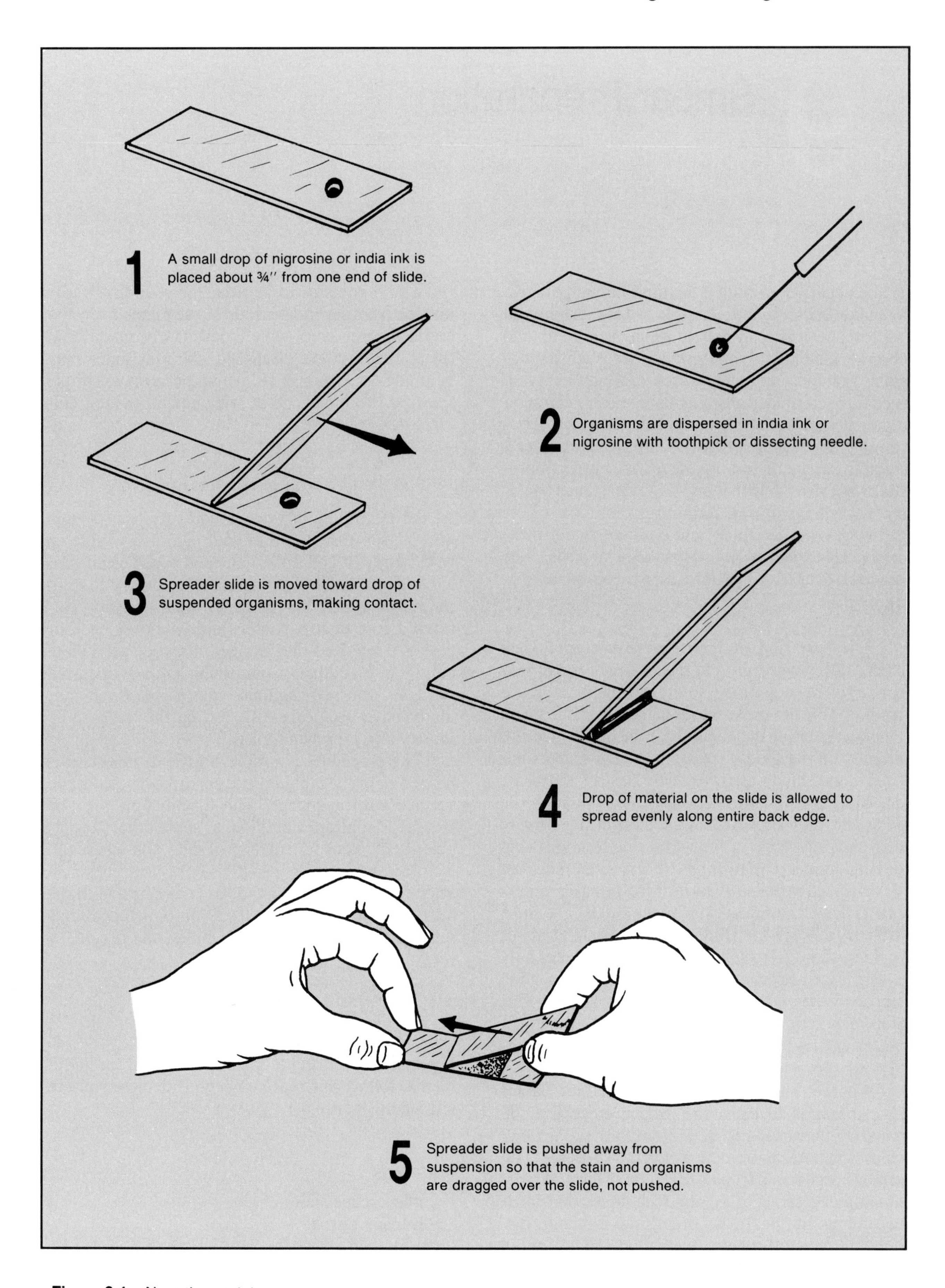

Figure 8.1 Negative staining procedure.

9 Smear Preparation

While negative staining is a simple enough process to make bacteria more visible with a brightfield microscope, it is of little help when one attempts to observe anatomical microstructures such as flagella, granules, and endospores. Only by applying specific bacteriological stains to organisms can such organelles be seen. However, success at bacterial staining depends first of all on the preparation of a suitable *smear* of the organisms. A properly prepared bacterial smear is one that withstands one or more washings during staining without loss of organisms; is not too thick; and does not result in excessive distortion due to shrinkage of cells. The procedure for making such a smear is illustrated in figure 9.1.

The first step in preparing a bacteriological smear differs according to the source of the organisms. If the organisms are growing in a liquid medium (broths, milk, saliva, urine, etc.), one starts by placing one or two loopfuls of the liquid medium directly on the slide. From solid media such as nutrient agar, blood agar, or some part of the body, one places one or two loopfuls of water first on the slide and then uses a straight inoculating wire to disperse organisms in the water. Bacteria growing on solid media tend to cling to each other and must be dispersed sufficiently by dilution in water; unless this is done, the smear will be too thick. The most difficult concept for students to understand about making slides from solid media is that it takes only a very small amount of material to make a good smear. When your instructor demonstrates this step, pay very careful attention to the amount of material that is placed on the slide.

Another hurdle to overcome in making bacterial smears is to learn the proper procedure for handling cultures. Figure 9.2 outlines the various steps that must become routine whenever you remove organisms from a test tube. This is the beginning of a series of aseptic techniques that must become as automatic as breathing. Tabletop disinfection must also be performed at the beginning of the period and at the close of the laboratory session.

The organisms to be used for your first slides may be from several sources. If the plates from Exercise 7 were saved, some slides may be made from them. If they were discarded, the first slides may be made for Exercise 10, which pertains to simple staining. Your instructor will indicate which cultures are to be used.

Tabletop Preparation

At the beginning of every laboratory period from now on, it is essential that your laboratory tabletop be cleared of all paraphernalia and wiped clean with a disinfectant. This procedure is necessary to minimize the possibility of contaminating your cultures. Loose dust particles (and bacteria) on your desk are removed that can get into your petri plate cultures. In addition, sanitizing your work area protects yourself against microorganisms that might have been carelessly left on the tabletop by someone in a previous class.

The procedure is simple: a little disinfectant is poured onto the tabletop and spread over the entire surface with a sponge. The disinfectant may be Roccal, Zephiran, Betadine, or some other acceptable agent. Your instructor will show you where the disinfectant and sponges are kept. As a consideration for those students that use your station in the next class it is expected that you will repeat this scrub-down procedure at the end of the period.

Procedure from Liquid Media

(from broths, saliva, milk, etc.)

If you are preparing a bacterial smear from liquid media, follow this routine which is depicted on the left side of figure 9.1.

Materials:

- microscope slides
- Bunsen burner
- wire loop
- china marking pencil
- slide holder (clothespin), optional

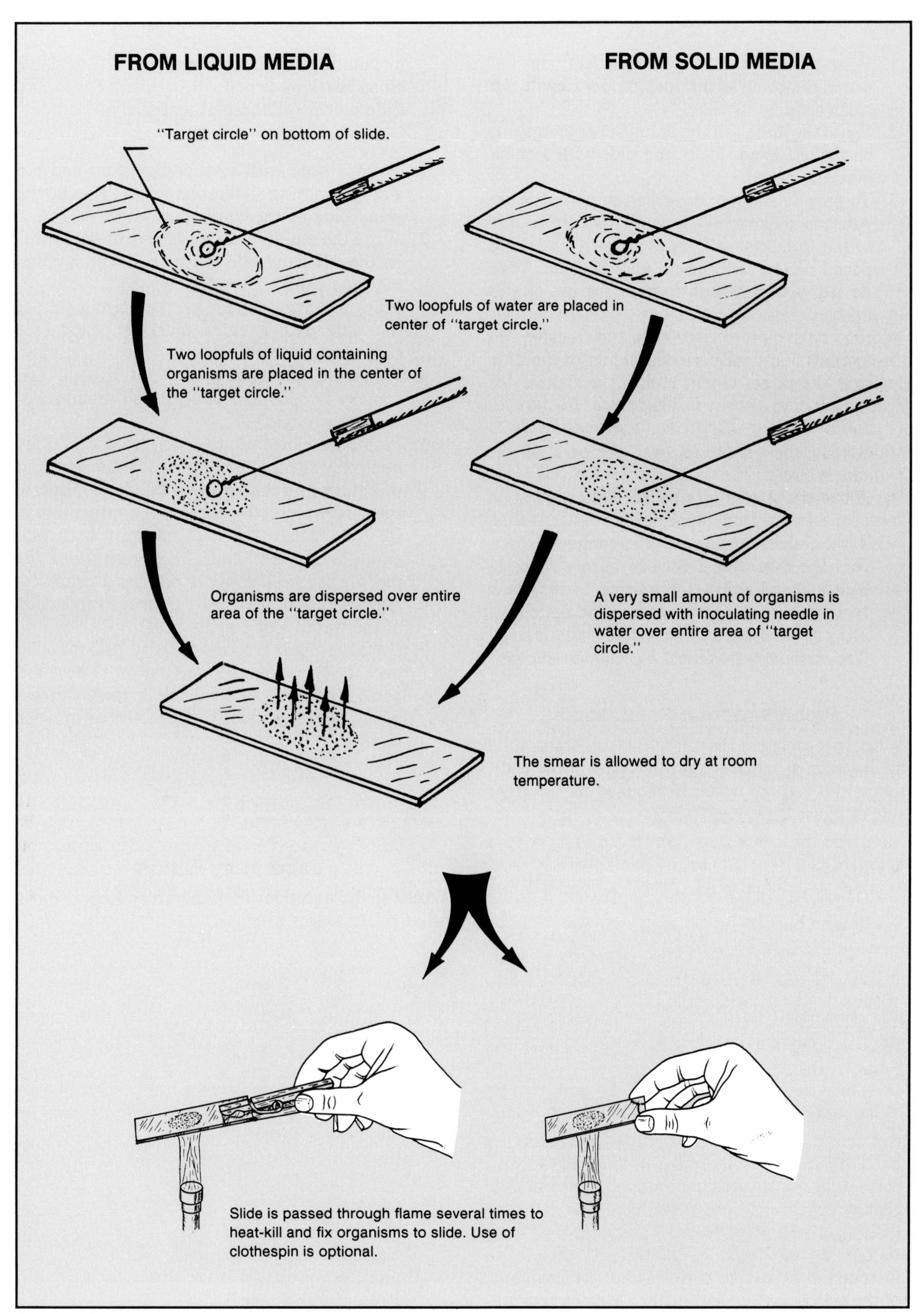

Figure 9.1 Procedure for making a bacterial smear.

1. Wash a slide with soap or Bon Ami and hot water, removing all dirt and grease. Handle the clean slide by its edges.
2. Write the initials of the organism or organisms on the left-hand side of the slide with a china marking pencil.
3. To provide a target on which to place the organisms, make a ½″ circle *on the bottom side of the slide,* centrally located, with a marking pencil. Later on, when you become more skilled, you may wish to omit the use of this "target circle."
4. Shake the culture vigorously and transfer two loopfuls of organisms to the center of the slide over the target circle. Follow the routine for inoculations shown in figure 9.2. *Be sure to flame the loop after it has touched the slide.*
5. Spread the organisms over the area of the target circle.
6. Allow the slide to dry by normal evaporation of the water. Don't apply heat.
7. After the smear has become completely dry, pass the slide over a Bunsen burner flame to heat-kill and fix the organisms to the slide. Note that in this step one has the options of using or not using a clothespin to hold the slide. *Use the option preferred by your instructor.*

Procedure from Solid Media

When preparing a bacterial smear from solid media, such as nutrient agar or a part of the body, follow this routine which is depicted on the right side of figure 9.1.

Materials:

microscope slides
Bunsen burner
inoculating needle and loop
china marking pencil
slide holder (clothespin), optional

1. Wash a slide with soap or Bon Ami and hot water, removing all dirt and grease. Handle the clean slide by the edges.
2. Write the initials of the organism or organisms on the left-hand side of the slide with a china marking pencil.
3. Mark a "target circle" on the bottom side of the slide with a china marking pencil (see comments in step 3 above).
4. Flame an inoculating loop, let it cool, and transfer two loopfuls of water to the slide over the target circle.
5. Flame an inoculating needle, let it cool, pick up a very small amount of the organisms, and mix them into the water on the slide. Disperse the mixture over the area of the target circle. Be certain that the organisms have been well emulsified in the liquid. *Be sure to flame the inoculating needle before placing it aside.*
6. Allow the slide to dry by normal evaporation of the water. Don't apply heat.
7. Once the smear is completely dry, pass the slide over the flame of a Bunsen burner to heat-kill and fix the organisms to the slide. Use a clothespin to hold the slide if it is preferred by your instructor.

Laboratory Report

Answer the questions on Laboratory Report 8–11 that relate to this exercise.

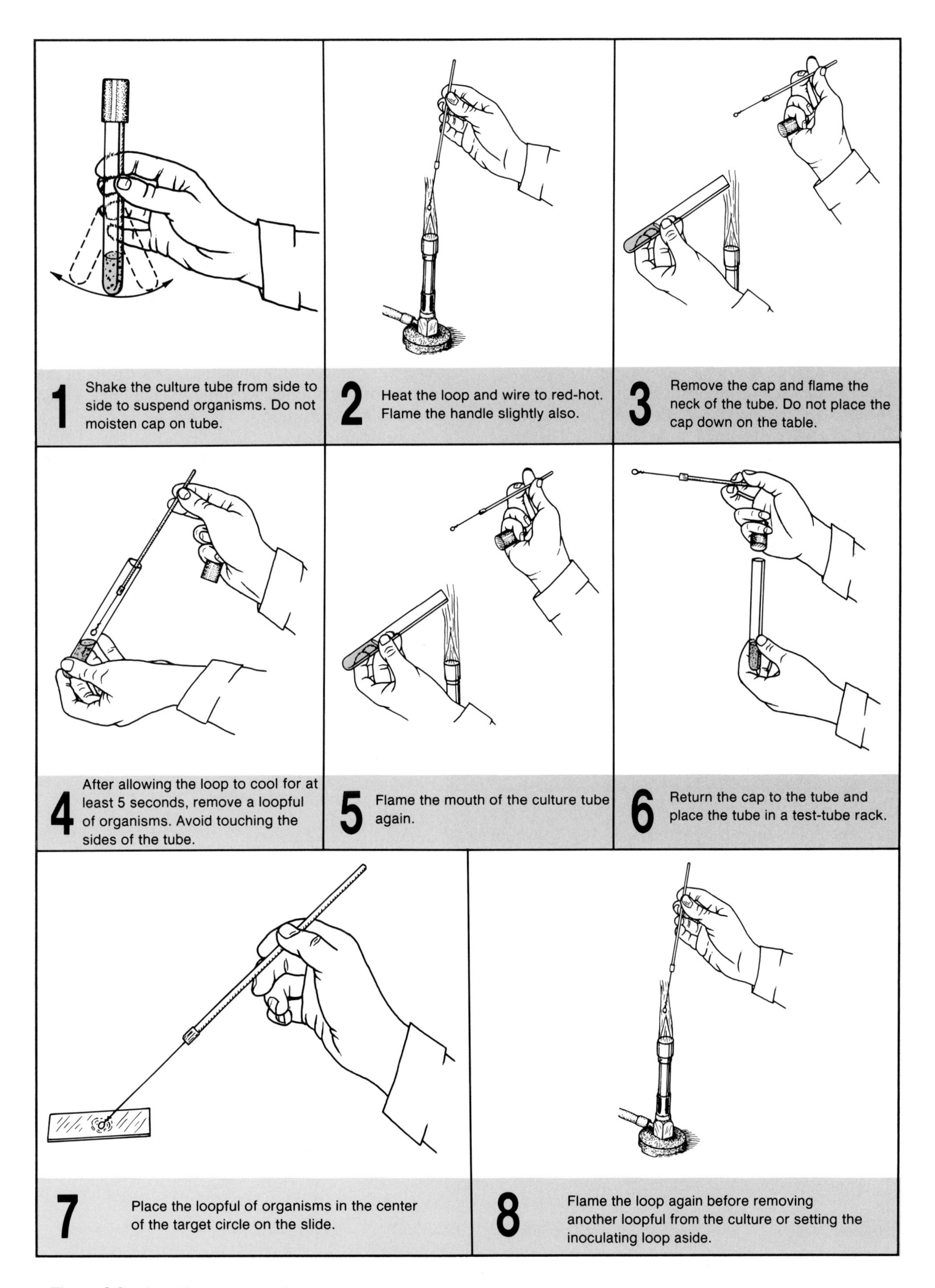

Figure 9.2 Aseptic procedure for organism removal.

10 Simple Staining

The use of a single stain to color a bacterial organism is commonly referred to as **simple staining.** Some of the most commonly used dyes for simple staining are methylene blue, basic fuchsin, and crystal violet. All of these dyes work well on bacteria because they have color-bearing ions (*chromophores*) that are positively charged (cationic). The fact that bacteria are slightly negatively charged produces a pronounced attraction between these cationic chromophores and the organism. Such dyes are classified as **basic dyes.** The basic dye methylene blue (methylene$^+$ chloride$^-$) will be used in this exercise. Those dyes that have anionic chromophores are called **acidic dyes.** Eosin (sodium$^+$ eosinate$^-$) is such a dye. The anionic chromophore, eosinate$^-$, will not stain bacteria because of the electrostatic repelling forces that are involved.

The staining times for most simple stains are relatively short, usually from 30 seconds to 2 minutes, depending on the dye affinity. After a smear has been stained for the required time, it is washed off gently, blotted dry, and examined directly under oil immersion. Such a slide is useful in determining basic morphology and the presence or absence of certain kinds of granules.

An avirulent strain of *Corynebacterium diphtheriae* will be used here for simple staining. In the pathogenic form, this organism is the cause of diphtheria, a very serious disease. One of the steps in identifying this pathogen is to do a simple stain of it to demonstrate the following unique characteristics: pleomorphism, metachromatic granules, and palisade arrangement of cells.

Pleomorphism pertains to *irregularity* of form: i.e., demonstrating several different shapes. While *C. diphtheriae* is basically rod-shaped, it also appears club-shaped, spermlike, or needle-shaped. *Bergey's Manual* uses the terms "pleomorphic" and "irregular" interchangeably.

Metachromatic granules are distinct reddish-purple granules within cells that show up when the organisms are stained with methylene blue. These granules are considered to be masses of volutin, a polymetaphosphate.

Palisade arrangement pertains to parallel arrangement of rod-shaped cells. This characteristic, also called "picket fence" arrangement, is common to many corynebacteria.

Procedure

Prepare a slide of *C. diphtheriae,* using the procedure outlined in figure 10.1. It will be necessary to refer back to Exercise 9 for the smear preparation procedure.

Materials:

- slant culture of avirulent strain of *C. diphtheriae*
- methylene blue (Loeffler's)
- wash bottle
- bibulous paper

After examining the slide, compare it with the photomicrograph in illustration 1, figure 12.3 (page 54). Record your observations on Laboratory Report 8–11.

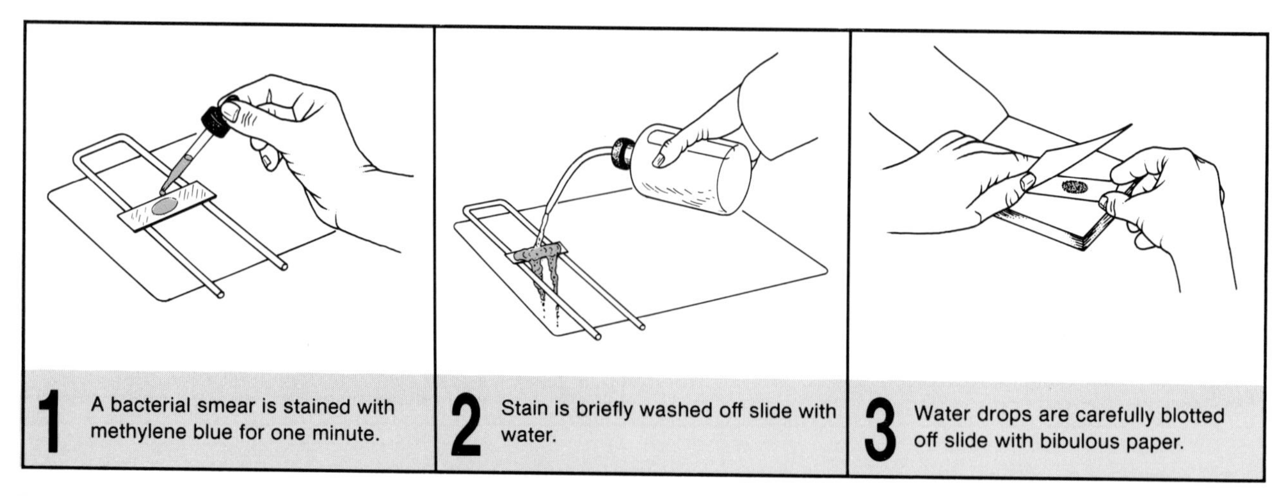

Figure 10.1 Procedure for simple staining.

Capsular Staining 11

Some bacterial cells are surrounded by a pronounced gelatinous or slimy layer called a **capsule.** There is considerable evidence to support the view that all bacteria have some amount of slime material surrounding the cells. In most instances, however, the layer is not of sufficient magnitude to be readily discernible. Although some capsules appear to be made of glycoprotein, others contain polypeptides. All appear to be water-soluble.

Staining the bacterial capsule cannot be accomplished by ordinary simple staining procedures. The problem with trying to stain capsules is that if you prepare a heat-fixed smear of the organism by ordinary methods, you will destroy the capsule; and if you do not heat-fix the slide, the organism will slide off the slide during washing. In most of our bacteriological studies our principal concern is simply to demonstrate the presence or absence of a capsule. This can be easily achieved by combining negative and simple staining techniques, as in figure 11.1. To learn about this technique prepare a slide of *Klebsiella pneumoniae,* using the procedure outlined in figure 11.1.

Materials:

36–48 hour milk culture of *Klebsiella pneumoniae*
india ink
crystal violet

Observation: Examine the slide under oil immersion and compare your slide with illustration 2, figure 12.3 (page 54). Record your results on Laboratory Report 8–11.

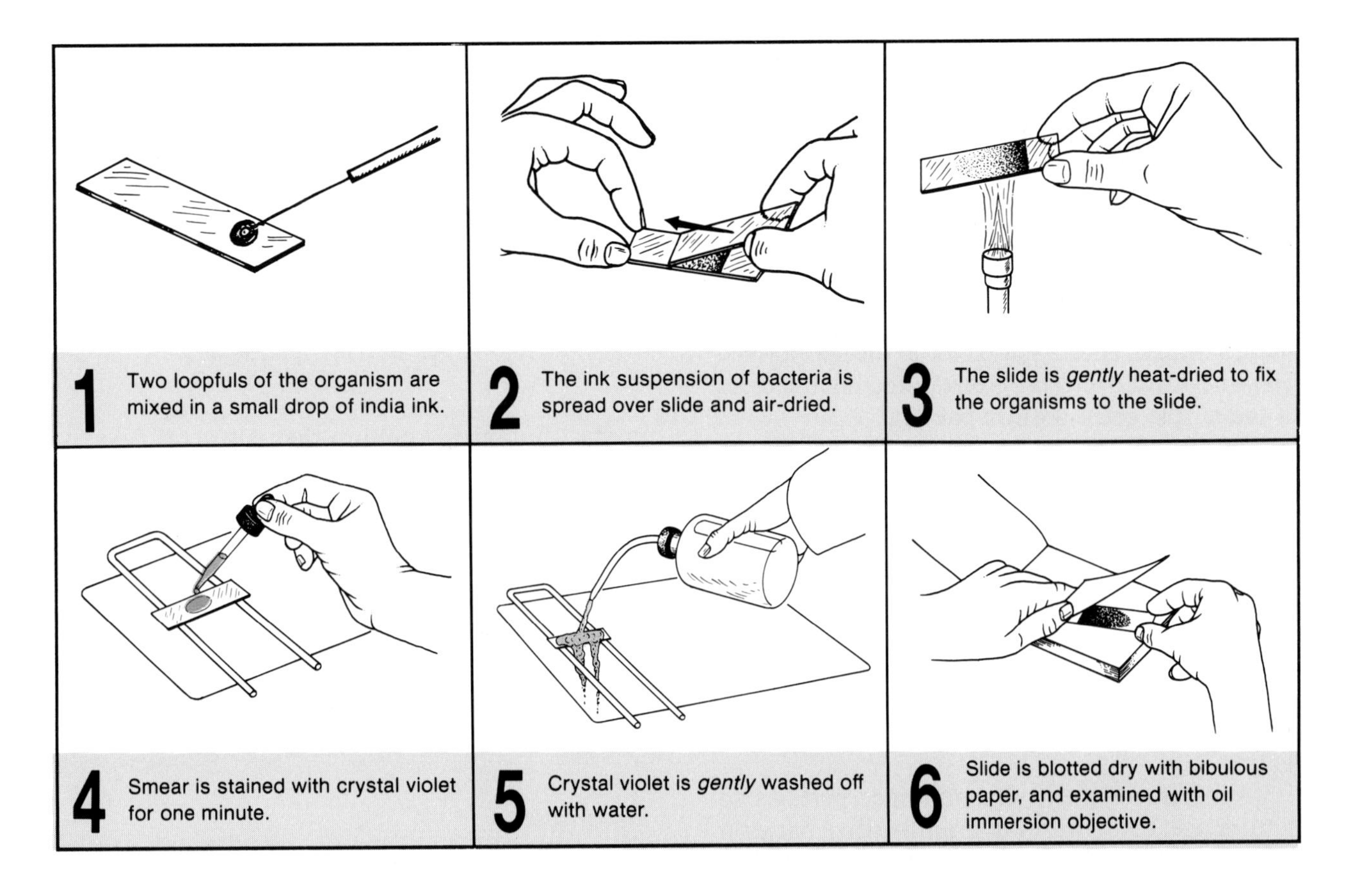

Figure 11.1 Procedure for demonstration of capsules.

12 Gram Staining

In 1884 the Danish bacteriologist Christian Gram developed a staining technique that separated bacteria into two groups: those that are gram-positive and those that are gram-negative. The procedure is based on the ability of microorganisms to retain the purple color of crystal violet during decolorization with alcohol. Gram-negative bacteria are decolorized by the alcohol, losing the purple color of crystal violet. Gram-positive bacteria are not decolorized and remain purple. After decolorization, safranin, a red counterstain, is used to impart a pink color to the decolorized gram-negative organisms.

Figure 12.1 illustrates the effects of the various reagents on bacterial cells at each stage in the process. Note that crystal violet, the **primary stain,** causes both gram-positive and gram-negative organisms to become purple after 20 seconds of staining. When Gram's iodine, the **mordant,** is applied to the cells for one minute, the color of gram-positive and gram-negative bacteria remains the same: purple. The function of the mordant here is to combine with crystal violet to form a relatively insoluble compound in the gram-positive bacteria, but not in the gram-negative bacteria. When the **decolorizing agent,** 95% ethanol, is added to the cells for 10–20 seconds, the gram-negative bacteria are leached colorless, but the gram-positive bacteria remain purple. In the final step a **counterstain,** safranin, adds a pink color to the decolorized gram-negative bacteria without affecting the color of the purple gram-positive bacteria.

Of all the staining techniques you will use in the identification of unknown bacteria, Gram staining is, undoubtedly, the most important tool you will use. Although this technique seems quite simple, performing it with a high degree of reliability is a goal that requires some practice and experience. Here are two suggestions that can be helpful: first, don't make your smears too thick, and second, pay particular attention to the comments in step 4 on the next page that pertains to decolorization.

Another comment of significance pertains to culture age: old cultures of gram-positive bacteria tend to decolorize more rapidly; thus, they might appear to be gram-negative. To get reliable results one should use cultures that are 18 to 24 hours old.

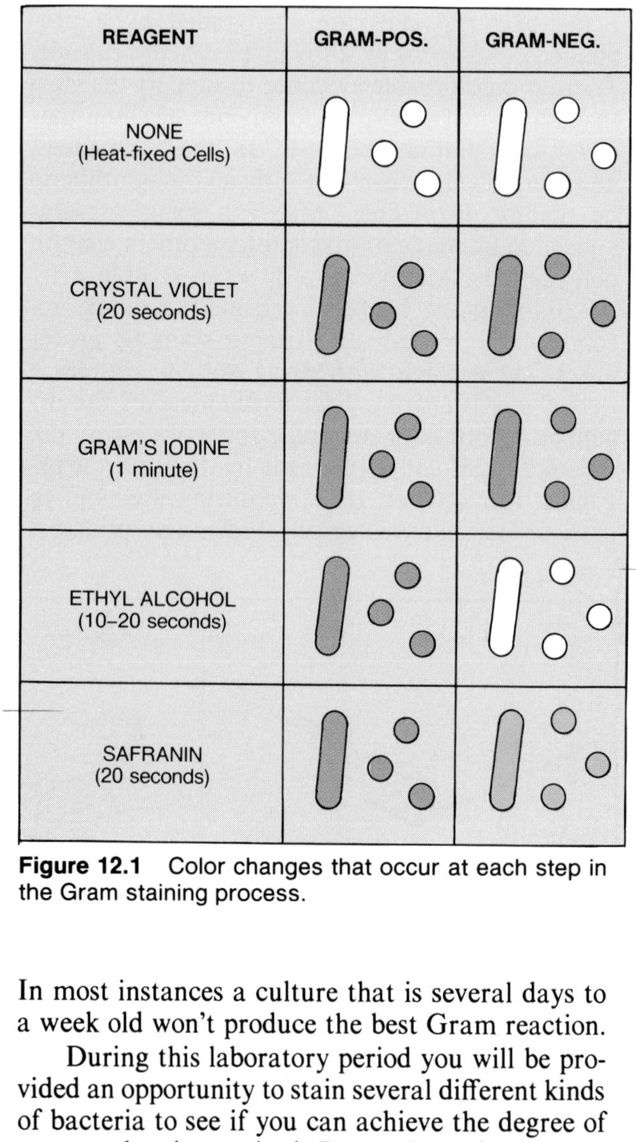

REAGENT	GRAM-POS.	GRAM-NEG.
NONE (Heat-fixed Cells)		
CRYSTAL VIOLET (20 seconds)		
GRAM'S IODINE (1 minute)		
ETHYL ALCOHOL (10–20 seconds)		
SAFRANIN (20 seconds)		

Figure 12.1 Color changes that occur at each step in the Gram staining process.

In most instances a culture that is several days to a week old won't produce the best Gram reaction.

During this laboratory period you will be provided an opportunity to stain several different kinds of bacteria to see if you can achieve the degree of success that is required. Remember, if you don't master this technique now, you will have difficulty with your unknowns later.

Staining Procedure

Materials:

slides with prepared smears
Gram staining kit and wash bottle
bibulous paper

1. Cover the smear with **crystal violet.** Let stand for **20 seconds.**
2. Briefly wash off the stain, using a wash bottle of distilled water. Drain off excess water.
3. Cover the smear with **Gram's iodine** solution and let stand for **one minute.** (Your instructor may prefer 30 to 60 seconds for this step.)
4. Pour off the Gram's iodine and flood the smear with **95% ethyl alcohol** for **10 to 20 seconds.** This step is critical. Thick smears will require more time than thin ones. *Decolorization has occurred when the solvent flows colorlessly from the slide.*
5. Stop the action of the alcohol by rinsing the slide with wash bottle for a few seconds.
6. Cover the smear with **safranin** for **20 seconds.**
7. Wash gently for a few seconds, blot with bibulous paper or paper toweling, and let dry at room temperature.
8. The slide may be examined under oil immersion immediately.

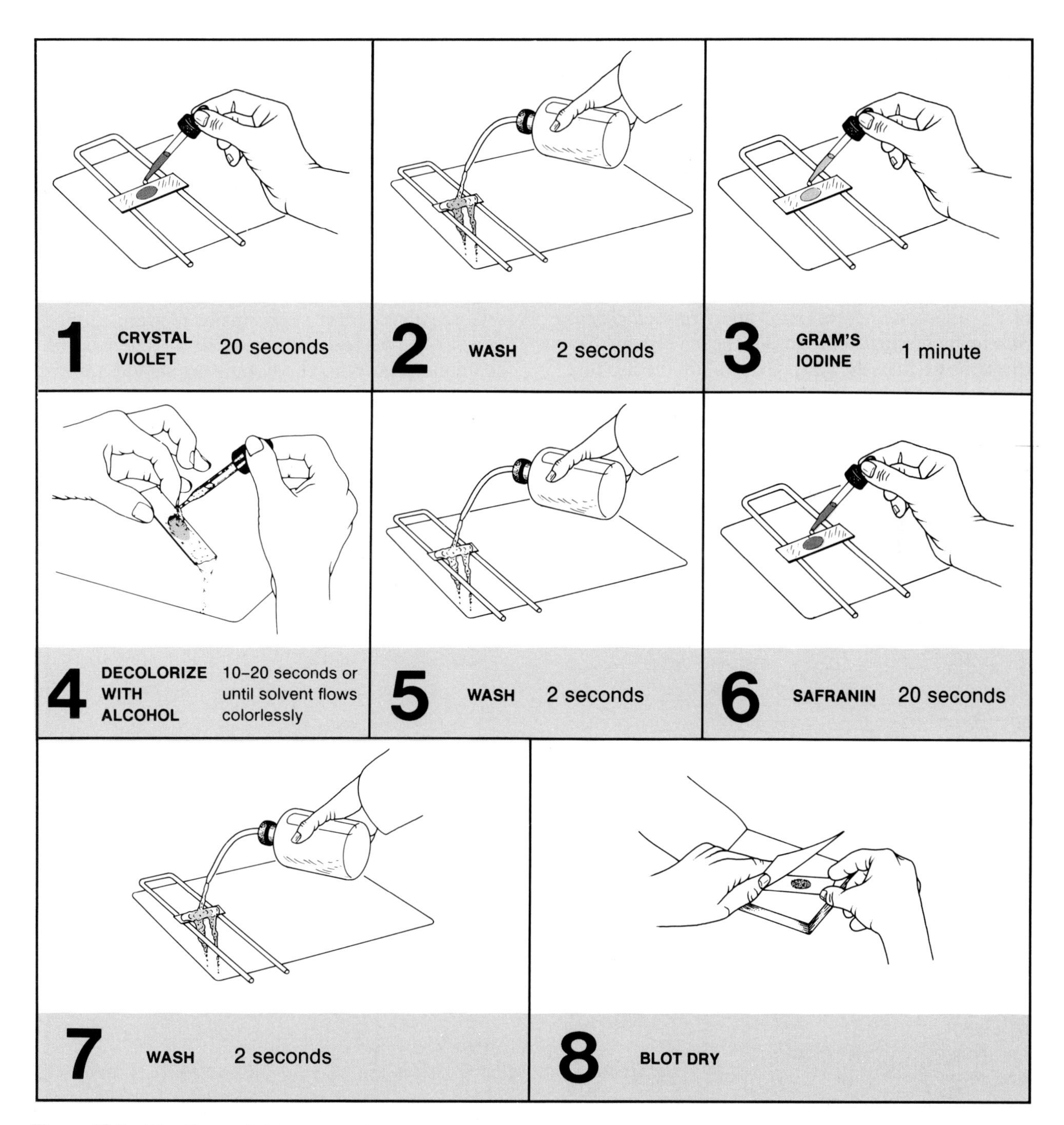

Figure 12.2 The Gram staining procedure.

Assignments

The organisms that will be used to prepare slides in this exercise represent a diversity of form and staining characteristics. Some of the rods and cocci are gram-positive; others are gram-negative. One rod-shaped organism is a spore-former and another is acid-fast. Note that two slides to be made are mixtures of two organisms. The challenge in making these slides is to demonstrate the color differences between gram-positive and gram-negative bacteria on the same slide.

Materials:

- broth cultures of *Pseudomonas aeruginosa, Staphylococcus aureus,* and *Moraxella (Branhamella) catarrhalis*
- nutrient agar slant cultures of *Bacillus megaterium* and *Mycobacterium smegmatis*

Mixed Organisms I Prepare a smear that consists of *Pseudomonas aeruginosa* and *Staphylococcus aureus* by mixing two loopfuls of each organism in the same area on the slide. Gram stain the mixture and examine under oil immersion. The slide should have pink rods and purple cocci, as indicated in illustration 3, figure 12.3.

Draw a few cells in the appropriate circle on Laboratory Report 12–14. Have your instructor look at this slide. If it is satisfactory, he or she will initial the drawing on your Laboratory Report sheet. If it is unsatisfactory, repeat the process until you produce a good slide.

Mixed Organisms II Make a gram-stained slide of a mixture of *Bacillus megaterium* and *Moraxella (Branhamella) catarrhalis.*

This mixture differs from your first slide in that the rods (*B. megaterium*) will be purple and the cocci (*M. B. catarrhalis*) will be large pink diplococci. See illustration 4, figure 12.3.

As you examine this slide look for clear areas on the rods which represent endospores. Since endospores are refractile and impermeable to crystal violet as it is applied in the Gram staining procedure, they will appear as transparent holes in the cells.

Draw a few cells in the appropriate circle on your Laboratory Report sheet.

Acid-Fast Bacteria To see how acid-fast mycobacteria react to Gram stain, make a gram-stained slide of *Mycobacterium smegmatis.* If your staining technique is correct, the organisms should appear gram-positive.

Draw a few cells in the appropriate circle on your Laboratory Report sheet.

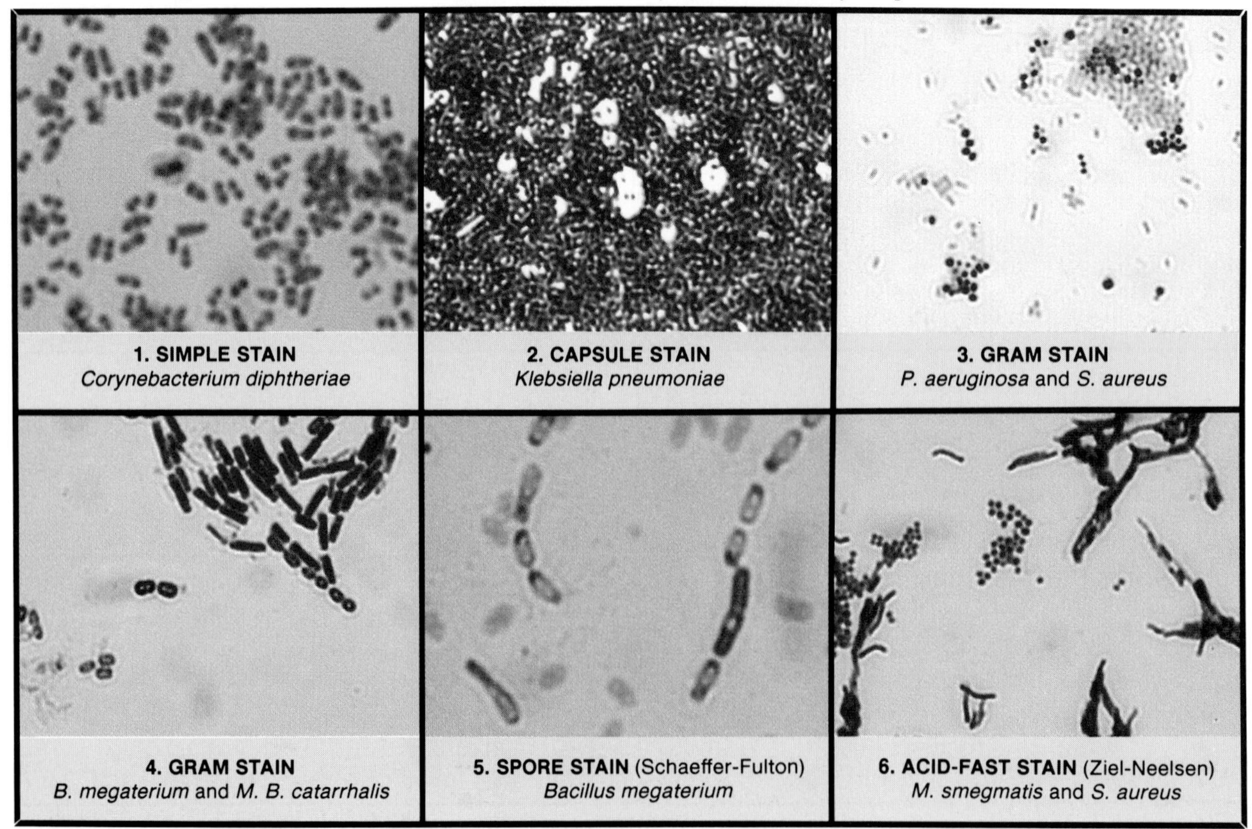

Figure 12.3 Photomicrographs of representative staining techniques (×8000).

Spore Staining (Two Methods) 13

Species of bacteria, belonging principally to the genera *Bacillus* and *Clostridium,* produce extremely heat-resistant structures called **endospores.** In addition to being heat-resistant, they are also very resistant to many chemicals that destroy non-spore-forming bacteria. This resistance to heat and chemicals is due primarily to a thick, tough spore coat.

It was observed in Exercise 12 that Gram staining will not stain endospores. Only if considerable heat is applied to a suitable stain can the stain penetrate the spore coat. Once the stain has entered the spore, however, it is not easily removed. Several methods are available that employ heat to provide stain penetration; however, since the Schaeffer-Fulton and Dorner methods are the principal ones used by most bacteriologists, both have been included in this exercise. Your instructor will indicate which procedure will be used.

Schaeffer-Fulton Method

This method, which is depicted in figure 13.1, utilizes malachite green to stain the endospore and safranin to stain the vegetative portion of the cell. Utilizing this technique, a properly stained spore-former will have a green endospore contained in a pink sporangium. Illustration 5, figure 12.3 on the opposite page, reveals what such a slide looks like.

After preparing a smear of *Bacillus megaterium* (Exercise 9), follow the steps outlined in figure 13.1 to stain the spores.

Materials:

24–36 hour nutrient agar slant culture of *Bacillus megaterium*
electric hot plate and small beaker of water
spore-staining kit consisting of a bottle each of 5% malachite green and safranin

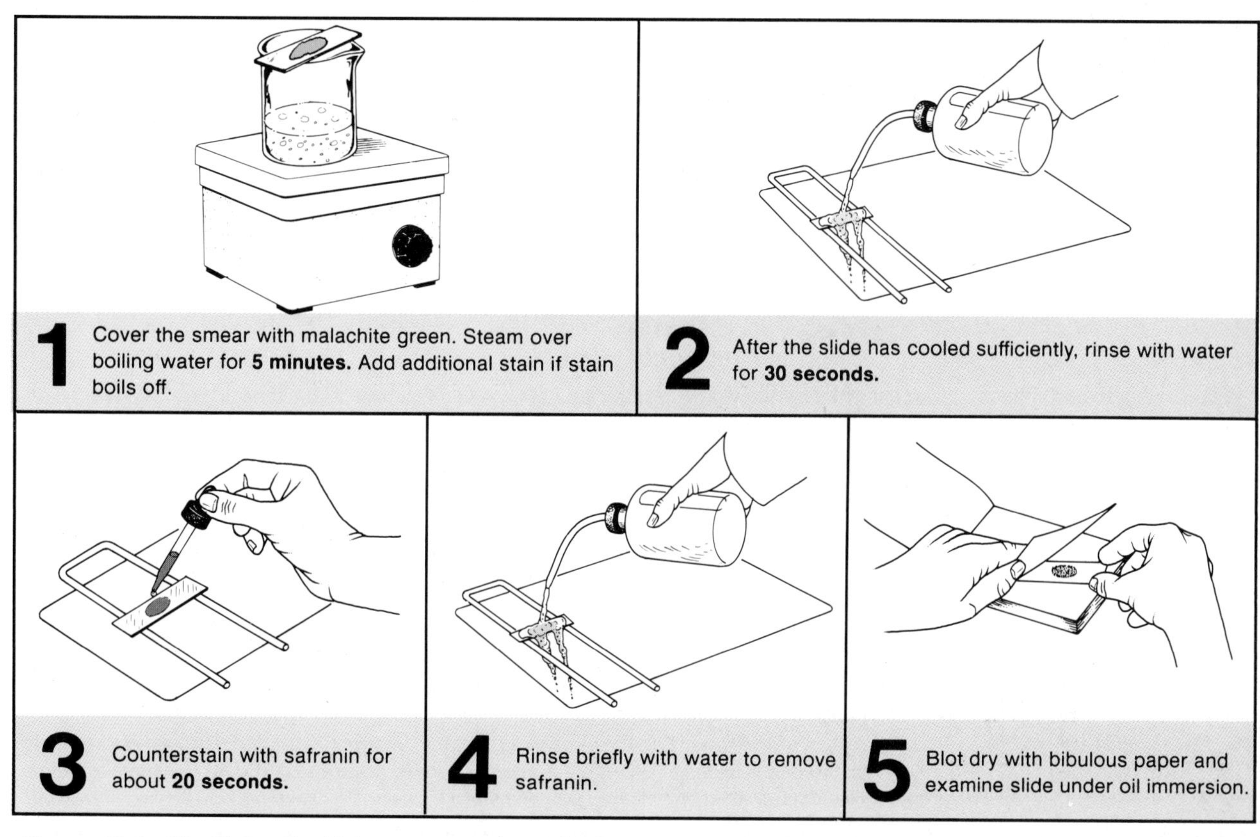

Figure 13.1 The Schaeffer-Fulton spore stain method.

Dorner Method

The Dorner method for staining endospores produces a red spore within a colorless sporangium. Nigrosine is used to provide a dark background for contrast. The six steps involved in this technique are shown in figure 13.2. Although both the sporangium and endospore are stained during boiling in step 3, the sporangium is decolorized by nigrosine in step 4.

Prepare a slide of *Bacillus megaterium* that utilizes the Dorner method. Follow the steps in figure 13.2.

Materials:

electric hot plate
carbolfuchsin stain (Ziehl's)
nigrosine
small beaker (25 ml size)
small test tube (10 × 75 mm size)
test-tube holder
24–36 hour nutrient agar slant culture of *Bacillus megaterium*

Laboratory Report

After examining the organisms under oil immersion, draw a few cells in the appropriate circles on Laboratory Report 12–14.

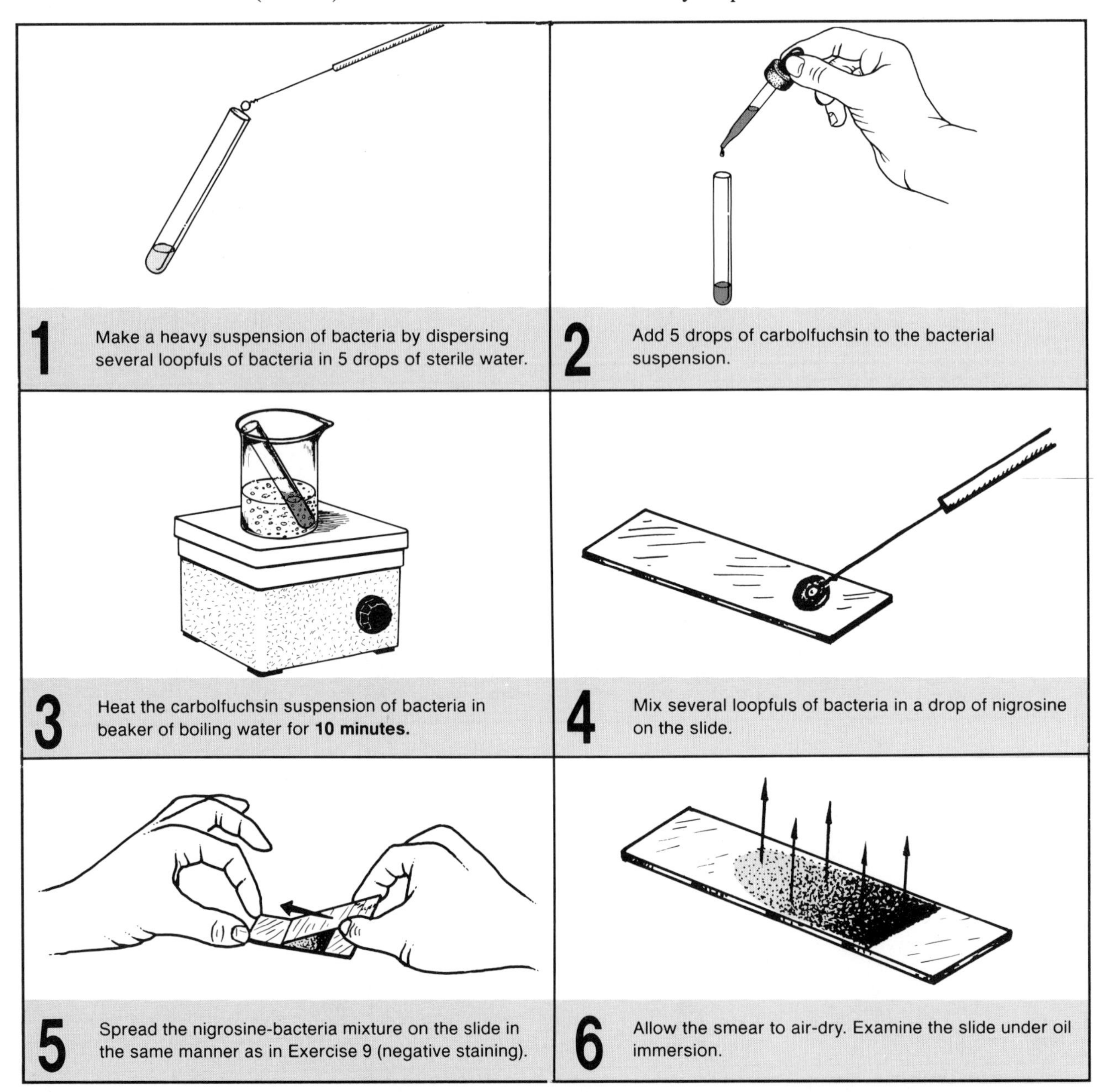

Figure 13.2 The Dorner spore stain method.

Acid-Fast Staining (Ziehl-Neelsen Method) 14

Bacteria which are not readily decolorized with acid-alcohol after staining with hot carbolfuchsin are said to be **acid-fast.** These microorganisms contain considerable quantities of waxlike lipoidal material which combines tenaciously with this red dye. This stain is used primarily in the identification of the tuberculosis bacillus, *Mycobacterium tuberculosis,* and the leprosy organism, *Mycobacterium leprae.* After decolorization, methylene blue is added to the organisms to counterstain any material that is not acid-fast; thus, a properly stained slide of a mixture of acid-fast organisms, tissue cells, and non-acid-fast bacteria will reveal red acid-fast rods with bluish tissue cells and bacteria (see illustration 6, figure 12.3).

The two organisms used in this staining exercise are *M. smegmatis,* a nonpathogenic acid-fast rod found in soil and on external genitalia, and *Staphylococcus aureus,* a non-acid-fast coccus.

Materials:

nutrient agar slant culture of *Mycobacterium smegmatis* (48 hour culture)
nutrient broth culture of *S. aureus*
electric hot plate and small beaker
acid-fast staining kit (carbolfuchsin, acid-alcohol, methylene blue)

Smear Preparation Prepare a mixed culture smear by placing two loopfuls of *S. aureus* on a slide and then transferring a very small amount of *M. smegmatis* to the broth on the slide with an inoculating needle. Since the latter organisms tend to cling to each other, break up the mass of organisms with the inoculating needle. After air-drying the smear, heat-fix it over a Bunsen burner flame.

Staining Follow the staining procedure outlined in figure 14.1.

Examination Examine under oil immersion and compare your slide with illustration 6, figure 12.3. Record your results on Laboratory Report 12–14.

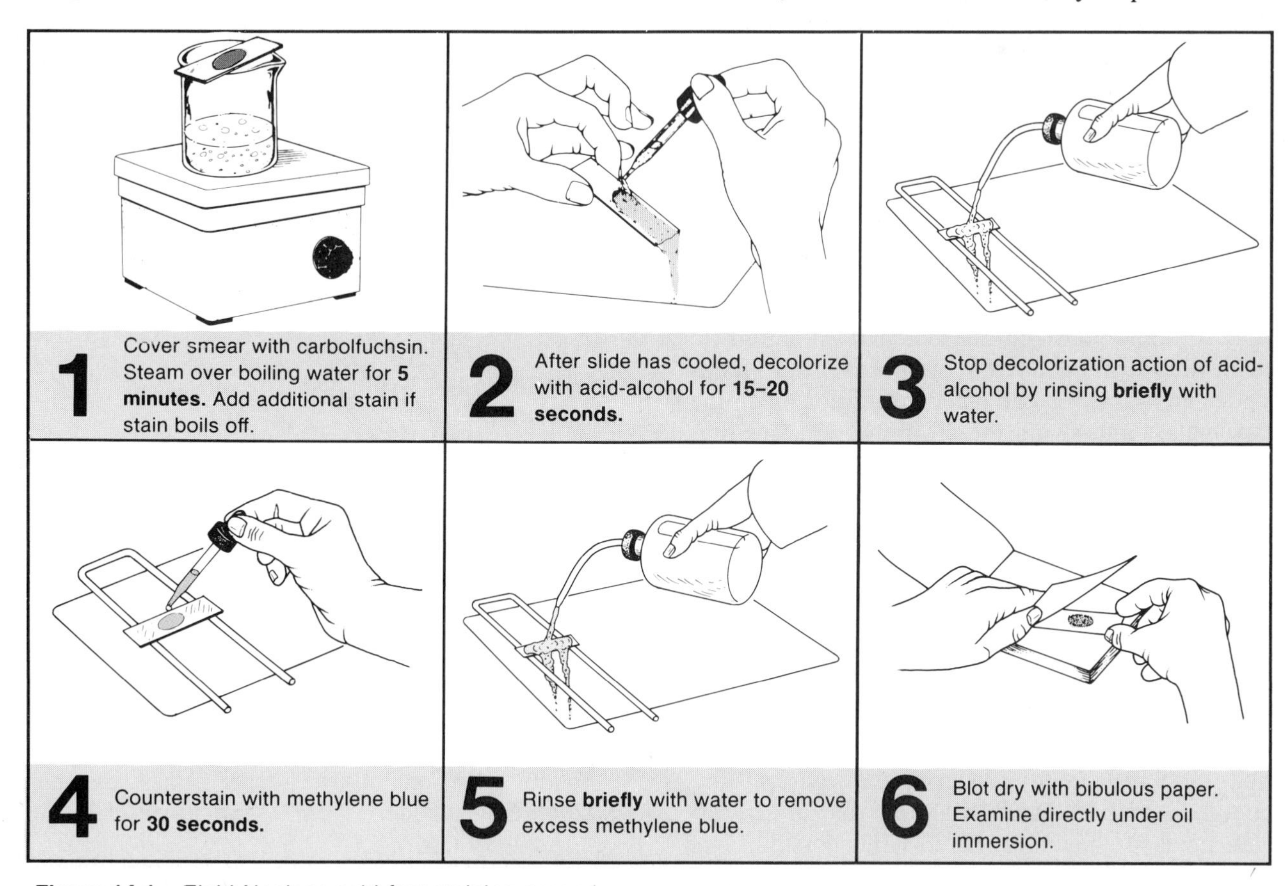

Figure 14.1 Ziehl-Neelsen acid-fast staining procedure.

15 Motility Determination

When attempting to identify an unknown bacterium it is often necessary to determine whether or not the microorganism is motile. There are several ways that one can use to make this determination. Three methods are described here. Each method has its advantages and disadvantages. When working with unknown bacteria, you will select the method that best suits your purposes. In this exercise you will try all three procedures to become familiar with all of them.

The Wet Mount

The simplest way to determine motility is to place a few loopfuls of the organism on a clean slide and cover with a cover glass. The best way to examine the slide is with a phase-contrast microscope, utilizing high-dry magnification. If the organisms are motile, you should see them darting about. In some instances, only a few of the organisms of a culture will exhibit motility. This is not unusual. Even if only a few move about, the organisms must be considered motile.

It is important that you be able to differentiate true motility from *Brownian movement.* When bacteria exhibit vibratory movement instead of actually moving from one place to another, you are observing Brownian movement. Brownian movement is motion caused by invisible molecules striking the bacteria, displacing them for short distances.

If a brightfield microscope is used, it is necessary to reduce the lighting to a bare minimum to achieve contrast. Once the organisms are in focus under high-dry, you can place immersion oil on the cover glass and swing the oil immersion lens into position.

The advantage of a wet mount is that it is the quickest means for determining motility. Such a preparation also is useful for determining cellular shape and arrangement. Whether or not the cells are arranged in irregular clusters, packets, pairs, or long chains is more readily determined by this method than by staining because heat-drying of a smear tends to cause organisms to coalesce.

One disadvantage of this method is that the slide quickly dries out, rendering the organisms immotile. Another disadvantage is that if the organism is pathogenic, there is the possibility of danger to the technician in handling viable organisms on a slide.

The Hanging Drop Method

When it is necessary to study viable organisms on a microscope slide for a longer period of time than is possible with a wet mount, one must resort to a hanging drop slide. As seen in illustration 4 of figure 15.1, organisms are observed in a drop that is suspended under a cover glass in a concave slide. Since the drop lies within an enclosed glass chamber, drying out occurs very slowly.

Although the preparation of a hanging drop slide presents no great difficulties, it is somewhat harder to examine than a wet mount. Since depression slides are often somewhat thicker than ordinary slides, it is necessary with some microscopes to employ a slightly different focusing method than would normally be used. Particular care must be taken to avoid breaking the cover glass since it is much more vulnerable when supported only around its edges.

In general, hanging drop slides usually are observed with a brightfield microscope, either with high-dry or oil immersion. Phase-contrast optics can be used, but not as satisfactorily as for wet mounts. Hanging drop slides yield the same information as wet mount slides: motility information, cell shape, and cell arrangement. The procedure is as follows:

Materials:

depression slides and cover glasses
inoculating loop
Vaseline and toothpicks
Bunsen burner
nutrient broth cultures of organisms (young cultures)

1. Prepare a clean depression slide and cover glass by washing with soap or Bon Ami and warm water. All grease must be removed.
2. With a china marking pencil, label the slide with the name of the organism.
3. Place a very small amount of Vaseline near each corner of a cover glass as shown in illustration 1 of figure 15.1. The Vaseline will provide adhesion of the cover glass to the depression slide. *Avoid using more than is shown in figure 15.1.*

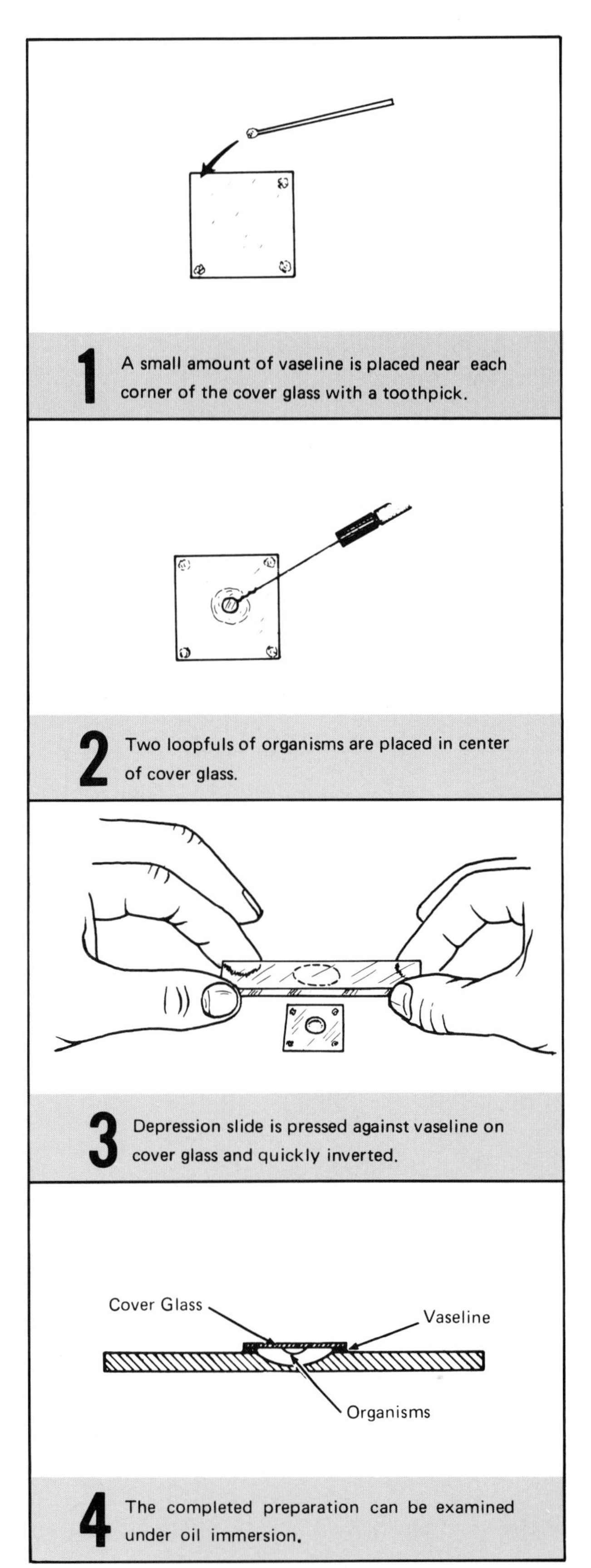

Figure 15.1 The hanging drop slide.

4. Shake the culture tube and transfer two loopfuls of the culture onto the cover glass. *Be sure to flame the loop prior to removing the second loopful from the culture.*

   If a hanging drop slide is to be made from organisms on agar or from some part of the body, they should be mixed with a drop of sterile water on the cover glass.
5. Place the slide in contact with the cover glass with the depression over the drop of suspended bacteria (see illustration 3).
6. Invert the slide quickly so that the drop cannot run off to one side.

Examination

1. Examine the slide first with the low-power objective. If your microscope is equipped with an automatic stop, it may be necessary to bring the image into focus by using the coarse adjustment knob. The greater thickness of some depression slides prevents one from being able to focus by conventional means. All you will be able to see under low power will be some very fine specks.
2. Once the image is seen under low power, swing the high-power objective into position and readjust the lighting. *Focus near the edge of the drop* where most organisms are drawn by surface tension.
3. If your microscope has phase-contrast optics, switch to high-dry phase. Although a hanging drop does not provide the shallow field desired for phase-contrast, you may find that it works fairly well.
4. If you wish to use oil immersion, simply rotate the high-power objective out of position and add immersion oil to the cover glass. Without moving the focusing knobs, rotate the oil immersion objective into place. Slight readjustment of the fine adjustment knob and diaphragm may be necessary. Be sure to keep lighting to a minimum. Bacteria are transparent phase objects that are difficult to see under brightfield microscopy.

A Culture Method
(Semisolid Medium)

When working with pathogenic microorganisms such as the typhoid bacillus, it is too dangerous to attempt to determine motility with slide techniques. A much safer method is to culture the organisms in a special medium that can demonstrate the presence of motility. The procedure is to inoculate a tube of semisolid medium with the organism by the *stab technique* as illustrated in figure 15.2. The inoculation is performed with a straight wire that is stabbed two-thirds of the way down into the medium. After the tube is incubated for 24 to 48 hours, it is examined for cloudiness. If the organism is motile, a cloudiness will exist around the stab due to the migration of motile bacteria from the point of inoculation. Semisolid medium differs from nutrient agar in that it contains less agar and allows motile bacteria to move through it.

Materials:

1 tube of semisolid medium per organism
(2 ml of medium in serological tube)
nutrient broth culture of organism
inoculating wire (straight)
Bunsen burner

1. Label a tube of semisolid medium with the name of the organism. Place your initials on the tube, also.
2. Flame and cool inoculating wire, and insert wire into culture after flaming neck of tube.
3. Remove plug from tube of semisolid medium, flame neck, and stab it ⅔ of the way down into the medium, as shown in figure 15.2. Flame neck of tube again before returning plug to tube.
4. Incubate at room temperature for 48 hours.

Assignment

Materials:

all materials listed above for hanging drop slide and culture method
nutrient broth cultures of *Micrococcus luteus* and *Proteus vulgaris*
slides and cover glasses

Prepare separate wet mounts, hanging drop slides, and semisolid medium inoculations of *Micrococcus luteus* and *Proteus vulgaris*. Record your observations of the slides on the Laboratory Report in this laboratory period. Record the results of the tube inoculations in the next laboratory period.

Laboratory Report

Complete the Laboratory Report for this exercise.

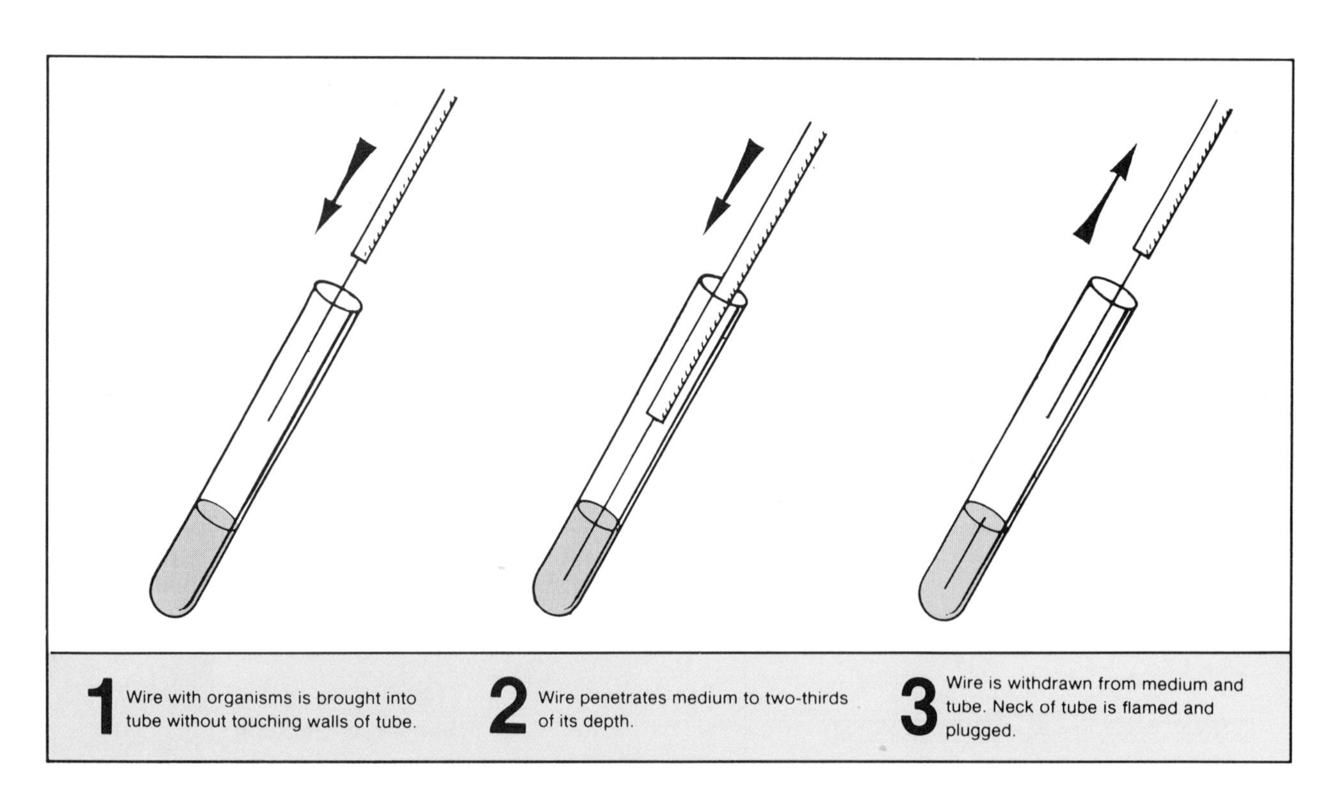

Figure 15.2 Stab technique for motility test.

Part 4 Culture Methods

The nutritional requirements of different microorganisms cover an extremely broad spectrum. Some organisms require highly enriched media; other thrive on simple inorganic substances. Some demand a lot of oxygen; others die in its presence. Because of this great diversity, a multiplicity of culture techniques have been developed. Each method is suited to a particular type of organism.

This unit contains seven exercises concerned with media preparation and various culture techniques that serve different purposes. Although no attempt has been made to include all the essential culture methods, most of the basic ones for an introductory course have been included.

16 Culture Media Preparation

From now on, most of the laboratory experiments in this manual will utilize bacteriological media. In most instances it will be provided for you. However, circumstances may occur when you will need some special medium that is not already prepared, and it will be up to you to put it together. It is for situations like this that the information in this exercise will be useful.

The first portion of this exercise pertains to the types of media and how they relate to the needs of microorganisms. The last part of the exercise pertains to the actual mechanics of making up a batch of medium. Whether or not you will be provided an opportunity to prepare some media during a designated laboratory period will depend on the availability of time and classroom needs. Your instructor will indicate how this exercise is to be used.

Media Consistency

Microbiological **media** (medium, singular) is the food that we use for culturing bacteria, molds, and other microorganisms. It exists in three consistencies: liquid, solid, and semisolid. If you have performed all of the exercises in Part 3, you are already familiar with all of them.

The **liquid media** include nutrient broth, citrate broth, glucose broth, litmus milk, etc. These media are used for the propagation of large numbers of organisms, fermentation studies, and various other tests.

If a little agar, gelatin, or silica gel is added to a liquid medium to impart a degree of firmness to it, it becomes a **solid medium.** Nutrient agar, blood agar, and Sabouraud's agar are examples of solid media that are used for developing surface colony growth of bacteria and molds. As we will see in the next exercise, the development of colonies on the surface of a medium is essential when trying to isolate organisms from mixed cultures.

Semisolid media fall in between liquid and solid media. Although they are similar to solid media in that they contain solidifying agents such as agar and gelatin, they are more jellylike due to lower percentages of these solidifiers. These media are particularly useful in determining whether or not certain bacteria are motile (Exercise 15).

Nutritional Needs of Bacteria

Before one can construct a medium that will achieve a desired result in the growth of organisms, one must understand their basic needs. Any medium that is to be suitable for a specific group of organisms must take into account the following seven factors: water, carbon, energy, nitrogen, minerals, growth factors, and pH. The role of each one of these follows:

Water Protoplasm is from 70% to 85% water. The water in a single-celled organism is continuous with the water of its environment, and the molecules pass freely in and out of the cell, providing a vehicle for nutrients inward, and secretions or excretions outward. All the enzymatically controlled chemical reactions that occur within the cell occur only in the presence of an adequate amount of water.

The quality of water used in preparing media is important. Hard tap water, high in calcium and magnesium ions, should not be used. Insoluble phosphates of calcium and magnesium may precipitate in the presence of peptones and beef extract. The best policy is to *always use distilled water.*

Carbon Organisms are divided into two groups with respect to their sources of carbon. Those that can utilize the carbon in carbon dioxide for the synthesis of all cell materials are called *autotrophs.* If they must have one or more organic compounds for their carbon source, they are called *heterotrophs.* In addition to organic sources of carbon, the heterotrophs are also dependent on carbon dioxide. If this gas is completely excluded from their environment, their growth is greatly retarded, particularly in the early stages of starting a culture.

Specific organic carbon needs are as diverse as the organisms themselves. Where one organism may require only a single simple compound such as acetic acid, another may require a dozen or more organic nutrients of various degrees of complexity.

Energy Organisms that have pigments that enable them to utilize solar energy are called *photoautotrophs* (photosynthetic autotrophs). Media for such organisms will not include components to provide energy.

Autotrophs that cannot utilize solar energy but are able to oxidize simple inorganic substances for energy are called *chemoautotrophs* (chemosynthetic autotrophs). The essential energy-yielding substance for these organisms may be as elemental as nitrite, nitrate, or sulfide.

Most bacteria fall into the category of *chemoheterotrophs* (chemosynthetic heterotrophs) that require an organic source of energy, such as glucose or amino acids. The amounts of energy-yielding ingredients in media for both chemosynthetic types is on the order of 0.5%.

A small number of bacteria are classified as *photoheterotrophs* (photosynthetic heterotrophs). These organisms have photosynthetic pigments enabling them to utilize sunlight for energy, but must have an organic source of carbon, such as alcohol.

Nitrogen Although autotrophic organisms can utilize inorganic sources of nitrogen, the heterotrophs get their nitrogen from amino acids and intermediate protein compounds such as peptides, proteoses, and peptones. Beef extract and peptone, as used in nutrient broth, provide the nitrogen needs for the heterotrophs grown on this medium.

Minerals All organisms require several metallic elements such as sodium, potassium, calcium, magnesium, manganese, iron, zinc, copper, phosphorus, and cobalt for normal growth. Bacteria are no exception. The amounts required are very small.

Growth Factors Any essential component of cell material that an organism is unable to synthesize from its basic carbon and nitrogen sources is classified as being a *growth factor.* This may include certain amino acids or vitamins. Many heterotrophs are satisfied by the growth factors present in beef extract of nutrient broth. More fastidious pathogens require enriched media such as blood agar for ample growth factors.

Hydrogen Ion Concentration The growth of organisms in a particular medium may be completely inhibited if the pH of the medium is not within certain limits. The enzymes of microorganisms are greatly affected by this factor. Since most bacteria grow best at around pH 7 or slightly lower, the pH of nutrient broth should be adjusted to pH 6.8. Pathogens, on the other hand, usually prefer a more alkaline pH. Trypticase soy broth, a suitable medium for the more fastidious organisms, should be adjusted to pH 7.3.

Synthetic and Nonsynthetic Media

Media can be prepared to exact specifications so that the exact composition is known. These media are generally made from chemical compounds that are highly purified and precisely defined. Such media are readily reproducible. They are known as **synthetic media.**

Media such as nutrient broth that contain ingredients of imprecise composition are called **nonsynthetic media.** Both the beef extract and peptone in nutrient broth are inexact in composition.

Dehydrated Media

Until around 1930, the laboratory worker had to spend a good deal of time preparing laboratory media from various raw materials. If a medium

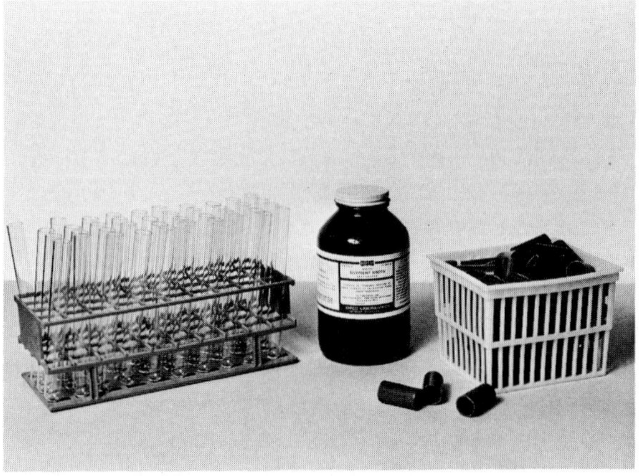

Figure 16.1 Basic supplies and equipment needed for making up a batch of medium.

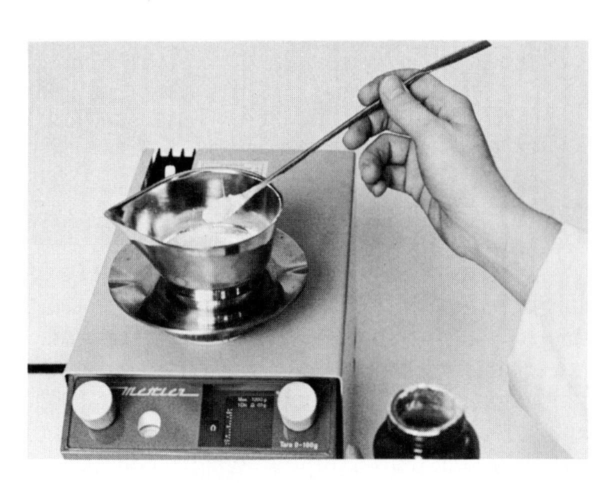

Figure 16.2 Correct amount of dehydrated medium is weighed on balance.

contained five or six ingredients, it was not only necessary to measure the various materials but also, in many instances, to fabricate some of the components such as beef extract or veal infusion by long tedious methods. Today, dehydrated media have revolutionized media preparation techniques in much the same way that commercial cake mixes have taken over in the modern kitchen. For most routine bacteriological work, media preparation has been simplified to the extent that all that is necessary is to dissolve a measured amount of dehydrated medium in water, adjust the pH, dispense into tubes, and sterilize.

Media Preparation Assignment

In this laboratory period you will work with your laboratory partner to prepare tubes of media that will be used in future laboratory experiments. Your instructor will indicate which media you are to prepare. Record in the space below the number of tubes of specific media that have been assigned to you and your partner.

nutrient broth ________

nutrient agar pours ________

nutrient agar slants ________

other ________

Several different sizes of test tubes are used for media, but the two sizes most generally used are either 16 mm or 20 mm diameter by 15 cm long. Select the correct size tubes first, according to these guidelines:

Large tubes (20 mm dia.): Use these test tubes for all *pours,* i.e., nutrient agar, Sabouraud's agar, EMB agar, etc. Pours are used for filling petri plates.

Small tubes (16 mm dia.): Use these tubes for all *broths, deeps,* and *slants.*

If the tubes are clean and have been protected from dust or other contamination, they can be used without cleaning. If they need cleaning, scrub them with warm water and detergent, using a test-tube brush. Rinse twice, first with tap water and finally with distilled water, to rid them of all traces of detergent. Place them in a wire basket or rack, inverted, so that they can drain. Do not dry with a towel.

Measurement and Mixing

The amount of medium you make for a batch of test tubes should be determined as precisely as possible to avoid shortage or excess.

Materials:

graduate, beaker, glass stirring rod
bottles of dehydrated media
Bunsen burner and tripod, or hot plate

1. The amount of distilled water you need will depend on the amount of medium that goes into each tube. The following volumes per tube should be used:

   pours .... 12 ml
   deeps .... 6 ml
   slants .... 4 ml
   broths .... 5 ml
   broths with fermentation tubes ...... 5–7 ml

Figure 16.3 Dehydrated medium is dissolved in a measured amount of distilled water.

Figure 16.4 If medium contains agar, it must be brought to a boil to bring agar into solution.

2. Consult the label on the bottle to determine how much powder is needed for 1000 ml and then determine by proportionate methods how much you need for the amount of water you are using. Weigh this amount on a balance and add it to the beaker of water. If the medium does not contain agar, the mixture usually goes into solution without heating.
3. **If the medium contains agar,** heat the mixture over a Bunsen burner (figure 16.4) or on an electric hot plate until it comes to a boil. To safeguard against water loss, *before heating, mark the level of the top of the medium on the side of the beaker with a china marking pencil.* As soon as it "froths up," turn off the heat. If an electric hot plate is used, the medium must be removed from the plate or it will boil over the sides of the container.

   Caution: Be sure to keep stirring from the bottom with a glass stirring rod so that the medium does not char on the bottom of the beaker.

   Check the level of the medium with the mark on the beaker to note if any water has been lost. Add sufficient distilled water as indicated.

   Keep the temperature of the medium at about 60° C to avoid solidification. The medium will solidify at around 42° C.

Adjusting the pH

Although dehydrated media contain buffering agents to keep the pH of the medium in a desired range, the hydrogen ion concentration of a freshly made batch of medium may differ from that stated on the label of the bottle. Before the medium is tubed, therefore, one must check the pH and make any necessary adjustments.

If a pH meter (figure 16.5) is available and already standardized, use it to adjust the pH of your media utilizing the bottles of HCl and NaOH in the materials list below. In most instances, however, pH test papers will work about as well.

Materials:

beaker of medium
acid and base kits (dropping bottles of 1N and .1N HCl and NaOH
glass stirring rod
pH test papers
pH meter (optional)

1. Dip a piece of pH test paper into the medium to determine the pH of the medium.
2. If the pH is too high, add a drop or two of HCl to lower the pH. For large batches use the 1N HCl. If the pH difference is slight use the .1N HCl. Use a glass stirring rod to mix the solution as the drops are added.
3. If the pH is too low, add NaOH a drop at a time. For slight pH differences, use .1N NaOH; for large pH differences use 1N NaOH. Use a glass stirring rod to mix the solution as the drops are added.

Filling the Test Tubes

Once the pH of the medium is adjusted it must be dispensed into test tubes. If an automatic pipetting machine is to be used, as shown in figure 16.6, it will have to be set up for you by your instructor. These machines can be adjusted to deliver any amount of medium at any desired speed. When

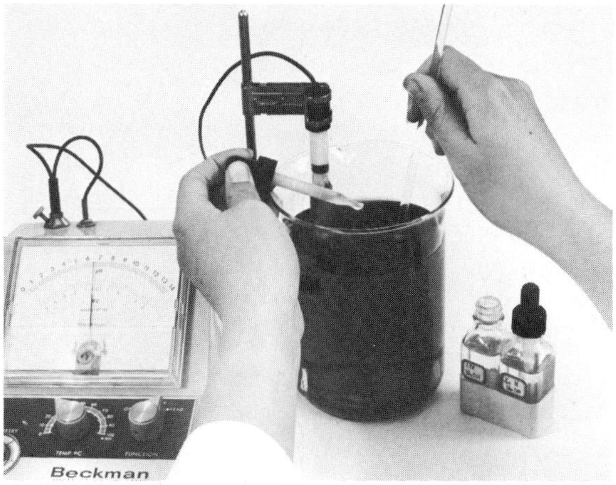

Figure 16.5 The hydrogen ion concentration of a medium must be adjusted to its recommended pH.

Figure 16.6 An automatic pipetting machine will deliver precise amounts of media at a controlled rate.

large numbers of tubes are to be filled, the automatic pipetting machine should be used. For smaller batches of tubes the funnel method shown in figure 16.7 is adequate. Use the following procedure when filling tubes with the funnel assembly.

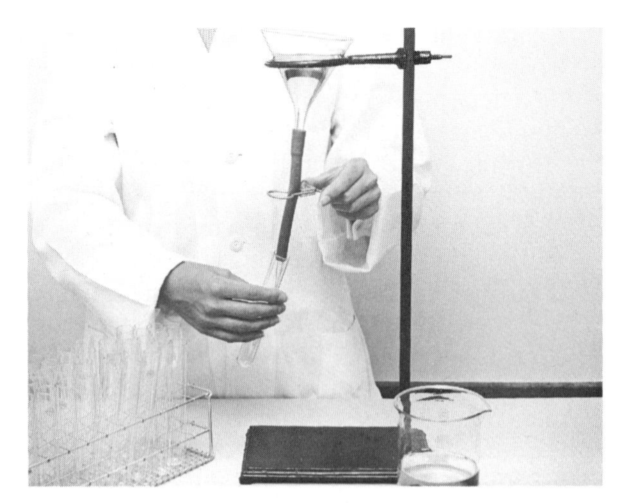

Figure 16.7 A glass funnel assembly and hose clamp is adequate for filling small batches of tubes.

Materials:

ring stand assembly
funnel assembly (glass funnel, rubber tubing, hose clamp, and glass tip)
graduate (small size)

1. Fill one test tube with a measured amount of medium. This tube will be your guide for filling the other tubes.
2. Fill the funnel and proceed to fill the test tubes to the proper level, holding the guide tube alongside of each empty tube to help you to determine the amount to allow in each tube.
3. Keep the beaker of medium over heat if it contains agar.
4. If fermentation tubes are to be used, add one to each tube at this time with the open end down.

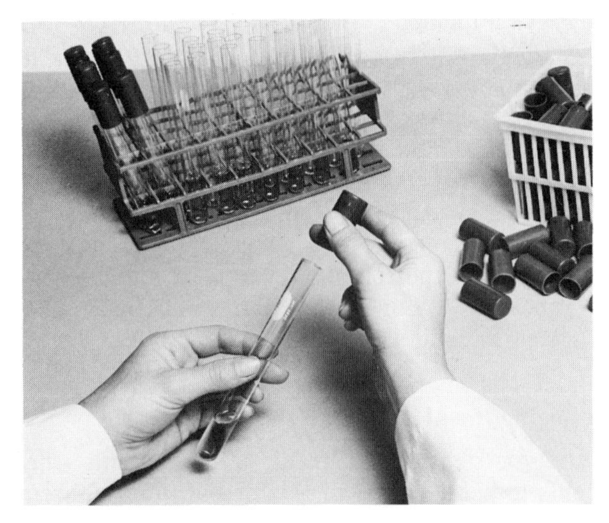

Figure 16.8 Once the medium has been dispensed to all the tubes, they are capped prior to sterilization.

Capping the Tubes

The last step before sterilization is to provide a closure for each tube. Plastic (polypropylene) caps are suitable in most cases. All caps that slip over the tube have inside ridges that grip the side of the tube and provide an air gap to allow steam to escape during sterilization. If you are using tubes with plastic screw-caps, the caps should not be screwed tightly before sterilization; each one must be left partly unscrewed.

If no slip-on caps of the correct size are available, it may be necessary to make up some cotton plugs. A properly made cotton plug should hold firmly in the tube so that it is not easily dislodged.

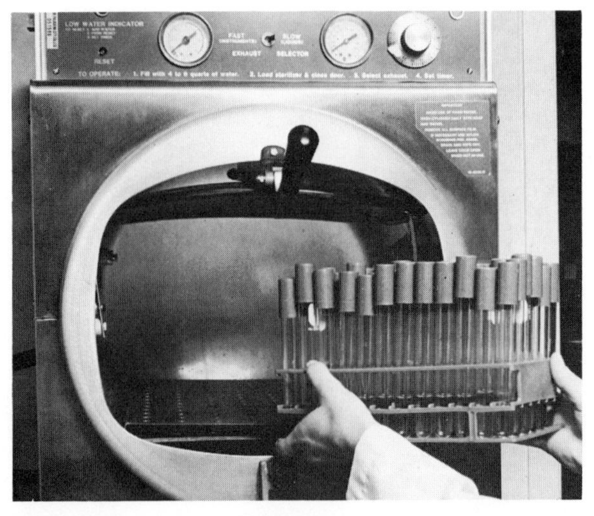

Figure 16.9 Tubes of media are sterilized in an autoclave for 20 to 30 minutes at 15 psi steam pressure.

Sterilization

As soon as the tubes of media have been stoppered they must be sterilized. Organisms on the walls of the test tubes, in the distilled water, and in the dehydrated medium will begin to grow within a short period of time at room temperature, destroying the medium.

Prior to sterilization the tubes of media should be placed in a wire basket with a label taped on the outside of the basket. The label should indicate the type of medium, the date, and your name.

Sterilization must be done in an autoclave. The following considerations are important in using an autoclave:

- Check with your instructor on the procedure to be used with your particular type of autoclave. Complete sterilization occurs at 250° F (121.6° C). To achieve this temperature the autoclave has to develop 15 pounds per square inch (psi) of steam pressure. To reach the correct temperature there must be some provision in the chamber for the escape of air. On some of the older units it is necessary to allow the steam to force air out through the door before closing it.
- Don't overload the chamber. One should not attempt to see how much media can be packed into it. Provide ample space between baskets of media to allow for circulation of steam.
- Adjust the time of sterilization to the size of load. Small loads may take only 10 to 15 minutes. An autoclave full of media may require 30 minutes for complete sterilization.

After Sterilization

Slants If you have a basket of tubes that are to be converted to slants, it is necessary to lay the tubes down in a near-horizontal manner as soon as they are removed from the autoclave. The easiest way to do this is to use a piece of rubber tubing (½" dia.) to support the capped end of the tube as it rests on the countertop. Solidification should occur in about 30 to 60 minutes.

Other Media Tubes of broth, agar deeps, nutrient gelatin, etc., should be allowed to cool to room temperature after removal from the autoclave. Once they have cooled down, place them in a refrigerator or cold storage room. When stored for long periods of time at room temperature media tend to lose moisture. At refrigerated temperatures media will keep for months.

Laboratory Report

Complete the Laboratory Report for this exercise.

17 Pure Culture Techniques

When we try to study the bacterial flora of the body, soil, water, food, or any other part of our environment, we soon discover that bacteria exist in mixed populations. It is only in very rare situations that they occur as a single species. To be able to study the cultural, morphological, and physiological characteristics of an individual species, it is essential, first of all, that the organism be separated from the other species normally found in its habitat. In other words, we must have a **pure culture** of the microorganism.

Several different methods of getting a pure culture from a mixed culture are available to us. The two most frequently used methods involve making a streak plate or a pour plate. Both plate techniques involve thinning the organisms so that one individual species can be selected from the others.

In this exercise you will have an opportunity to use both methods in an attempt to separate three distinct species from a tube that contains a mixture. The principal difference between the three organisms will be their colors: *Serratia marcescens* is red, *Micrococcus luteus* is yellow, and *Escherichia coli* is white.

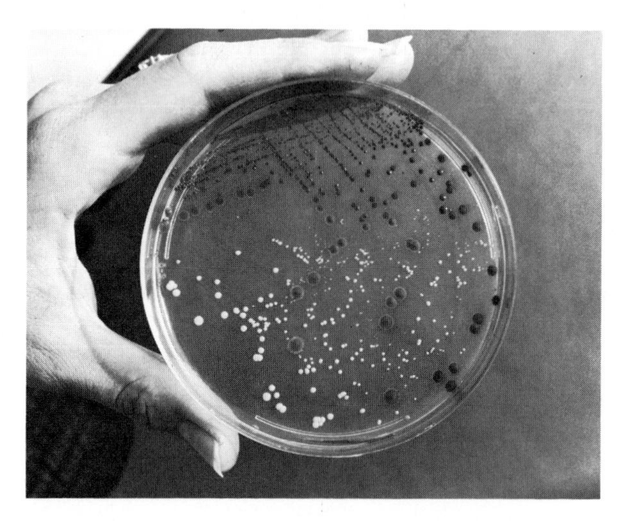

Figure 17.1 Isolated colonies on a streak plate.

Streak Plate Method

For economy of materials and time, this method is best. It requires a certain amount of skill, however, which is forthcoming with experience. A properly executed streak plate will give as good an isolation as is desired for most work. Figure 17.1 shows what a properly made streak plate looks like after incubation. Note how well spaced the colonies are.

Materials:

- electric hot plate (or tripod and wire gauze)
- Bunsen burner
- beaker of water
- wire loop, thermometer, and china marking pencil
- 1 nutrient agar pour
- 1 sterile petri plate
- 1 mixed culture of *Serratia marcescens, Micrococcus luteus,* and *Escherichia coli*

1. Prepare your tabletop by disinfecting its surface with the disinfectant that is available in the laboratory (Roccal, Zephiran, Betadine, etc.). Use a sponge to scrub it clean.
2. Label the bottom surface of a sterile petri plate with your name and date. Use a china marking pencil.
3. Liquefy a tube of nutrient agar, cool to 50° C, and pour the medium into the bottom of the plate, following the procedure illustrated in figure 17.2. Be sure to flame the neck of the tube prior to pouring to destroy any bacteria around the end of the tube.

   After pouring the medium into the plate, gently rotate the plate so that it becomes evenly distributed, but do not splash any medium up over the sides.

   Agar, the solidifying agent in this medium, becomes liquid when boiled and resolidifies at around 42° C. Failure to cool it prior to pouring into the plate will result in condensation of moisture on the cover. Any moisture on the cover is undesirable because if it drops down on the colonies, the organisms of one colony can spread to other colonies, defeating the entire isolation technique.

4. Streak the plate by one of the methods shown in figure 17.4. Your instructor will indicate which technique you should use.

Caution: Be sure to follow the routine in figure 17.3 for getting the organism out of culture.

5. Incubate the plate in an *inverted position* at 37° C for 24 to 48 hours. By incubating plates upside down, the problem of moisture on the cover is minimized.

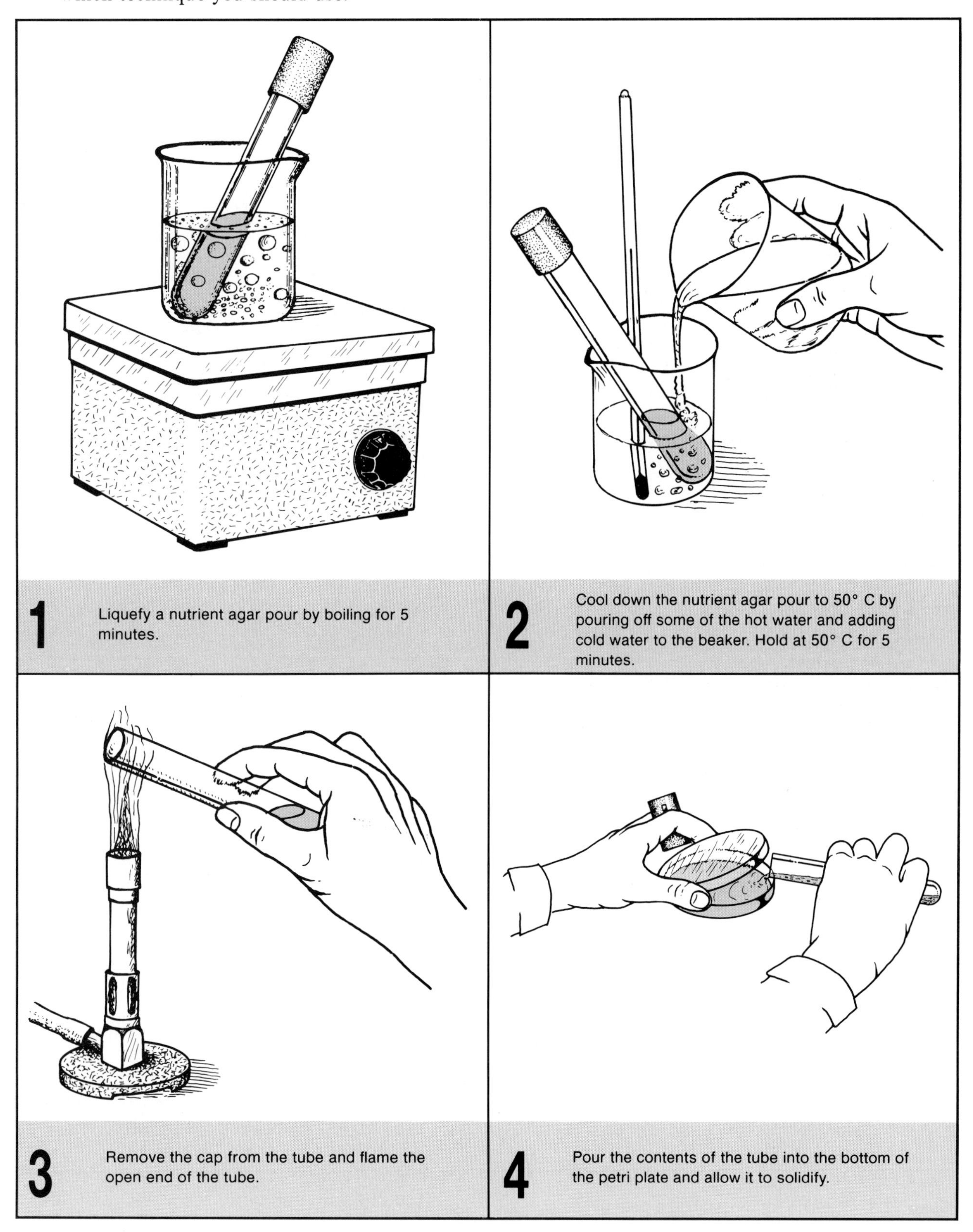

Figure 17.2 Procedure for pouring an agar plate for streaking.

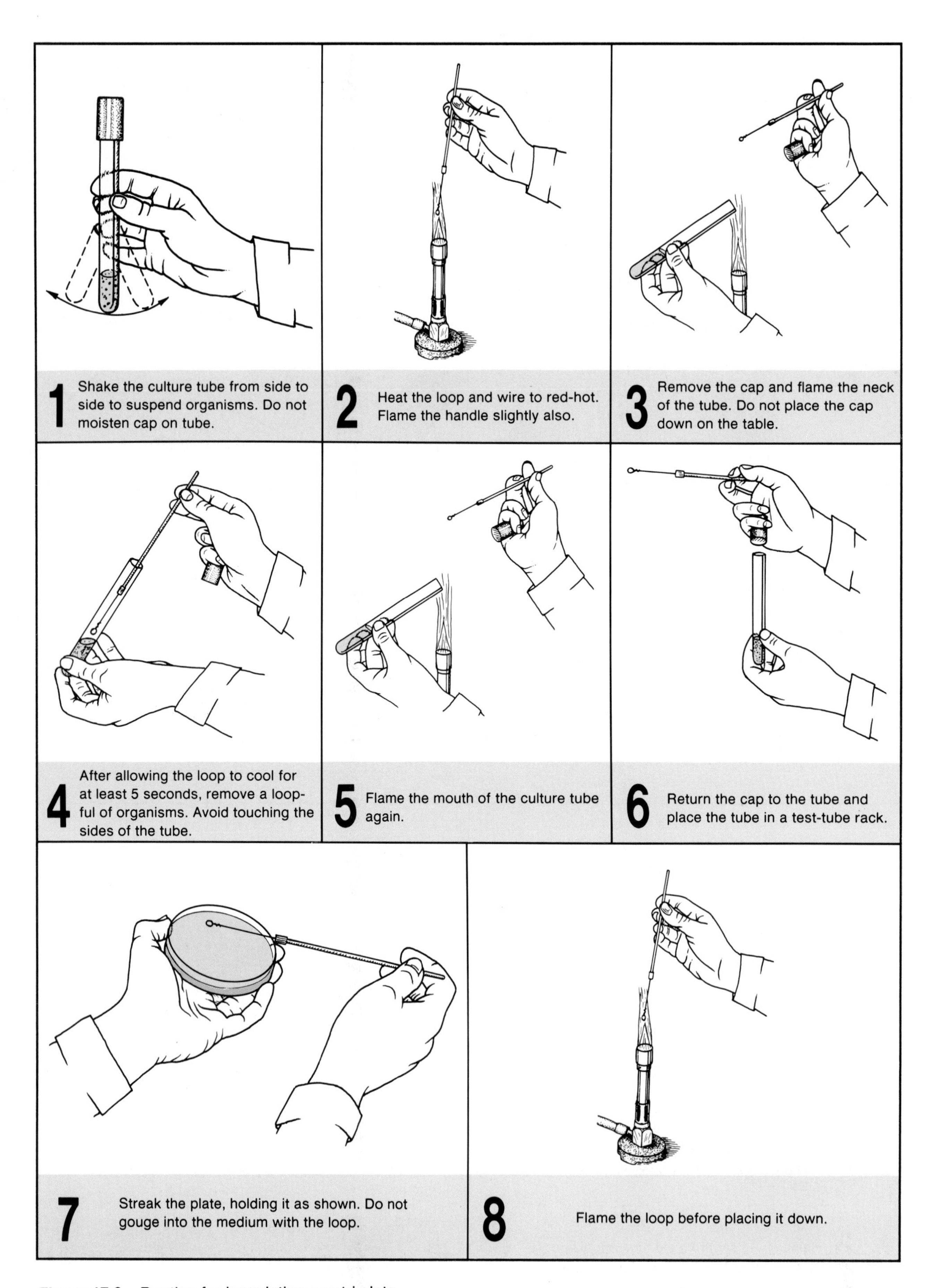

Figure 17.3 Routine for inoculating a petri plate.

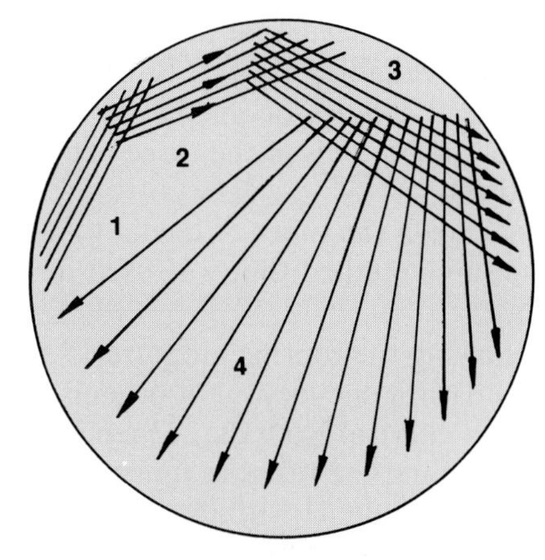

QUADRANT STREAK

1. Streak one loopful of organisms over Area 1 near edge of the plate. ***Apply the loop lightly.*** Don't gouge into the medium.

2. Flame the loop, cool 5 seconds, and make 5 or 6 streaks from Area 1 through Area 2.

3. Flame the loop again, cool it, and make 6 or 7 streaks from Area 2 through Area 3.

4. Flame the loop again and make as many streaks as possible from Area 3 into Area 4, using up the remainder of the plate surface.

5. Flame the loop before putting it aside.

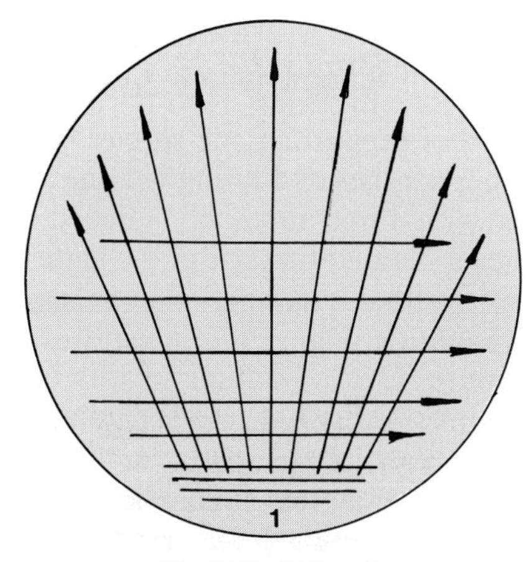

RADIANT STREAK

1. Spread a loopful of organisms in small area near edge of plate **(Area 1).**

2. Flame the loop and allow it to cool for 5 seconds.

3. **From the edge** of Area 1 make seven or eight straight streaks to the opposite side of the plate.

4. Flame the loop again and cross-streak over the last streaks, **starting near Area 1.**

5. Flame the loop again before putting it aside.

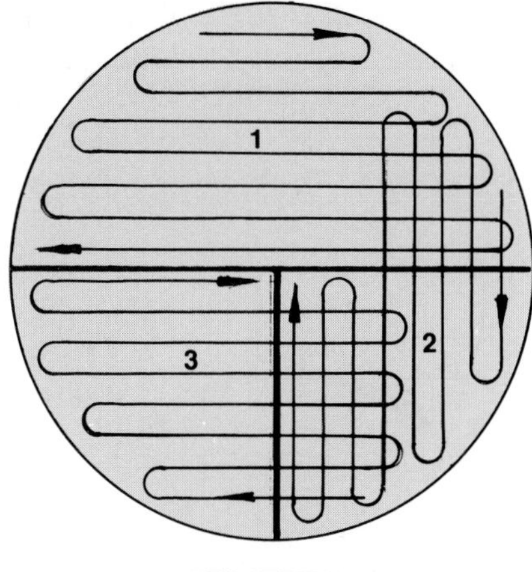

"T" STREAK

1. With marking pencil draw a "T" on bottom of plate, dividing the plate into one half and two quarters.

2. Inoculate the half portion with one loopful of organisms in a continuous line from edge of the plate to the midline of the plate. Don't gouge the medium.

3. After flaming the loop, cross-streak from Area 1 into Area 2 with one continuous streak, filling Area 2.

4. After flaming loop again, cross-streak from Area 2 to Area 3 with a single continuous streak.

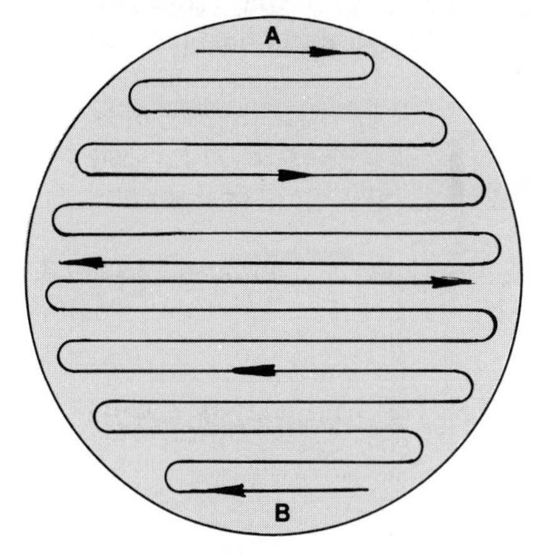

CONTINUOUS STREAK

1. Starting at the edge of the plate (Area A) with a loopful of organisms, spread the organisms in a single continuous movement to the center of the plate.

2. Rotate the plate 180 degrees so that the uninoculated portion of the plate is away from you.

3. Without flaming loop, and using the same face of the loop, continue streaking the other half of the plate by starting at Area B and working toward the center.

4. Flame your loop before setting it aside.

Figure 17.4 Four ways to make a streak plate.

Pour Plate Method

(Loop Dilution)

This method of separating one species of bacteria from another consists of diluting out one loopful of organisms with three tubes of liquefied nutrient agar in such a manner that one of the plates poured will have an optimum number of organisms to provide good isolation. Figure 17.5 illustrates the general procedure. One advantage of this method is that it requires somewhat less skill than that required for a good streak plate; a disadvantage, however, is that it requires more media, tubes, and plates. Proceed as follows to make three dilution pour plates, using the same mixed culture you used for your streak plate.

Materials:

mixed culture of bacteria
3 nutrient agar pours
3 sterile petri plates
electric hot plate
beaker of water
thermometer
inoculating loop and china marking pencil

1. Label the three nutrient agar pours **I, II,** and **III** with a marking pencil and place them in a beaker of water on an electric hot plate to be liquefied. To save time start with hot tap water if it is available.
2. While the tubes of media are being heated, label the bottoms of the three petri plates **I, II,** and **III.**
3. Cool down the tubes of media to 50° C, using the same method that was used for the streak plate.
4. Following the routine in figure 17.5, inoculate tube I with one loopful of organisms from the mixed culture. Note the sequence and manner of handling the tubes in figure 17.6.
5. Inoculate tube II with one loopful from tube I after thoroughly mixing the organisms in tube I by shaking the tube from side to side or by rolling the tube vigorously between the palms of both hands. ***Do not splash any of the medium up onto the tube closure.*** Return tube I to the water bath.
6. Agitate tube II to completely disperse the organisms, then inoculate tube III with one loopful from tube II. Return tube II to the water bath.
7. Agitate tube III, flame its neck, and pour its contents into plate III.
8. Flame the necks of tubes I and II and pour their contents into their respective plates.
9. After the medium has completely solidified, incubate the *inverted* plates at 37° C for 24 to 48 hours.

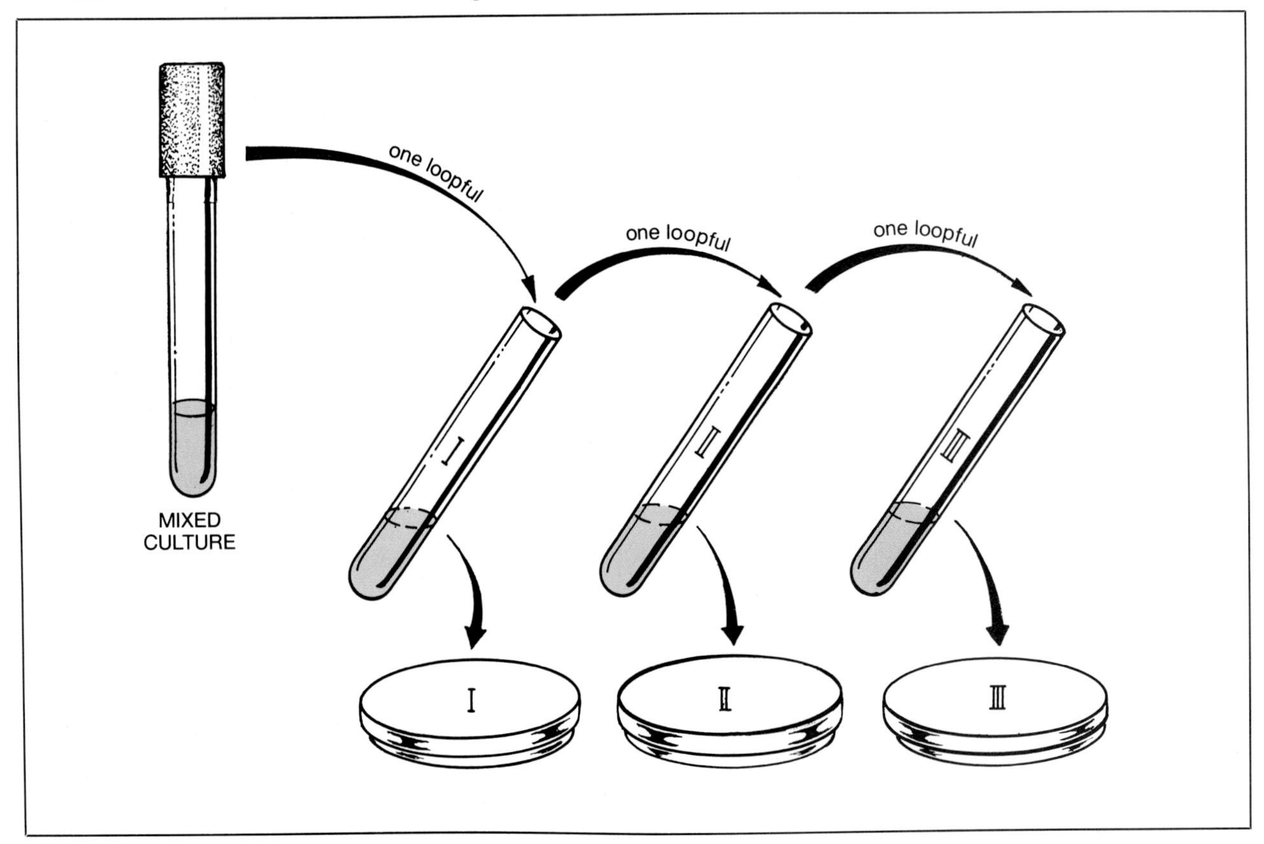

Figure 17.5 Three steps in the loop dilution technique for separating out organisms.

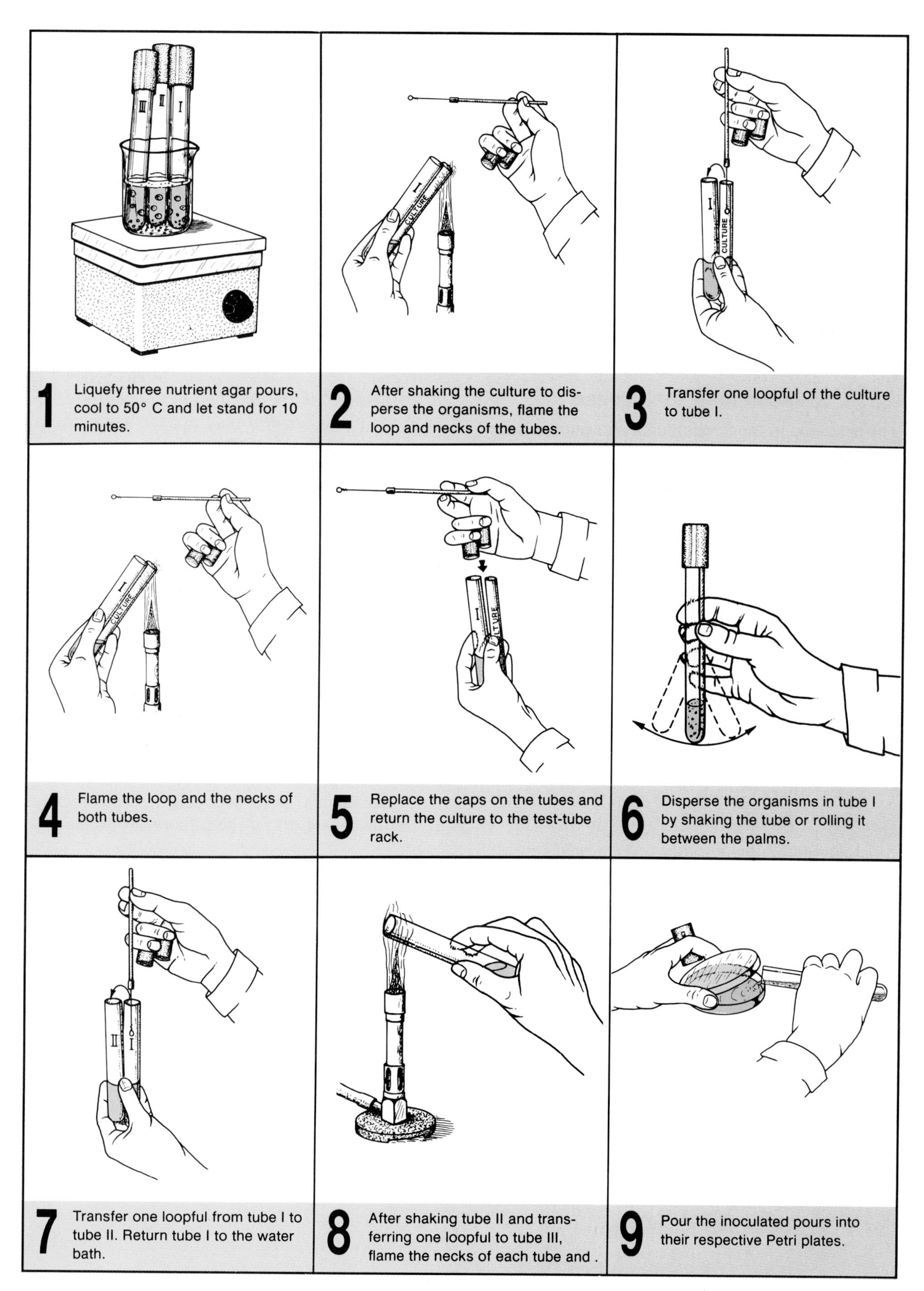

Figure 17.6 Tube handling procedure in making inoculations for pour plates.

Evaluation of Isolation Techniques

After 24 to 48 hours of incubation examine all four petri plates. Look for colonies that are well isolated from others. Note how crowded the colonies appear on pour plate I as compared with plates II and III. Plate I will be unusable. Either plate II or III will have the most favorable isolation of colonies. Can you pick out three well-isolated colonies on your best pour plate and streak plate that are white, yellow, and red?

Draw the appearance of your streak plate and pour plates on the Laboratory Report. Answer the questions pertaining to this portion of the experiment.

Subculturing Techniques

The next step in the development of a pure culture is to transfer the organisms from the petri plate to a tube of nutrient broth or a slant of nutrient agar. After this subculture has been incubated for 24 hours, a stained slide of the culture can be made to determine if a pure culture has been achieved. When transferring the organisms from the plate, an inoculating needle (straight wire) is used instead of the wire loop. The needle is inserted into the center of the colony where there is a greater probability of getting only one species of organisms. Use the following routine in subculturing out the three different organisms.

Materials:

3 nutrient agar slants
inoculating needle
Bunsen burner

1. Label one tube *Serratia marcescens,* another *Micrococcus luteus,* and the third *Escherichia coli.*
2. Select a well-isolated red colony on either the streak plate or pour plate for your first transfer. Insert the inoculating needle into the center of the colony.
3. In the tube labeled *Serratia marcescens* streak the slant by placing the needle near the bottom of the slant and drawing it up over its surface. One streak is sufficient.
4. Repeat this inoculating procedure on the other two slants for a yellow colony (*Micrococcus luteus*) and a white colony (*E. coli*).
5. Incubate for 24 to 48 hours at 37° C.

Evaluation of Slants

After incubation, examine the slants. Is *Serratia marcescens* red? Is *Micrococcus luteus* yellow? Is *E. coli* white? If the incubation temperature has been too high, *Serratia* may appear white due to the fact that the red pigment forms only at moderate temperatures. Draw the appearance of the slants with colored pencils on the Laboratory Report.

Although the colors of the growths on the slants may lead you to think that you have pure cultures, you cannot be absolutely certain until you have made a microscopic examination of each culture. For example, it is entirely possible that the yellow slant (*Micrococcus luteus*) may have some *E. coli* present that are masked by the yellow pigment.

To find out if you have a pure culture on each slant, make a gram-stained slide from each slant. Knowing that *Serratia marcescens* and *E. coli* are gram-negative rods and *Micrococcus luteus* is a gram-positive coccus, you should be able to evaluate your slants more precisely microscopically. Draw the organisms on the Laboratory Report.

Laboratory Report

Complete the Laboratory Report for this exercise.

Cultivation of Anaerobes 18

The procedures used to grow bacteria in the last exercise work well only if the organisms are aerobes or facultative anaerobes. Growing strict anaerobes, on the other hand, requires special equipment and media to exclude oxygen. The degree of oxygen tolerance of anaerobes extends over a considerable range; while some anaerobes will exhibit some growth when exposed to air, others will grow only in the complete absence of oxygen.

It is not the purpose of this exercise to dwell on the physiology of the different kinds of anaerobes at this time. However, when we get to the oxygen needs of unknowns in Exercise 36, we will explore this subject in greater detail.

In this laboratory session we will employ two popular methods that are used for growing anaerobic bacteria: the candle method and the GasPak system. Each method has its particular application. A brief mention will also be made of a few systems that have been used in the past, but are less frequently used today.

Older Methods

The Wright's tube, Brewer's petri dish, and the vacuum and gas displacement methods are worthy of mention here because under certain circumstances they might still be useful.

Wright's Tube Method Figure 18.1 illustrates the procedure that can be used to provide anaerobic conditions for a culture that has been streaked on an agar slant. Note that the cut portion of a cotton plug is pushed down to a point above the agar slant to provide a chamber into which equal amounts of pyrogallol and sodium hydroxide are added. A rubber stopper must be used to close the open end of the tube so that the tube can be incubated in an inverted position. Pyrogallol is used because it is a powerful reducing agent that is activated by NaOH. This combination results in the removal of oxygen in the tube. The greatest drawback of this system is its messiness. Removal of organisms from the tube is hampered to some extent by the wetness of the inside of the tube.

Brewer's Petri Dish Method Figure 18.2 illustrates how a thick Brewer dish cover provides an air space between itself and anaerobic agar on which organisms can grow. Good growth of anaerobes is possible because the anaerobic agar contains thioglycollate, a reducing agent that removes the oxygen in the small space. One of the problems with this method is that considerable moisture accumulates between the cover and the medium. If the organisms exist in mixed cultures it is difficult to separate the different species.

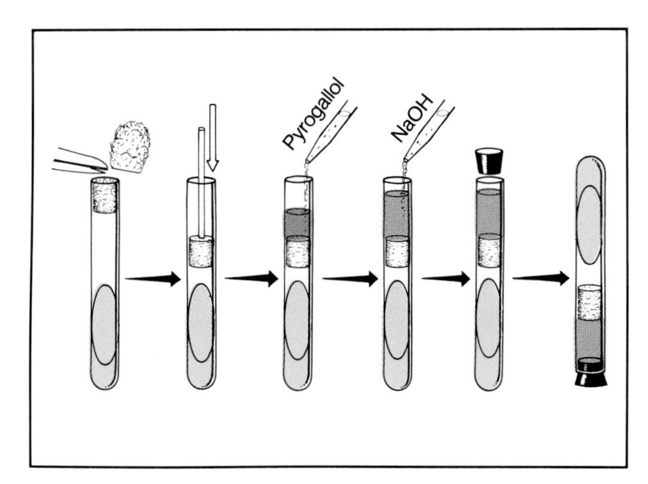

Figure 18.1 Wright's tube method: Anaerobic conditions for growth are achieved by adding equal amounts of pyrogallol and NaOH to chamber in culture tube.

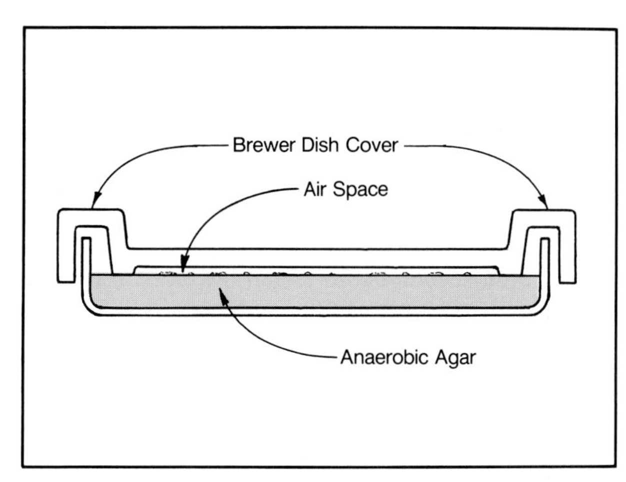

Figure 18.2 Brewer's petri dish method: Anaerobic organisms are provided an oxygen-free environment in shallow space between Brewer cover and anaerobic agar medium.

Vacuum and Gas Displacement Method This system (figure 18.3) requires considerable equipment: a vacuum pump, an anaerobic jar, and a tank that contains 95% nitrogen and 5% carbon dioxide. The procedure involves pumping the air out of the anaerobic jar so that it can be filled with the N_2-CO_2 mixture. Although this system works well, it has been replaced by the GasPak system since the latter is so much faster to set up.

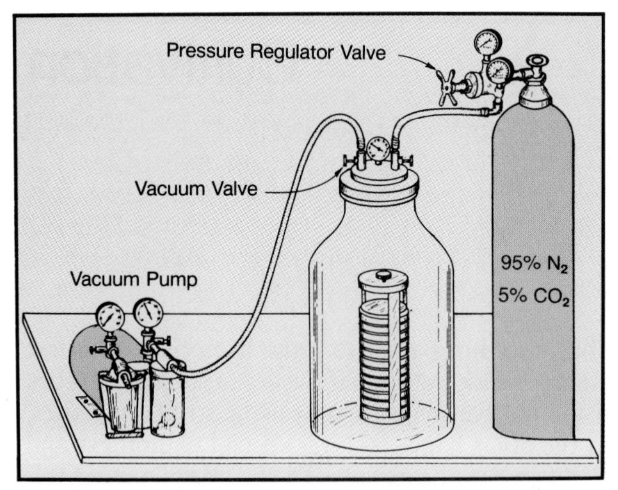

Figure 18.3 Vacuum and gas displacement method: Anaerobic jar is filled with N_2-CO_2 mixture after it is emptied of air with vacuum pump.

The Candle Method

If a lighted candle is placed in a container that is tightly sealed (figure 18.4) the oxygen will be used up, CO_2 will be produced, and the candle will quit burning once the oxygen has been exhausted. Although a setup like this will not produce strict anaerobic conditions, it works very well for the growth of some bacteria such as *Neisseria gonorrhoeae.* The enclosure need not be as sophisticated as a bell jar; a coffee can, if large enough, may suffice. To provide an airtight seal for the bell jar, one should use Vaseline between it and a glass plate. For a coffee can one should have a tight lid and use tape around the edge of the lid.

Although this method is an old one, it is still used for culturing *Neisseria.* For many organisms, however, it is considered too toxic because of the amount of carbon monoxide that is produced by the burning candle.

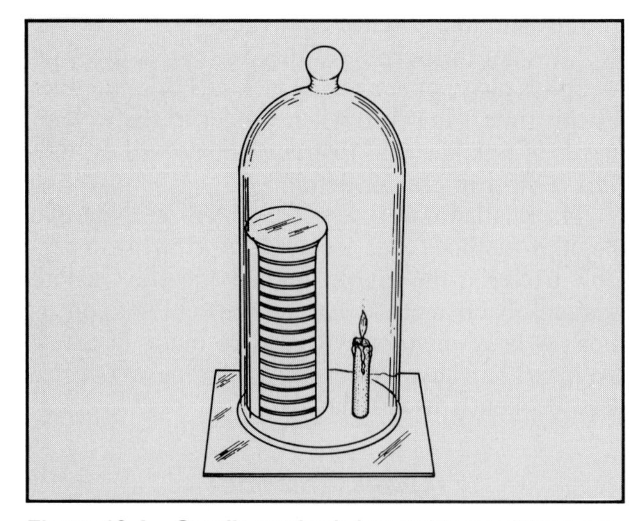

Figure 18.4 Candle method: Anaerobic conditions are achieved by burning candle in sealed container. Considered a good method for growing *Neisseria gonorrhoeae.*

The GasPak Anaerobic Jar

Because of its effectiveness and ease of use, the GasPak jar (figure 18.6) is undoubtedly the most widely used system for culturing anaerobic bacteria at the present time.

Note in the illustration that hydrogen is generated in the jar which removes the oxygen by forming water. Palladium pellets catalyze the reaction at room temperature. The generation of hydrogen is achieved by adding water to a plastic envelope of chemicals. Note also that CO_2 is produced, which is a requirement for the growth of many fastidious bacteria.

To make certain that anaerobic conditions actually exist in the jar, an indicator strip of methylene blue becomes colorless in the total absence of oxygen. If the strip is not reduced (decolorized) within two hours, the jar has not been sealed properly or the chemical reaction has failed to occur.

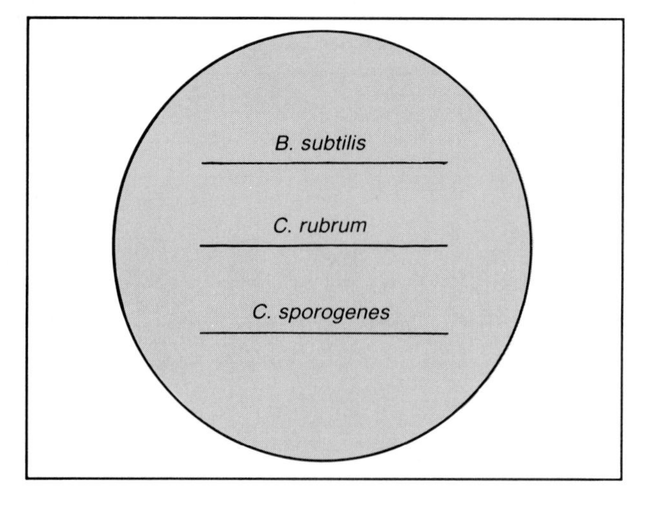

Figure 18.5 Three petri plates of Brewer's anaerobic agar are streaked in this manner for the experimental evaluation of candle and GasPak methods.

Experimental Procedures

Three petri plates of Brewer's anaerobic agar will be streaked with three organisms in the manner illustrated in figure 18.5. One plate will be used in a candle setup, another one will be placed in a GasPak jar, and the third one will be a control that is incubated without placing in any special anaerobic enviornment. The three organisms have differing oxygen requirements. It will be the purpose of this experiment to determine the oxygen needs of each species, and to learn how each method is performed. Proceed as follows:

Materials:

3 petri plates of Brewer's anaerobic agar
broth cultures of *Bacillus subtilis, Clostridium rubrum,* and *Clostridium sporogenes*
bell jar, glass plate, candle, Vaseline
GasPak anaerobic jar, 3 GasPak generator envelopes, 1 GasPak anaerobic generator strip, scissors, and one 10-ml pipette

1. With a china marking pencil draw three lines on the bottom of each petri plate in the manner illustrated in figure 18.5. Label one line ***B. subtilis,*** another ***C. rubrum,*** and the third one ***C. sporogenes.***

   In addition, label one plate **Control,** another **GasPak,** and the third **Candle.** Finally, put your initials and date somewhere on the bottom of each plate.
2. Inoculate each plate with three straight line streaks of the organisms along their respective lines.
3. Place your plates in the petri dish holders that are set aside for the three different methods.
4. After the plates are placed in their respective containers, proceed as follows if it is your responsibility to complete the setups.

Candle Method

1. Place the petri dishes in the center of the glass plate and light the candle which has been affixed to the glass plate with melted candle wax.
2. Apply a liberal amount of Vaseline around bottom of bell jar and place it over the petri plates and candle. Make certain that the jar-to-glass plate seal is tight by rotating the jar a little against the glass plate.
3. Once the candle flame expires place the entire assembly into a 37° C incubator.
4. Incubate for 24 to 48 hours.

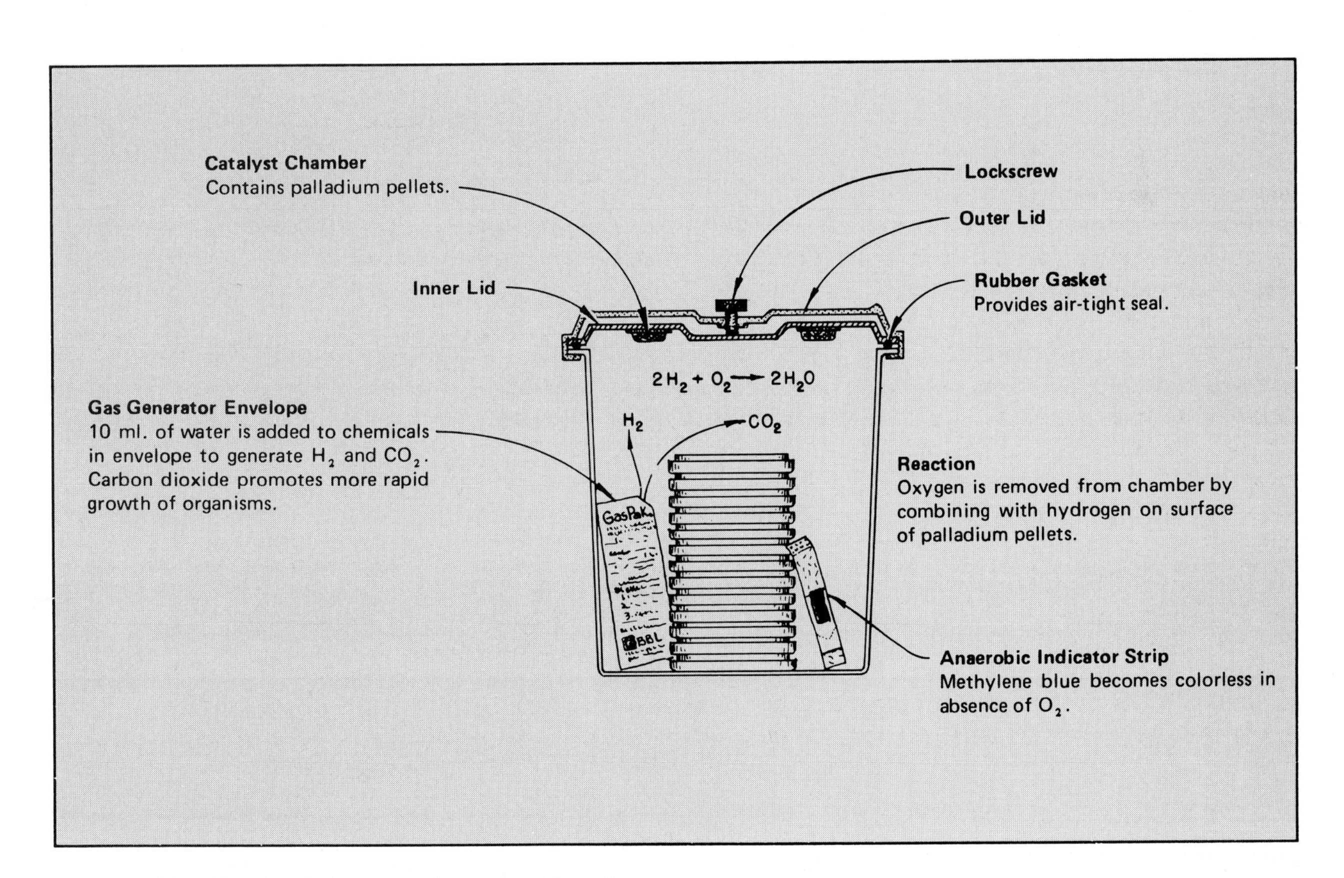

Figure 18.6 The GasPak system for anaerobic cultures.

GasPak Method

1. Place the plates into the GasPak jar.
2. Peel apart the foil at one end of a GasPak indicator strip, and pull it halfway down. The indicator will turn blue on exposure to the air. Place the indicator strip in the jar so that the wick is visible.
3. Cut off the corner of each of three GasPak gas generator envelopes with a pair of scissors. Place them in the jar in an upright position.
4. Pipette 10 ml of tap or distilled water into the open corner of each envelope. Avoid forcing the pipette into the envelope.
5. Place the inner section of the lid on the jar making certain it is centered on top of the jar. Do not use grease or other sealant on the rim of the jar since the O-ring gasket provides an effective seal when pressed down on a clean surface.
6. Unscrew the thumbscrew of the outer lid until the exposed end is completely withdrawn into the threaded hole. Unless this is done, it will be impossible to engage the lugs of the jar with the outer lid.
7. Place the outer lid on the jar directly over the inner lid and rotate the lid slightly to allow it to drop in place. Now rotate the lid firmly to engage the lugs. The lid may be rotated in either direction.
8. Tighten the thumbscrew by turning clockwise. If the outer lid raises up, the lugs are not properly engaged.
9. Place the jar in a 37° C incubator.
10. After two or three hours check the jar to note if the indicator strip has lost its blue color. If decolorization has not occurred, replace the palladium pellets and repeat the entire process.
11. After 24 to 48 hours incubation, remove the lid. If a vacuum holds the inner lid firmly to the jar, break the vacuum by sliding the lid to the edge.

Plate Evaluation and Spore Study

Plates: Examine the three plates and record your observations on the Laboratory Report.

Spore Study: Make a combined slide with three separate smears of the three spore-formers, using either one of the two spore-staining methods in Exercise 13. Draw the organisms in the circles provided on the Laboratory Report.

Bacterial Population Counts 19

Many bacteriological studies require that we be able to determine the number of organisms that are present in a given unit of volume. Several different methods are available to us for such population counts. The method one uses is determined by the purpose of the study.

To get by with a minimum of equipment, it is possible to do a population count by diluting out the organisms and counting the organisms in a number of microscopic fields on a slide. Direct examination of milk samples with this technique can be performed very quickly, and the results obtained are quite reliable. A technique similar to this can be performed on a Petrof-Hauser counting chamber.

Bacterial counts of gas-forming bacteria can be made by inoculating a series of tubes of lactose broth and using statistical probability tables to estimate bacterial numbers. This method, which we will use in Exercise 46 to estimate numbers of coliform bacteria in water samples, is easy to use, works well in water testing, but is limited to water, milk, and food testing.

In this exercise we will use **quantitative plating** (standard plate count, or SPC) and **turbidity measurements** to determine the number of bacteria in a culture sample. Although the two methods are somewhat parallel in the results they yield, there are distinct differences. For one thing, the SPC reveals information only as related to viable organisms; that is, colonies that are seen on the plates after incubation represent only living organisms, not dead ones. Turbidimetry results, on the other hand, reflect the presence of all organisms in a culture, dead and living.

Quantitative Plating Method
(Standard Plate Count)

In determining the number of organisms present in water, milk, and food, the **standard plate count** (SPC) is universally used. It is relatively easy to perform and gives excellent results. We can also use this basic technique to calculate the number of organisms in a bacterial culture. It is in this respect that this assignment is set up.

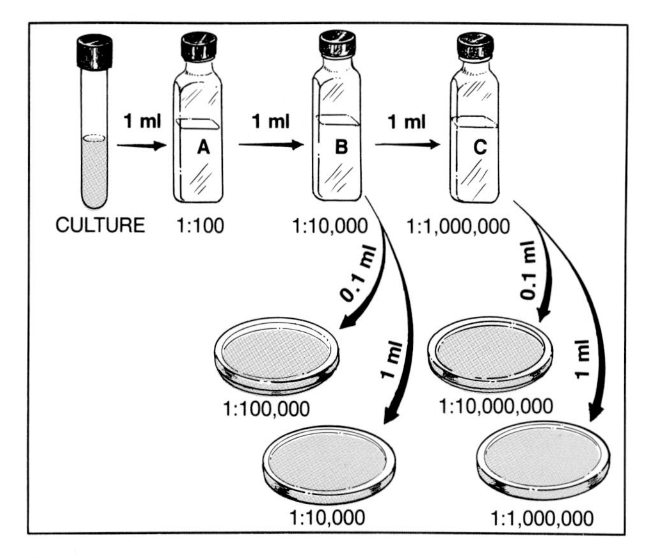

Figure 19.1 Quantitative plating procedure.

The procedure consists of diluting the organisms with a series of sterile water blanks as illustrated in figure 19.1. Generally, only three bottles are needed, but more could be used if necessary. By using the dilution procedure indicated below, a final dilution of 1:1,000,000 occurs in blank C. From blanks B and C, measured amounts of the diluted organisms are transferred into empty petri plates. Nutrient agar, cooled to 50° C, is then poured into each plate. After the nutrient agar has solidified, the plates are incubated for 24 to 48 hours and examined. A plate which has between 30 and 300 colonies is selected for counting. From the count it is a simple matter to calculate the number of organisms per milliliter of the original culture. It should be pointed out that greater accuracy can be achieved by pouring two plates for each dilution and averaging the counts. Duplicate plating, however, has been avoided for obvious economic reasons.

Pipette Handling

Success in this experiment depends considerably on proper pipetting techniques. Pipettes may be available to you in metal cannisters or in individual envelopes; they may be disposable or reusable. In some laboratories pipetting must be done only with a mechanical pipetter; in other laboratories, delivery by mouth is preferred. Your instructor will indicate the techniques that will prevail in this laboratory. If this is the first time that you have used sterile pipettes, consult figure 19.2, keeping the following points in mind.

- When removing a sterile pipette from a cannister, do so without contaminating the ends of the other pipettes with your fingers. This can be accomplished by *gently* moving the cannister from side to side in an attempt to isolate one pipette from the rest. The obvious reason for avoiding the other pipettes is to keep your germs off the ends of pipettes that someone else is going to have to put into his or her mouth.
- After removing your pipette, replace the cover on the cannister to maintain sterility of the remaining pipettes.
- Don't touch the body of the pipette with your fingers or lay the pipette down on the table before or after you use it. **Keep that pipette sterile** until you have used it, and don't contaminate the table or yourself with it after you have used it.
- If a mechanical pipetting device is to be used, utilize your thumb, as shown in figure 19.3, to regulate the filling and delivery.
- If the pipette is to be filled by mouth suction, draw the fluid up into the pipette just past the mark you desire, stopping the upward flow with your tongue. With the index fingertip placed over the end of the pipette, allow the fluid to flow downward

Figure 19.2 Pipette handling techniques.

slowly to the desired mark. **Don't use your thumb to regulate the downward flow.** Use of the index finger to control flow requires a little practice before it becomes easy to perform; however, once learned, the technique is more controllable than can be achieved with the thumb.

• Remove and use only one pipette at a time; if you need three pipettes for the whole experiment and remove all three of them at once, there is no way that you will be able to keep two of them sterile while you are using the first one.

• When finished with a pipette, place it in the *discard cannister.* The discard cannister will have a disinfectant in it. At the end of the period the pipettes will be washed and sterilized by the laboratory assistant. Students have been known to absentmindedly return used pipettes to the original sterile cannister, and, occasionally, even toss them into the wastebasket. We are certain that no one in this laboratory would *ever* do that!

Diluting and Plating Procedure

Proceed as follows to dilute out a culture of *E. coli* and pour four plates, as illustrated in figure 19.1.

Materials:

per four students:
 1 bottle (35 ml) broth culture of *E. coli*
per student:
 1 bottle (50 ml) nutrient agar
 4 petri plates
 1.1-ml pipettes
 3 sterile 99-ml water blanks
discard cannister on demonstration table

1. Liquefy a bottle of nutrient agar. While it is being heated, label three 99-ml sterile water blanks A, B, and C. Also, label the four petri plates 1:10,000, 1:100,000, 1:1,000,000, and 1:10,000,000.
2. Shake the culture of *E. coli* and transfer 1 ml of the organisms to blank A, using a sterile 1.1-ml pipette. After using the pipette, place it in the discard cannister.
3. Shake water blank A 25 times, as illustrated in figure 19.4. Forceful shaking not only brings about good distribution, but also breaks up clumps of bacteria.
4. With a fresh 1.1-ml pipette, transfer 1 ml from blank A to blank B.
5. Shake water blank B 25 times and, with another sterile pipette, transfer 0.1 ml to the 1:100,000 plate and 1.0 ml to the 1:10,000 plate. With the same pipette, transfer 1.0 ml to blank C.
6. Shake blank C 25 times and, with another sterile pipette, transfer 0.1 ml to the 1:10,000,000 plate and 1.0 ml to the 1:1,000,000 plate.
7. After the bottle of nutrient agar has boiled for 8 minutes, cool in a water bath at 50° C for 10 minutes.

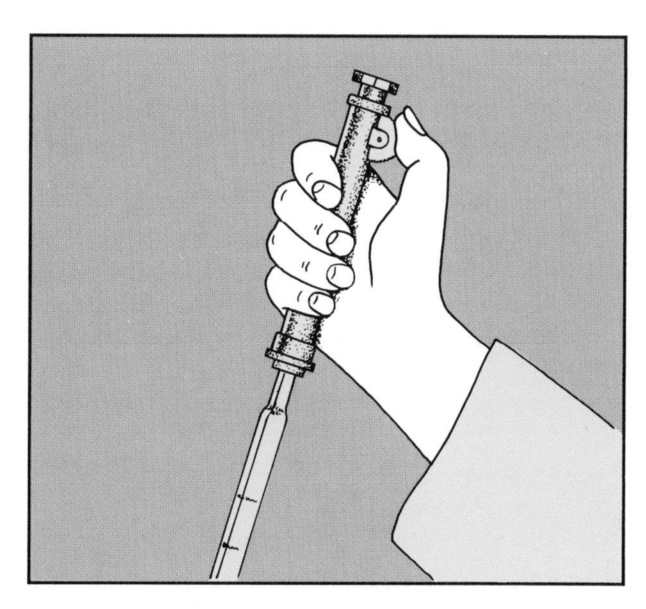

Figure 19.3 The thumb provides good control of delivery when using a mechanical pipetting device.

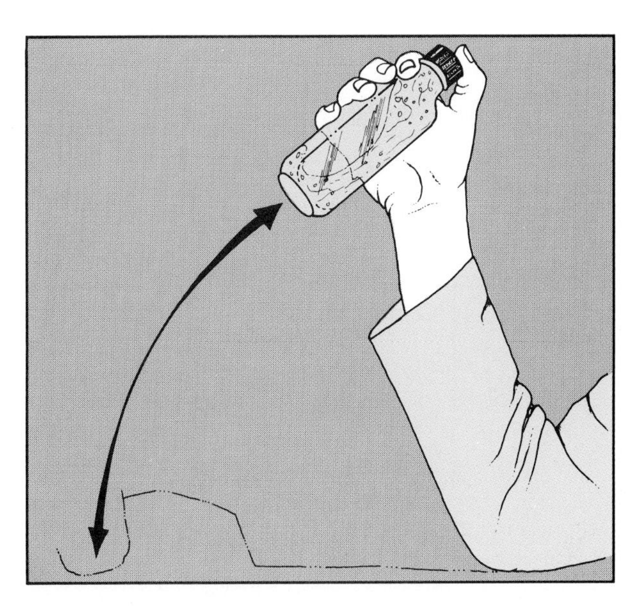

Figure 19.4 The standard procedure for shaking water blanks requires that the elbow remain fixed on the table.

8. Pour one-fourth of the nutrient agar into each of the four plates. Rotate the plates gently to get adequate mixing of medium and organisms.
9. After the medium has cooled completely, incubate at 35° C for 48 hours, inverted.

Counting and Calculations

Materials:

4 culture plates
Quebec colony counter
mechanical hand counter

1. Lay out the plates on the table in order of dilution and compare them. *Select the plate that has no fewer than 30 nor more than 300 colonies for your count.* Plates with less than 30 or more than 300 colonies are statistically unreliable.
2. Place the plate on the Quebec colony counter with the lid removed. See figure 19.5. Start counting at the top of the plate, using the grid lines to prevent counting the same colony twice. Use a mechanical hand counter. Count every colony, regardless of how small or insignificant.
3. Calculate the number of organisms per ml of culture by multiplying the number of colonies counted by the dilution factor. For example, if you counted 220 colonies on the 1:1,000,000 dilution plate,

   $220 \times 1{,}000{,}000 = 220{,}000{,}000$ bacteria per ml.

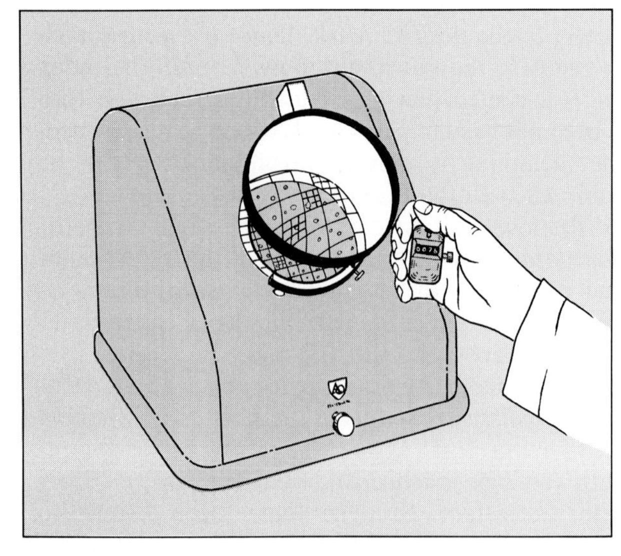

Figure 19.5 Colony counts are made on Quebec colony counter, using a mechanical hand tally.

   Use only two significant figures. If your count, for example, was 227, your number of organisms would be 230,000,000.
4. Record your results on the Laboratory Report.

Turbidimetry Determinations

When it is necessary to make bacteriological counts on large numbers of cultures, the quantitative plate count method becomes a rather cumbersome tool. It not only takes a considerable amount of glassware and media, but it is also time-consuming. A much faster method is to measure the turbidity of

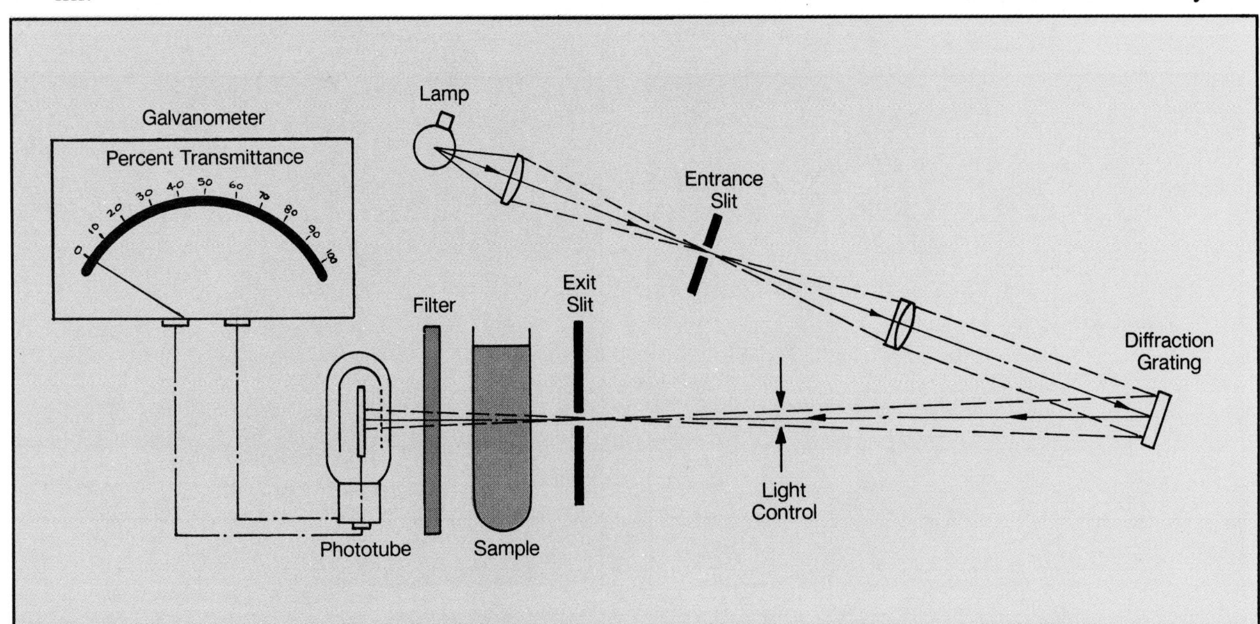

Figure 19.6 Schematic of a photocolorimeter.

the culture with a photocolorimeter and translate this into the number of organisms. To accomplish this, however, the plate count method is used to establish the count for one culture of known turbidity.

To understand how a photocolorimeter works, it is necessary, first, to recognize the fact that a culture of bacteria acts as a colloidal suspension which will intercept light as it passes through. Within certain limits the amount of light that is absorbed is directly proportional to the concentration of cells. A photocolorimeter is an instrument which has a photocell that can measure the amount of light that passes through the culture. Figure 19.6 illustrates the path of light through such an instrument. A light source in the instrument transmits a beam of white light through two lenses and an entrance slit into a diffraction grating which disperses the light into horizontal beams of all colors of the spectrum. Short wavelengths (violet and ultraviolet) are at one end and long wavelengths (red and infrared) are at the other end. The spectrum of light falls on a dark screen with a slit (exit slit) cut in it. Only that portion of the spectrum which happens to fall on the slit goes through into the sample. It will be a monochromatic beam of light. By turning a wavelength control knob on the instrument, the diffraction grating can be reoriented to allow different wavelengths to pass through the slit. The light that gets through the culture activates a phototube, which, in turn, registers **percent transmittance** on a galvanometer. The higher the percent transmittance, the fewer are the cells in suspension.

Before the turbidity can be determined, however, the instrument has to be calibrated with a tube of sterile nutrient broth to establish 100% transmittance. After it is calibrated, turbidity, in terms of transmittance, is read by inserting a cuvette (tube) of the culture into the sample holder of the instrument.

To illustrate the direct proportional relationship between the concentration of bacterial cells and the absorbance of light (**optical density,** O. D.) you will measure the percent transmittance of various dilutions of the culture you used for the plate count. These values will be converted to optical density and plotted on a graph. Figure 19.8 illustrates the general procedure of setting up the dilutions.

Materials:

broth culture of *E. coli* (same one as used for plate count)

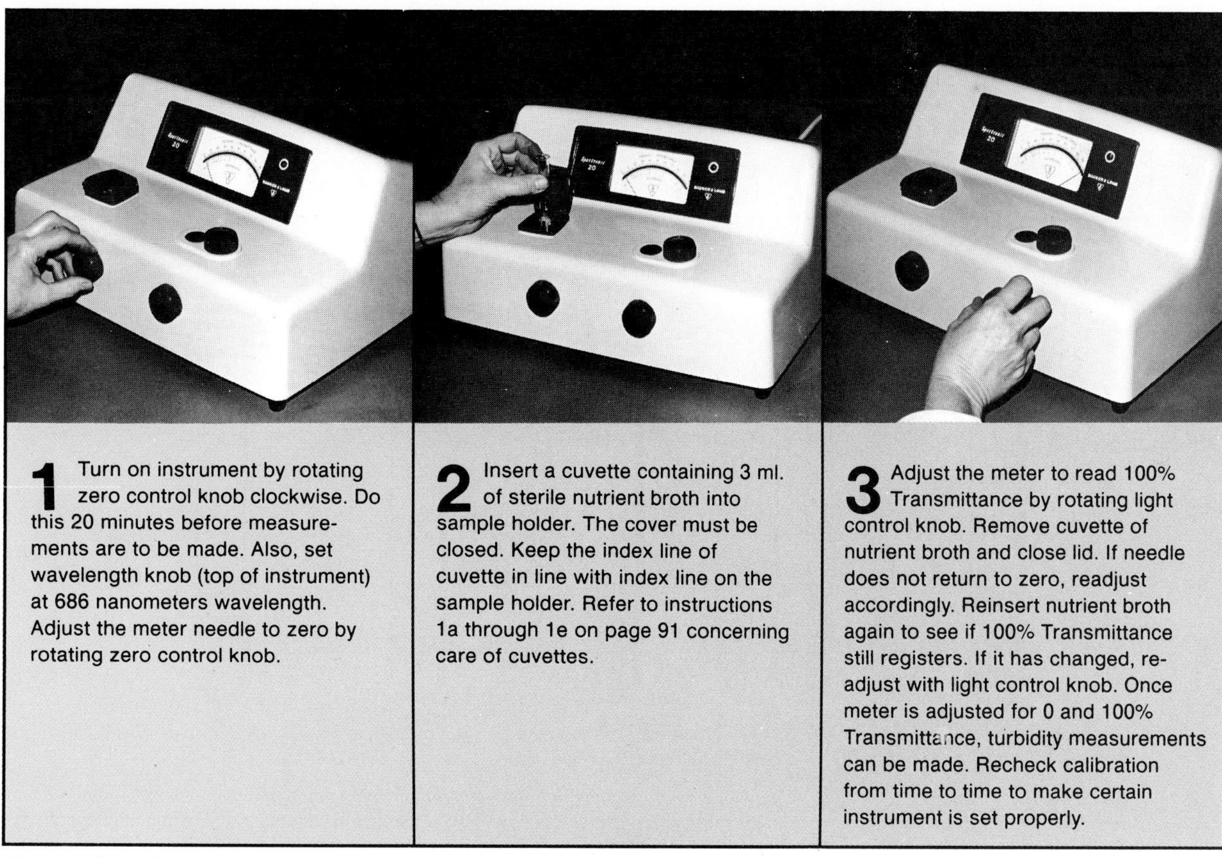

Figure 19.7 Calibration procedure for B & L Spectronic 20.

photocolorimeter cuvettes (5 per student)
5-ml pipettes
bottle of sterile nutrient broth (20 ml per student)

1. Calibrate the photocolorimeter, using the procedure described in figure 19.7. These instructions are specifically for the Bausch and Lomb Spectronic 20. In handling the cuvettes, keep the following points in mind:
   a. Rinse the cuvette several times with distilled water to get it clean before using.
   b. Keep the lower part of the cuvette spotlessly clean by keeping it free of liquids, smudges, and fingerprints. Wipe it clean with Kimwipes or some other lint-free tissue. Don't wipe cuvettes with towels or handkerchiefs.
   c. Insert the cuvette into the sample holder, with its index line registered with the index line on the holder.
   d. After the cuvette is seated, line up the index lines exactly.
   e. Handle these tubes with great care. They are expensive.
2. Label the five cuvettes near the top: 1:1, 1:2, 1:4, 1:8, 1:16.
3. With a 5-ml pipette, dispense 3 ml of sterile nutrient broth into tubes 1:2, 1:4, 1:8, and 1:16.
4. Shake the culture of *E. coli* vigorously to suspend the organisms. With the 5-ml pipette,

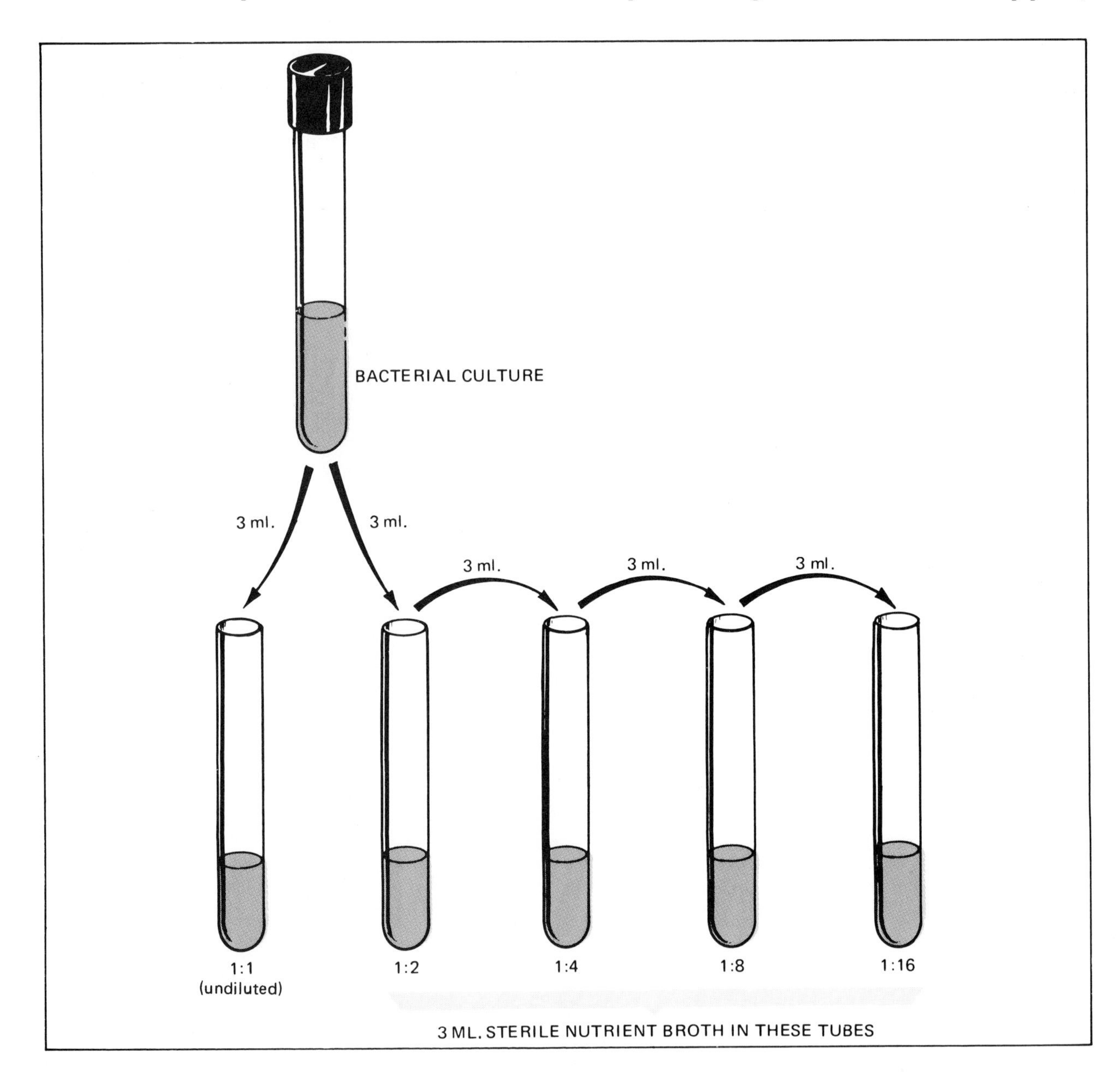

Figure 19.8 Dilution procedure for cuvettes.

transfer 3 ml to the 1:1 cuvette and 3 ml to the 1:2 cuvette.

5. Mix the contents in the 1:2 tube by drawing the mixture up into the pipette and discharging it into the tube three times.
6. Transfer 3 ml from the 1:2 tube to the 1:4 tube, mix three times, and go on to the other tubes in a similar manner. Tube 1:16 will have 6 ml of diluted organisms.
7. Measure the percent transmittance of these five tubes by inserting each one into the sample holder and closing the lid. Record the readings on the Laboratory Report.
8. Convert the percent transmittance values to optical density (O.D.), using the following formula:

   O.D. = 2 − log of percent transmittance.

   Example: If the percent transmittance of one of your dilutions is 53.5, you would solve the problem in this way:

   $$\begin{aligned} \text{O.D.} &= 2 - \log \text{ of } 53.5 \\ &= 2 - 1.7284 \\ &= .272. \end{aligned}$$

   Refer to table II of appendix A for logarithms.

Logarithm Refresher: In case you have forgotten how to use logarithms, recall these facts:

Mantissa: The value you find in the log table (.7284 in the previous example) is the mantissa.

Characteristic: The number to the left of the decimal (1 in the example) is the characteristic. This figure (the characteristic) is always one number less than the number of digits of the figure you are looking up.

Examples:

number	*characteristic*	*mantissa*
5.31	0	.7251
531	2	.7251

Although the galvanometer may show absorbance (O.D.) values, greater accuracy will result from calculating them from percent transmittance.

9. Record the O.D. values in the table of the Laboratory Report.
10. Plot the O.D. values on the graph of the Laboratory Report.

20 Slide Culture: Molds

The isolation, culture, and microscopic examination of molds require the use of suitable selective media and special microscopic slide techniques. If simple wet mount slides of molds were attempted in Exercise 6, it became apparent that wet mount slides made from mold colonies usually don't reveal the arrangement of spores that is so necessary in identification. The process of merely transferring hyphae to a slide breaks up the hyphae and sporangiophores in such a way that identification becomes very difficult. In this exercise a slide culture method will be used to prepare stained slides of molds. The method is superior to wet mounts in that the hyphae, sporangiophores, and spores remain more or less intact when stained.

When molds are collected from the environment, as in Exercise 6, Sabouraud's agar is most frequently used. It is a simple medium consisting of 1% peptone, 4% glucose, and 2% agar. The pH of the medium is adjusted to 5.6 to inhibit bacterial growth.

Unfortunately, for some molds the pH of Sabouraud's agar is too low and the glucose content is too high. A better medium for these organisms is one suggested by C. W. Emmons that contains only 2% glucose with 1% neopeptone, and an adjusted pH of 6.8–7.0. To inhibit bacterial growth, 40 mg of chloramphenical is added to one liter of the medium.

In addition to the above two media, cornmeal agar, Czapek solution agar, and others are available for special applications in culturing molds.

Figure 20.2 illustrates the procedure that will be used to produce a mold culture on a slide that can be stained directly on the slide. Note that a sterile cube of Sabouraud's agar is inoculated on two sides with spores from a mold colony. Figure 20.1 illustrates how the cube is held with a scalpel blade as inoculation takes place. The cube is placed in the center of a microscope slide with one of the inoculated surfaces placed against the slide. On the other inoculated surface of the cube is placed a cover glass. The assembled slide is incubated at room temperature for 48 hours in a moist chamber (petri dish with a small amount of water). After incubation the cube of medium is carefully separated from the slide and discarded.

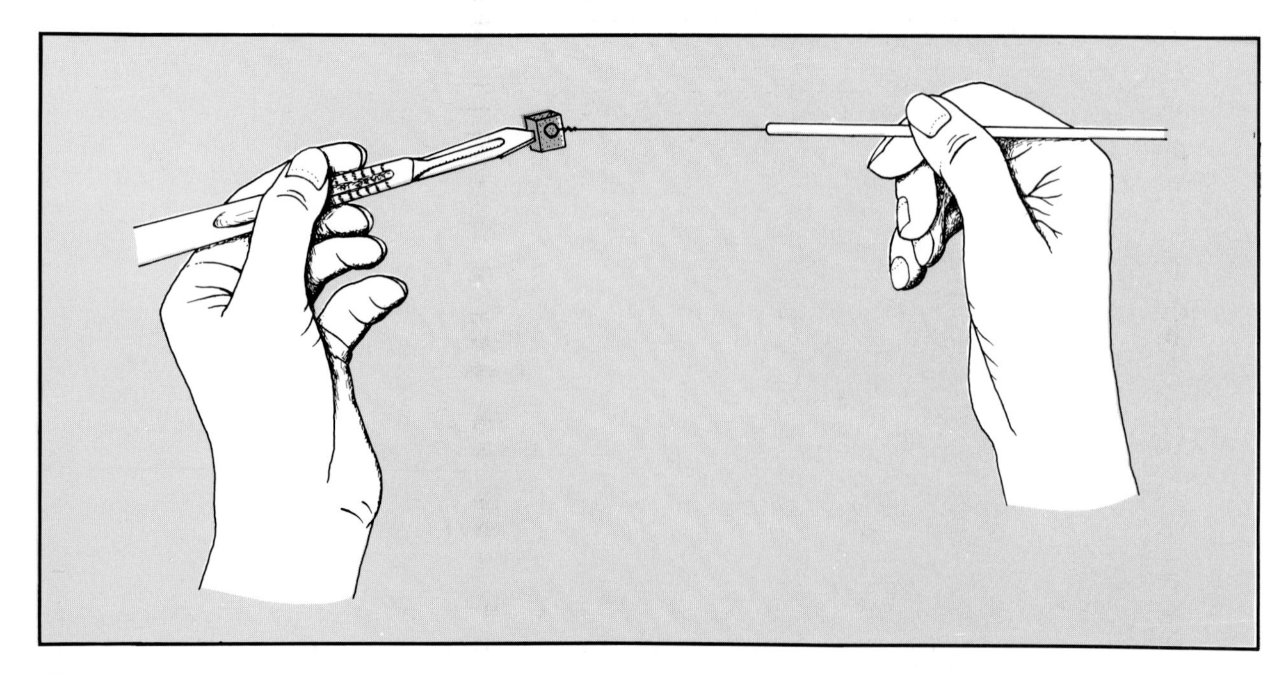

Figure 20.1 Inoculation technique.

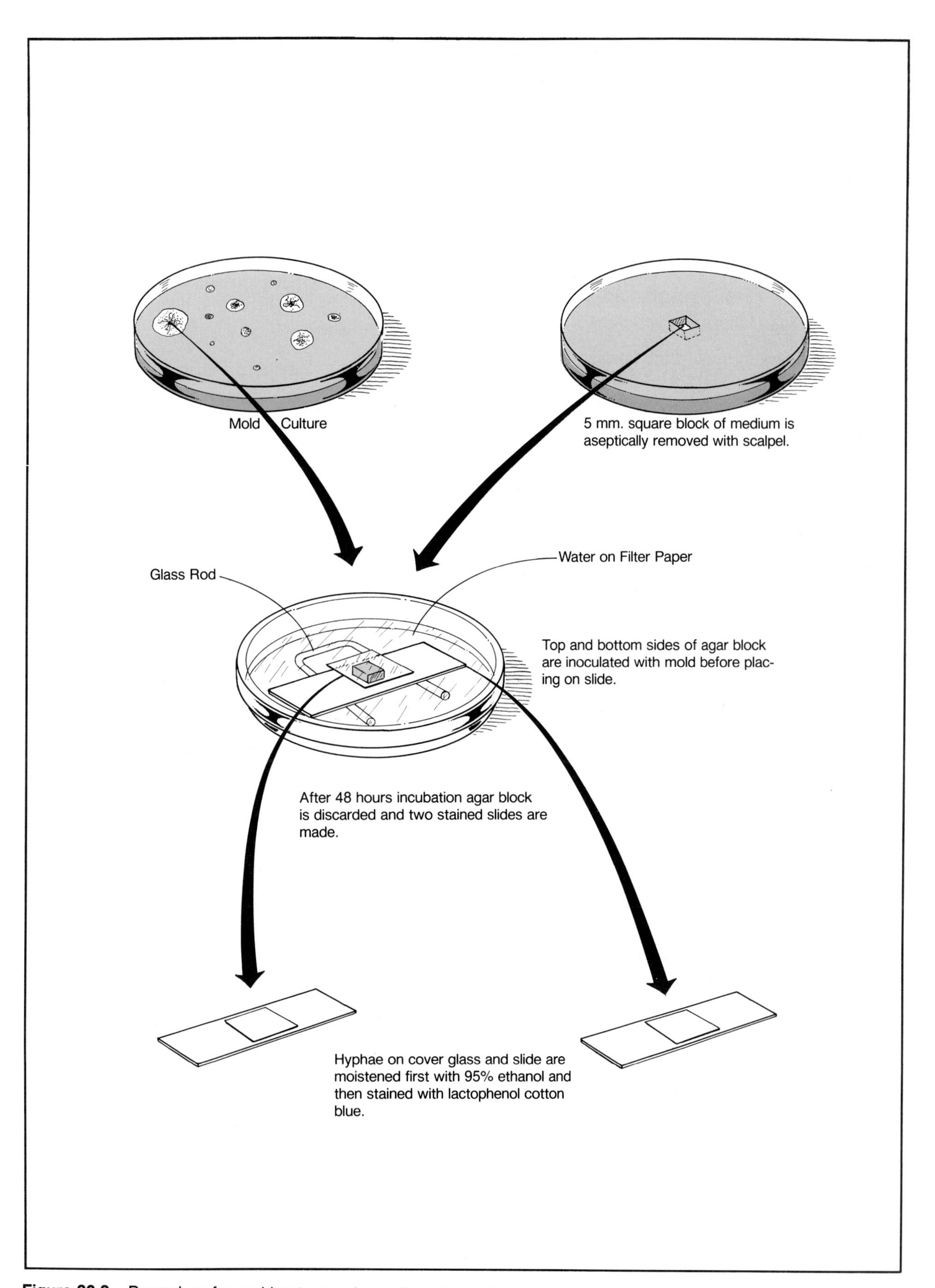

Figure 20.2 Procedure for making two stained slides from slide culture.

During incubation the mold will grow over the glass surfaces of the slide and cover glass. By adding a little stain to the slide, a semipermanent slide can be made by placing a cover glass over it. The cover glass can also be used to make another slide by placing it on another clean slide with a drop of stain on it. Before the stain (lactophenol cotton blue) is used, it is desirable to add a drop of alcohol to the hyphae. This acts as a wetting agent.

First Period

(Slide Culture Preparation)

Proceed as follows to make slide cultures of one or more mold colonies.

Materials:

petri dishes, glass, sterile
filter paper (9 cm dia., sterile)
glass U-shaped rods
mold culture plate (mixture)
1 petri plate of Sabouraud's agar or Emmons' medium for 4 students
scalpels
inoculating loop
sterile water
microscope slides and cover glasses (sterile)
forceps

1. Aseptically, with a pair of forceps, place a sheet of sterile filter paper into bottom of petri dish.
2. Place a sterile U-shaped glass rod on the filter paper. (Rod can be sterilized by flaming if held with forceps.)
3. Pour enough sterile water (about 4 ml) on filter paper to completely moisten it.
4. With a forceps, place a sterile slide on the U-shaped rod.
5. *Gently* flame a scalpel to sterilize, and cut a 5 mm square block of the medium from the plate of Sabouraud's agar or Emmons' medium.
6. Pick up the block of agar by inserting the scalpel into one side as illustrated in figure 20.1. Inoculate both top and bottom surfaces of the agar block with spores from the mold colony. Be sure to flame and cool the loop prior to picking up spores.
7. Place the inoculated block of agar in the center of a microscope slide. Be sure to place one of the inoculated surfaces down.
8. Aseptically, place a sterile cover glass on the upper inoculated surface of the agar cube.
9. Place the cover on the petri dish and incubate at room temperature for 48 hours.
10. After 48 hours examine the slide under low power. If growth has occurred you should see hyphae and spores. If growth is inadequate and spores are not evident, allow the mold to grow another 24 to 48 hours before making the stained slides.

Second Period

(Application of Stain)

As soon as there is evidence of spores on the slide, prepare two stained slides from the slide culture, using the following procedure:

Materials:

microscope slides and cover glasses
95% ethanol
lactophenol cotton blue stain
forceps

1. Place a drop of lactophenol cotton blue stain on a clean microscope slide.
2. Remove the cover glass from the slide culture and discard the block of agar.
3. Add a drop of 95% ethanol to the hyphae on the cover glass. As soon as most of the alcohol has evaporated place the cover glass, mold side down, on the drop of lactophenol cotton blue stain on the slide. This slide is ready for examination.
4. Remove the slide from the petri dish, add a drop of 95% ethanol to the hyphae and follow this up with a drop of lactophenol cotton blue stain. Cover the entire preparation with a clean cover glass.
5. Compare both stained slides under the microscope; one slide may be better than the other one.

Laboratory Report

There is no Laboratory Report for this exercise.

Slide Culture: Autotrophs 21

There is probably no single medium or method that one can use to do a comprehensive population count of all living microorganisms in a specific biosphere. Those media that we categorize as being "general purpose" will, for various reasons, inhibit the growth of many organisms. To make comparative studies of free-living organisms in freshwater lakes, A. T. Henrici, in 1932, devised an immersed slide technique that revealed the presence of many organisms that did not show up by other methods. Although his original concern was with algal populations, the technique works as well for bacteria and other microorganisms. His method consists of suspending glass microscope slides in the body of water for a specified period of time. Microorganisms in the water adhere to the glass and multiply to form small colonies that are observable under the microscope.

In this exercise we will suspend slides in aquaria or other containers of water known to have a stable microbial flora. After the slides have been immersed for about a week, a study will be made of stained and living forms to see what types predominate.

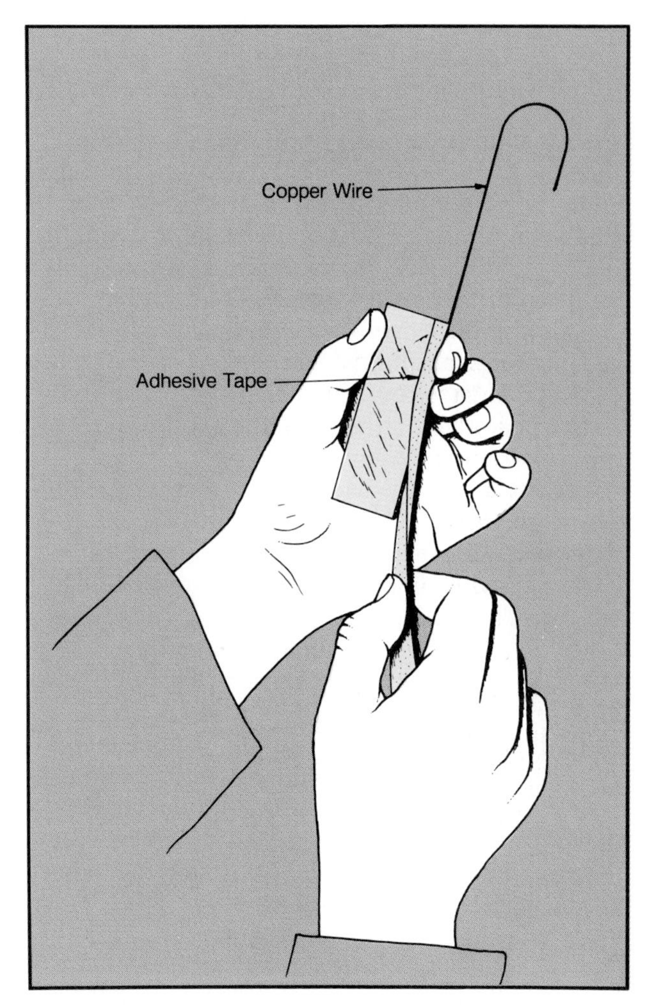

Figure 21.1 Preparation of slide for immersion.

Materials:

adhesive tape (½″ width)
microscope slides
copper wire
gummed labels
acid-alcohol

1. Clean two microscope slides as follows:
   a. Scrub with green soap or Bon Ami.
   b. Dip them in acid-alcohol for one minute and dry with tissue.
   c. Place them in a beaker of distilled water for five minutes to allow any residual solvent to dissipate.
2. Tape a piece of wire to one edge as illustrated in figure 21.1. Hold the slides back to back by their edges. Do not touch the flat surfaces with your fingers. Wrap all four edges with tape. For identification attach a gummed label with your name to the wire.
3. Suspend the slide in an aquarium or container of water that is known to have a stabilized natural flora of bacteria.
4. After one week remove the binding from the slides. Prepare one slide with Gram stain and place a drop of water and cover glass on the other one.
5. Examine both slides under oil immersion and record your observations on the Laboratory Report.

Reference: Henrici, A. T. 1933. Studies of Fresh Water Bacteria. *J. Bact.* 25 (3): 277–286.

22 Bacteriophage: Isolation and Culture

Viruses differ from bacteria in being much smaller than bacteria, lacking cellular structure and existing as intracellular parasites. Because they lack complete enzyme systems, they can only exist inside of cells that provide the materials they need.

Viruses that parasite bacteria are called *bacteriophage,* or *phage.* If phage virions and bacterial host cells are cultured together on nutrient agar, evidence of bacterial destruction (lysis) shows up on the agar plate in the form of clear areas called *plaques* (Figure 22.1).

Although many bacteria are parasitized by phage, we know more about the phage of *E. coli* than any other organism because more research has been focused upon them. The phage of these bacteria are, collectively, referred to as *coliphages.* Since *E. coli* is the predominant organism in the lower intestines, these phage can readily be isolated from sewage. It is from this source that we will attempt to isolate and culture bacteriophage.

Figure 22.2 illustrates how a typical coliphage invades an *E. coli* cell and within 45 minutes produces a whole family of new individuals by reorienting the bacterial cell physiology to viral production.

The phage illustrated in Figure 22.2 has a "head," or *capsid,* of protein and a hollow "tail." The capsid provides a protective shell for the DNA of the organism. The extreme end of the tail has the ability to become attached to specific receptor sites on the surface of phage-sensitive bacteria. Once the tail becomes attached to the cell, it literally digests its way through the host cell wall.

In this experiment we shall attempt to isolate phage from sewage. Three steps are involved: enrichment, filtration, and seeding. Figures 22.3 and 22.4 illustrate the procedures to be followed. Note in figure 22.3 that *enrichment* of phage numbers is accomplished with a special medium before *filtration* is used to separate the phage from the bacteria. As illustrated in figure 22.4, the filtrate is then used for *seeding* a "lawn" of bacteria with phage in the filtrated. Proceed as follows:

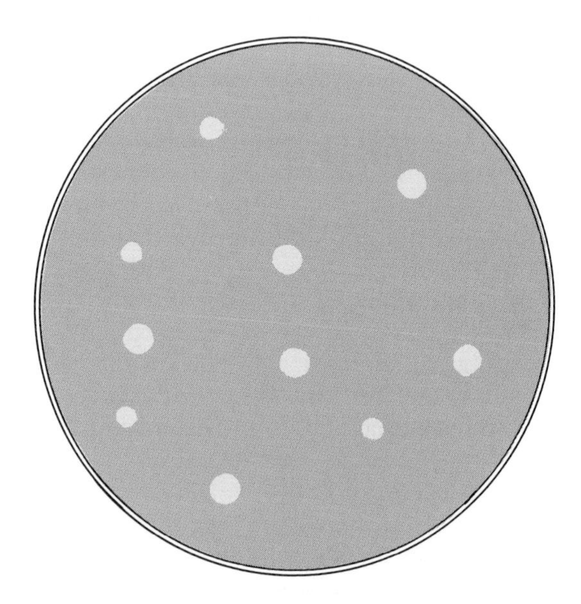

Figure 22.1 Plaques on cultures of bacteria are evidence of the lytic action of bacteriophage.

Enrichment

To increase the number of phage virions in a raw sewage sample, it is necessary to add 5 ml of deca strength phage broth (DSPB) and 5 ml of *E. coli* to 45 ml of raw sewage as in illustration 1 of figure 22.3. The DSPB medium is ten times as strong as ordinary broth to accommodate dilution with 45 ml of sewage. This mixture is incubated at 37° C for 24 hours.

Materials:

graduate
flask of raw sewage
1 Erlenmeyer flask (125 ml size)
5 ml of DSPB medium
nutrient broth culture of *E. coli* (strain B)
5 ml pipettes

1. With a graduate, measure out 45 ml of raw sewage and decant into an Erlenmeyer flask.
2. Pour 5 ml of DSPB medium and 5 ml of *E. coli* into the flask of sewage. If these constit-

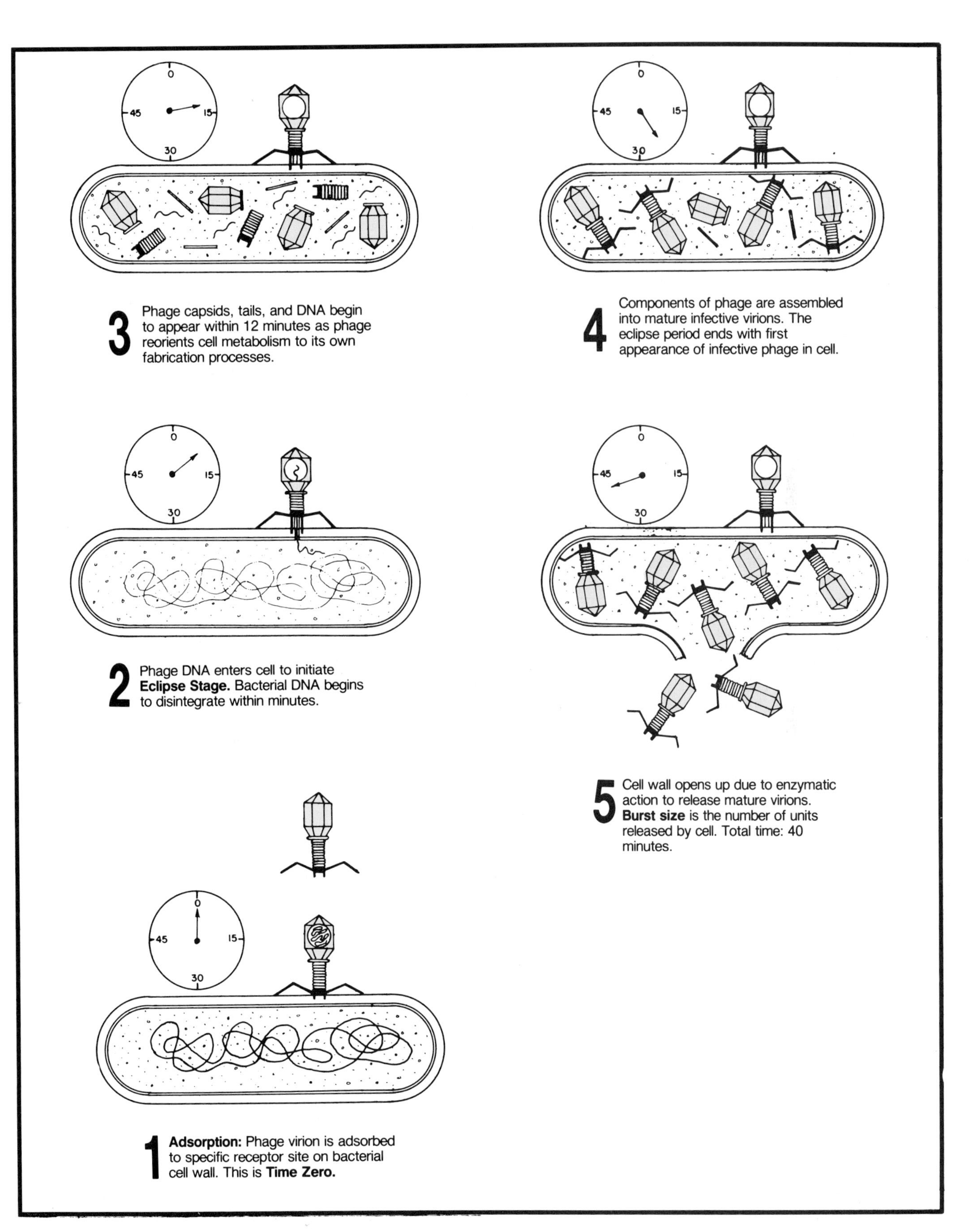

Figure 22.2 The lytic cycle of a virulent bacteriophage.

uents are not premeasured, use a 5 ml pipette. If the medium is pipetted first, the same pipette can be used for pipetting the *E. coli*.

3. Place the flask in the 37° C incubator for 24 hours.

Filtration

Rapid filtration to separate the phage from *E. coli* in the enrichment mixture requires adequate centrifugation first. If centrifugation is incomplete, the membrane filter clogs quickly and filtration will progress slowly. To minimize filter clogging, a triple centrifugation procedure will be used. To save time in the event filter clogging does occur, an extra filter assembly and an adequate supply of membrane filters should be available. These membrane filters have a pore size of 0.45 μm and less, which holds back all bacteria, allowing only the phage virions to pass through.

Materials:

centrifuge and centrifuge tubes (6–12)
2 sterile membrane filter assemblies (funnel, glass base, clamp, and vacuum flask)
package of sterile membrane filters
sterile Erlenmeyer flask with cotton plug (125 ml size)
forceps and Bunsen burner
vacuum pump and rubber hose

1. Into six or eight centrifuge tubes, dispense the sewage-*E. coli* mixture, filling each tube to within ½ inch of the top. Place the tubes in the centrifuge so that the load is balanced. Centrifuge the tubes at 2500 r.p.m. for 10 minutes.

   Without disturbing the material in the bottom of the tubes, decant all material from the tubes to within 1 inch of the bottom into another set of tubes.

   Centrifuge this second set of tubes at 2500 r.p.m. for another 10 minutes. While centrifugation is taking place, rinse out the first set of tubes.

   When the second centrifugation is complete, pour off the top two-thirds of each tube into the clean set of tubes and centrifuge again in the same manner.

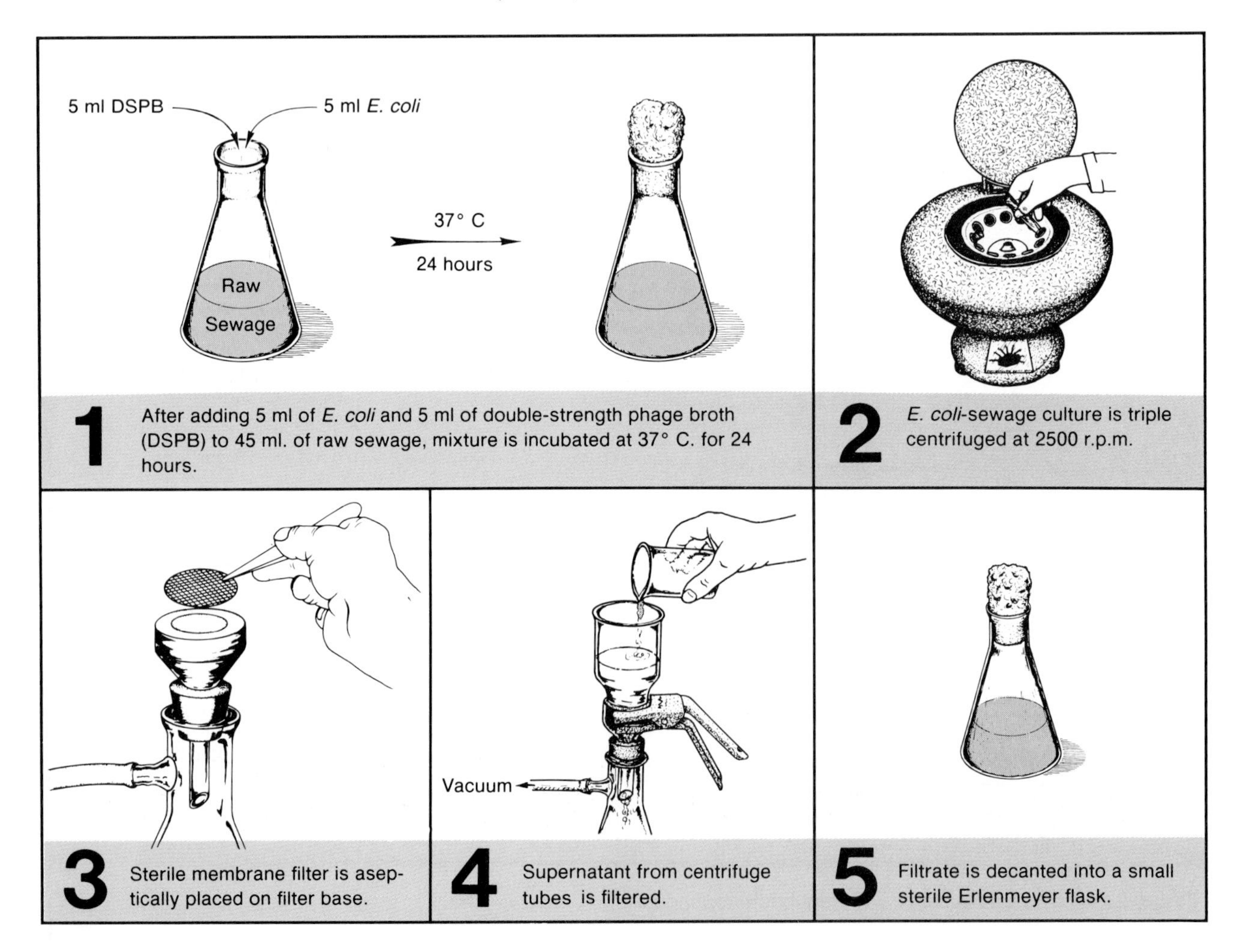

Figure 22.3 Enrichment and separation of phage from sewage.

2. While the third centrifugation is taking place, aseptically place a membrane filter on the glass base of a sterile filter assembly (illustration 3, figure 22.3). Use flamed forceps. Note that the filter is a thin sheet with grid lines on it.
3. Place the glass funnel over the filter and fix the clamp in place.
4. Hook up a rubber hose between the vacuum flask and pump.
5. Carefully decant the top three-fourths of each tube into the filter funnel. **Do not disturb the material in the bottom of the tube.**
6. Turn on the vacuum pump. If centrifugation has removed all bacteria, filtration will occur almost instantly. If the filter becomes clogged

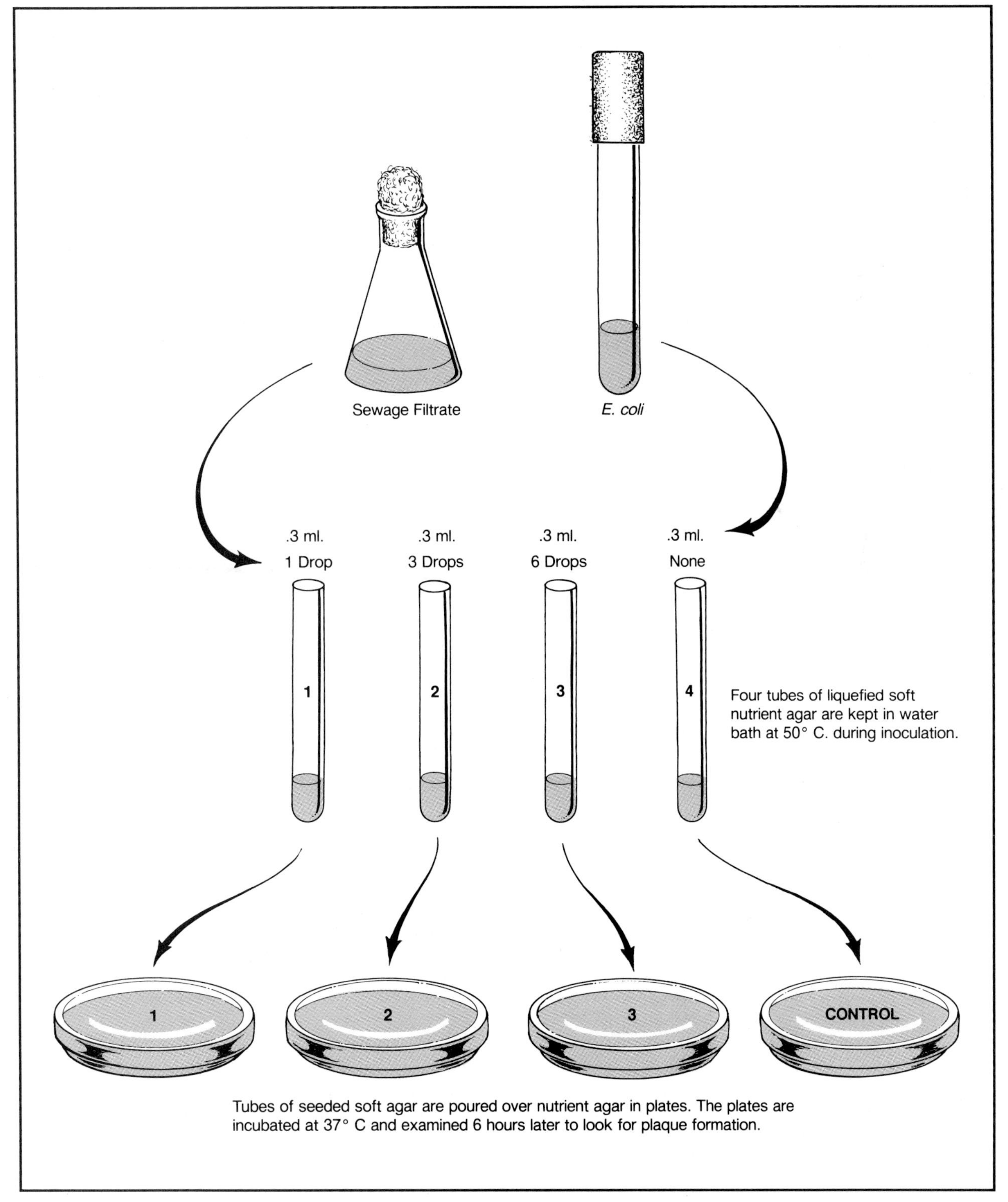

Figure 22.4 Overlay method of seeding *E. coli* cultures with phage.

and you have enough filtrate to complete the experiment, go on to step 7. (If this filtrate is to be used by the entire class, you will need 25–50 ml.)

If the filter clogs before you have enough filtrate, pour the unfiltered material from the funnel back into another set of centrifuge tubes and recentrifuge for 10 minutes at 2500 r.p.m. While centrifugation is taking place, set up the other filter assembly and pour whatever filtrate you have from the first flask into the funnel of the new setup. After centrifugation, decant the top three-fourths of material from each tube into the funnel and turn on the vacuum pump. Filtration should take place rapidly now.

7. Aseptically transfer the final filtrate from the vacuum flask to a sterile 125 ml Erlenmeyer flask that has a sterile cotton plug. Putting the filtrate in a small flask is necessary to facilitate pipetting. Be sure to flame the necks of both flasks while pouring from one to the other.

Seeding

Evidence of phage in the filtrate is produced by providing a "lawn" of *E. coli* and phage. The medium used is *soft nutrient agar.* Its jellylike consistency allows for better development of plaques. The soft agar is poured over the top of hard nutrient agar. Figure 22.4 illustrates the general procedure.

Materials:

nutrient broth culture of *E. coli,* strain B
flask of enriched sewage filtrate
4 metal-capped tubes of soft nutrient agar (5 ml per tube)
4 Petri plates of nutrient agar (15 ml per plate)
1 ml serological pipettes

1. Liquefy 4 tubes of soft nutrient agar and cool to 50° C. Keep the tubes in 50° C water bath to prevent solidification.
2. Label the tubes 1, 2, 3, and 4. Label the plates 1, 2, 3 and Control.
3. With a 1 ml pipette, transfer 1 drop of filtrate to tube 1, 3 drops to tube 2, and 6 drops to tube 3. **Don't put any filtrate into tube 4.**
4. With a fresh 1 ml pipette, transfer 0.3 ml of *E. coli* to each of the four tubes of soft agar.
5. After flaming the necks of each of the soft agar tubes, pour the contents of each tube over the hard agar of similarly numbered agar plates. Note that tube 4 is poured over the Control plate.
6. Once the agar is cooled completely, put the plates, inverted, into a 37° C incubator. If possible, examine the plates 6 hours later to look for plaque formation. If some are visible, measure them and record their diameters on the Laboratory Report. Plaque size should be checked every 2 hours for changes.
7. Complete the Laboratory Report.

Part 5 Environmental Influences and Control of Microbial Growth

The eleven exercises of this unit are concerned with two aspects of microbial growth: promotion and control. On the one hand, the microbiologist is concerned with providing optimum growth conditions to favor maximization of growth. The physician, nurse, and other members of the medical arts profession, on the other hand, are concerned with the limitation of microbial populations in disease prevention and treatment. An understanding of one of these facets of microbial existence enhances the other.

In Part 4 we were primarily concerned with providing media for microbial growth which contain all the essential nutritional needs. Very little emphasis was placed on other limiting factors such as temperature, oxygen, or hydrogen ion concentration. An organism provided with all its nutritional needs may fail to grow if one or more of these essentials are not provided. The total environment must be sustained to achieve the desired growth of microorganisms.

Microbial control by chemical and physical means involves the use of antiseptics, disinfectants, antibiotics, ultraviolet light, and many other agents. The exercises of this unit that relate to these aspects are intended, primarily, to demonstrate methods of measurement; no attempt has been made to make in-depth evaluation.

23 Temperature: Effects on Growth

Temperature is one of the most important factors influencing the activity of bacterial enzymes. Unlike warm-blooded animals, the bacteria lack mechanisms which conserve or dissipate heat generated by metabolism, and consequently their enzyme systems are directly affected by ambient temperatures. Enzymes have minimal, optimal, and maximal temperatures. At the **optimum** temperature the enzymatic reactions progress at maximum speed. Below the **minimum** and above the **maximum** temperatures the enzymes become inactive. At some point above the maximum temperature, destruction of a specific enzyme will occur. Low temperatures are less deleterious in most cases.

In this experiment we will attempt to measure the effects of various temperatures on two physiological reactions: pigment production and growth rate. *Serratia marcescens* will be grown on nutrient agar slants at 25° and 38° C to see which temperature favors the enzymatic reactions for pigment formation. This organism and *Escherichia coli* will also be grown in nutrient broth at various temperatures (5°, 25°, 38°, and 42° C) to determine, approximately, their optimum growth temperatures. More precise determinations of optimum growth temperatures would necessitate growth at a greater number of temperatures, however. In the evaluation of amount of growth, turbidity determinations will be made.

Inoculations

Materials:

nutrient broth cultures of *Serratia marcescens* and *Escherichia coli*
2 nutrient agar slants
4 nutrient broths

1. Label the tubes as follows:
   slants: label one 25° C and the other 38° C
   broths: 5° C, 25° C, 38° C, and 42° C
2. Inoculate the tubes with a wire loop as follows:
   odd-numbered students:
   all tubes: *S. marcescens*
   even-numbered students:
   slants: *S. marcescens*
   broths: *E. coli*
3. Place the tubes in the appropriately labeled baskets that are on the demonstration table. The instructor will see that they are incubated at the proper temperature for 24 hours.

Evaluation of Results

Materials:

slants and broth cultures that have been incubated at various temperatures
photocolorimeter
sterile nutrient broth
cuvettes

Slants Compare the nutrient agar slants of *Serratia marcescens*. Using colored pencils, draw the appearance of the growths on the Laboratory Report.

Broths Shake the broth cultures and compare them, noting the differences in turbidity. Those tubes which appear to have no growth should be compared with a tube of sterile nutrient broth. If no photocolorimeter is available, record turbidity by visual observation. If a photocolorimeter is available, it will not be necessary to determine the turbidity by visual means.

Visual Method Record your interpretation of these tubes on the Laboratory Report in the following manner:

no growth .... none
least growth .... +
more growth .... ++
most growth .... +++

Photocolorimetric Method If a photocolorimeter is available, determine the turbidity of each tube as follows:

1. Calibrate the photocolorimeter with sterile nutrient broth according to the instructions on page 83.
2. Label four cuvettes 5°, 25°, 38°, and 42°.
3. Shake each culture sufficiently to completely disperse the organisms and pour them into their respective cuvettes.
4. Refer to page 84 to review the techniques concerned with the handling of cuvettes.
5. Measure the percent transmittance of each of your cultures and record these values on the Laboratory Report.
6. Calculate the O.D. values (formula on page 85) for each culture. Record these figures on the Laboratory Report.
7. Plot the O.D. values for each culture on the graph of the Laboratory Report. Use different colored pencils for the two curves, with the names of the organisms written on the curves.

Temperature: Lethal Effects 24

In attempting to compare the susceptibility of different organisms to elevated temperatures, it is necessary to use some yardstick of measure. Two methods are available: the TDP and TDT. The **thermal death point,** or TDP, is the temperature at which an organism is killed in ten minutes. The **thermal death time,** or TDT, is the time required to kill a suspension of cells or spores at a given temperature. Various factors such as pH, moisture, composition of medium, and age of cells will greatly influence these values and consequently must be clearly stated.

In this exercise we will subject cultures of three different organisms to temperatures of 60°, 70°, 80°, 90°, and 100°C. At intervals of ten minutes, organisms will be removed and plated out to test their viability. The spore-former *Bacillus megaterium* will be compared with the non-spore-formers *Staphylococcus aureus* and *Escherichia coli.*

Due to the large number of plates that have to be inoculated to perform the entire experiment, it will be necessary for each member of the class to be assigned a specific temperature and organism to work with. A table on the next page provides assignments by student number. After the plates have been incubated, each student's results will be tabulated on the chalkboard so that all students will be provided with the necessary results.

Figure 24.1 illustrates the overall procedure. Note that before the culture is heated a control plate is incubated with 0.1 ml of the organism. When the culture is placed in the water bath, a tube of nutrient broth with a thermometer inserted into it is placed in the bath at the same time. Timing

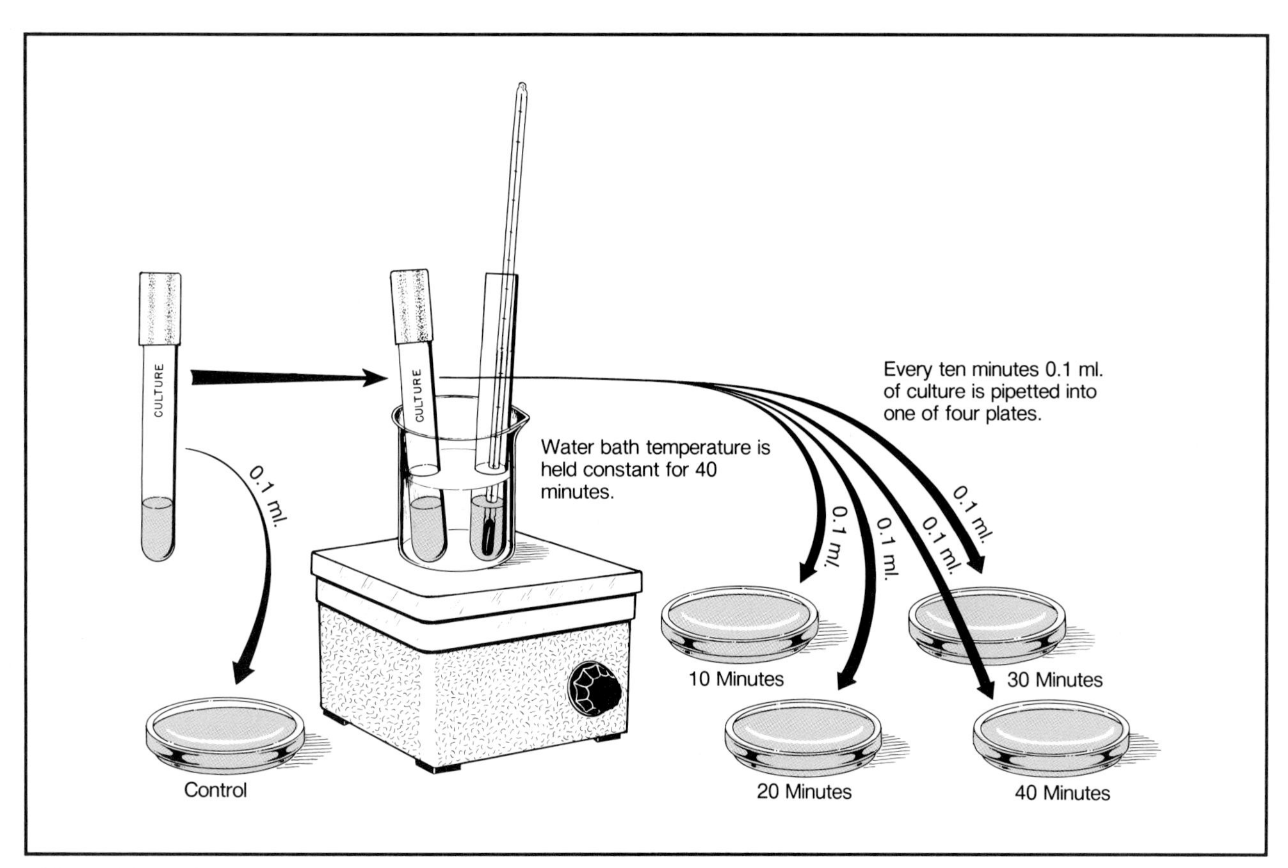

Figure 24.1 Procedure for determining thermal endurance.

of the experiment starts when the thermometer temperature reaches the test temperature.

Materials:

per student:
- 5 petri plates
- 5 pipettes (1 ml size)
- 1 tube of nutrient broth
- 1 bottle of nutrient agar (60 ml)
- 1 culture of organisms

class equipment:
- water baths set up at 60°, 70°, 80°, and 90° C

cultures:
- broth cultures of *Staphylococcus aureus* (⅓ of class or minimum of 5 cultures)
- broth cultures of *Escherichia coli* (⅓ of class or minimum of 5 cultures)
- saline suspension of slant cultures of *Bacillus megaterium* (⅓ of class or minimum of 5 tubes)

1. Consult the table below to determine what organism and temperature has been assigned to you. If several thermostatically controlled water baths have been provided in the lab, locate the one that you will use. If a bath is not available for your temperature, set up a bath on an electric hot plate or over a tripod and Bunsen burner. If your temperature is 100° C, a hot plate and beaker of water is the only way to go.
2. Liquefy a bottle of 60 ml of nutrient agar and cool to 50° C. This can be done while the rest of the experiment is in progress.
3. Label five petri plates: **control, 10 min, 20 min, 30 min,** and **40 min.**
4. Shake the culture of organisms and transfer 0.1 ml of organisms with a 1-ml pipette to the control plate.
5. Place the culture and a tube of sterile nutrient broth into the water bath. Remove the cap from a tube of nutrient broth and insert a thermometer into the tube. *Don't make the mistake of inserting the thermometer into the culture of organisms!*
6. As soon as the temperature of the nutrient broth reaches the desired temperature, record the time here: ______

   Watch the temperature carefully to make sure it does not vary appreciably.
7. After 10 minutes have elapsed, transfer 0.1 ml from the culture to the 10-minute plate with a *fresh* 1-ml pipette. Repeat this operation at 10-minute intervals until all the plates have been inoculated. *Use fresh pipettes each time and be sure to shake the culture before each delivery.*
8. Pour liquefied nutrient agar (50° C) into each plate, rotate, and cool.
9. Incubate at 37° C for 24 to 48 hours. After evaluating your plates, record your results on the chalkboard and Laboratory Report.

Organism	Student Number				
	60° C	70° C	80° C	90° C	100° C
Staphylococcus aureus	1, 16	4, 19	7, 22	10, 25	13, 28
Escherichia coli	2, 17	5, 20	8, 23	11, 26	14, 29
Bacillus megaterium	3, 18	6, 21	9, 24	12, 27	15, 30

Osmotic Pressure and Bacterial Growth 25

Growth of bacteria can be profoundly affected by the amount of water entering or leaving the cell. When the medium surrounding an organism is **hypotonic** (low solute content), a resultant higher osmotic pressure occurs in the cell. Except for some marine forms, this situation is not harmful to most bacteria. The cell wall structure of most bacteria is so strong and rigid that even slight cellular swelling is generally inapparent. In the reverse situation, however, when bacteria are placed in a **hypertonic** solution (high solute content), their growth may be considerably inhibited. The degree of inhibition will depend on the type of solute and the nature of the organism. In media of growth-inhibiting osmotic pressure, the cytoplasm becomes dehydrated and shrinks away from the cell wall. Such **plasmolyzed** cells are often simply inhibited in the absence of sufficient cellular water and return to normal when placed in an **isotonic** solution. In other instances, the organisms are irreversibly affected due to permanent inactivation of enzyme systems.

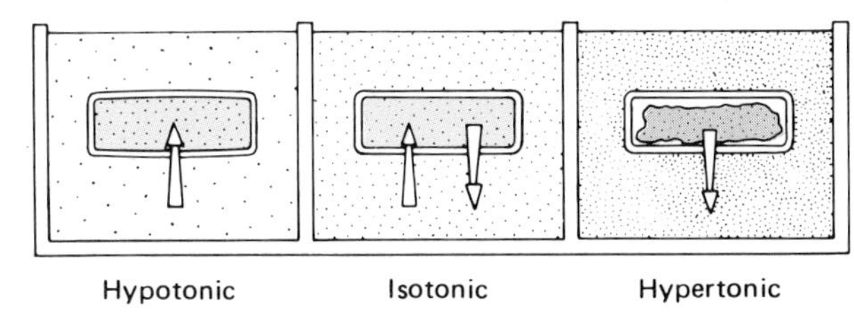

Figure 25.1 Osmotic variabilities.

In this exercise we will test the degree of inhibition of organisms that results with media containing different concentrations of sodium chloride. To accomplish this, you will streak three different organisms on four plates of media. The specific organisms used differ in their tolerance of salt concentrations. The salt concentrations will be 0.5, 5, 10, and 15%. After incubation for 48 hours, a comparison will be made of the degree of growth to determine the differences between the organisms.

Materials:

per student:
- 1 petri plate nutrient agar (0.5% NaCl)
- 1 petri plate nutrient agar (5% NaCl)
- 1 petri plate nutrient agar (10% NaCl)
- 1 petri plate milk salt agar (15% NaCl)

cultures:
- *Escherichia coli* (nutrient broth)
- *Staphylococcus aureus* (nutrient broth)
- *Halobacterium salinarium* (slant culture)

1. Mark the bottoms of the four petri plates as indicated in figure 25.2.
2. Streak each organism in a straight line on the agar, using a wire loop.
3. Incubate all of the plates for 48 hours at room temperature with exposure to light (the pigmentation of *H. salinarium* requires light to develop.) Record your results on the Laboratory Report.
4. Continue to incubate the milk salt agar plate for another 48 hours in the same manner, and record your results again on the Laboratory Report.

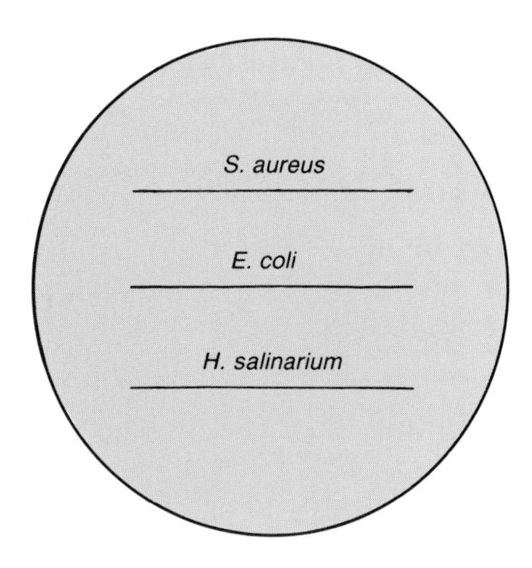

Figure 25.2 Streak pattern.

pH and Microbial Growth 26

Aside from temperature, the hydrogen ion concentration of an organism's environment exerts the greatest influence on its growth. The concentration of hydrogen ions, which is customarily designated by the term pH ($\log 1/H^+$), limits the activity of enzymes with which an organism is able to synthesize new protoplasm. As in the case of temperature, there exists for each organism an optimum concentration of hydrogen ions in which it grows best. The pH values above and below which an organism fails to grow are respectively referred to as the minimum and maximum hydrogen ion concentrations. These values hold only when other environmental factors remain constant. If the composition of the medium, incubation temperature, or osmotic pressure is varied, the hydrogen ion requirements become different.

In this exercise we will test the degree of inhibition of organisms that results from media containing different concentrations of hydrogen ions.

Materials:

per student:

- 1 tube of nutrient broth of pH 5.0
- 1 tube of nutrient broth of pH 7.0
- 1 tube of nutrient broth of pH 9.0

class material:

- broth cultures of *Escherichia coli*
- broth cultures of *Sporosarcina ureae**
- broth cultures of *Saccharomyces cerevisiae***
- broth cultures of *Staphylococcus aureus*
- photocolorimeter and cuvettes (second period)

1. Inoculate a tube of each of the above broths with one organism. Use the organism following your assigned seat number.

Student number	Organism
1,5,9,13,17,21,25,29,33	*Escherichia coli*
2,6,10,14,18,22,26,30,34	*Staphylococcus aureus*
3,7,11,15,19,23,27,31,35	*Sporosarcina ureae**
4,8,12,16,20,24,28,32,36	*Saccharomyces cerevisiae***

2. Incubate the tubes of *S. aureus* and *E. coli* at 37° C for 48 hours. Incubate the tubes of *Sporosarcina ureae* and *S. cerevisiae* at room temperature for 48 to 72 hours.
3. Measure the turbidities and record the data on the Laboratory Report.

**Alcaligenes* spp. may be substituted if *Sporosarcina ureae* is unavailable. The *Alcaligenes* cultures should be incubated at 37° C.

***Candida glabrata* is a good substitute for *Saccharomyces cerevisiae.*

27 Ultraviolet Light: Lethal Effects

Except for the photosynthetic bacteria, most bacteria are harmed by ultraviolet radiation. Those that contain photosynthetic pigments require exposure to sunlight in order to synthesize substances needed in their metabolism. Although sunlight contains the complete spectrum of short to long wavelengths of light, it is only the short, invisible ultraviolet wavelengths that are injurious to the nonphotosynthetic bacteria.

Wavelengths of light may be expressed in nanometers (nm) or angstrom units (A). The angstrom unit is equal to 10^{-8} cm. In terms of nanometers, 10 A equal one nanometer. Thus, a wavelength of 4500×10^{-8} cm would be expressed as 4500 A, 450 nm, or 0.45 μm.

Figure 27.1 illustrates the relationship of ultraviolet to other types of radiations. By definition, ultraviolet light includes electromagnetic radiations that fall in the wavelength band between 4000 and 40 A. It bridges the gap between the X rays and the shortest wavelengths of light visible to the human eye. The visible range is approximately between 4000 and 7800 A. Actually, the practical range of ultraviolet, as far as we are concerned, lies between 2000 and 4000 A. The "extreme" range (2000–40 A) includes radiations that are absorbed by air and consequently function only in a vacuum. This region is also referred to as *vacuum ultraviolet.*

Ultraviolet is not a single entity, but is a very wide band of wavelengths. This fact is often not realized. Extending from 4000 to 40 A, it encompasses a span of 100:1; visible wavelengths (7800–4000 A), on the other hand, represent only a two-fold spread.

The germicidal effects of ultraviolet are limited to only a specific region of the ultraviolet spectrum. As indicated in figure 27.1, the most effective wavelength is 2650 A. Low-pressure mercury vapor lamps, which have a high output (90%) of 2537 A, make very effective bactericidal lamps.

In this exercise, organisms that have been streaked on nutrient agar will be exposed to ultraviolet radiation for various lengths of time to determine the minimum amount of exposure required to effect a 100% kill. One half of each plate will be shielded from the radiation to provide a control comparison. *Bacillus megaterium,* a spore-former, and *Staphylococcus aureus,* a non-spore-former, will be used to provide a comparison of the relative resistances of vegetative and spore types.

Your instructor will assign you a specific organism and exposure times to be used. The number of exposures assigned will depend on class size and laboratory time.

Materials:

14 petri plates of nutrient agar
ultraviolet lamp
timers (bell type)
nutrient broth cultures of *S. aureus*
saline suspensions of *Bacillus megaterium*

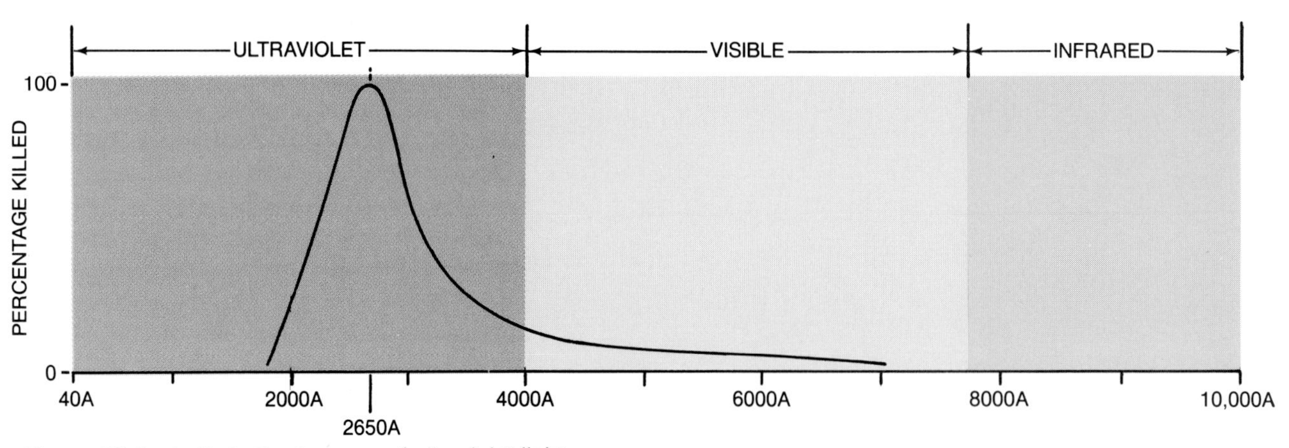

Figure 27.1 Lethal effectiveness of ultraviolet light.

1. After your instructor has informed you as to the organism you are to use and the length of time you are to expose the organisms, record your assignment in the table below.
2. Completely streak the surface of each plate, using several loopfuls of organisms. Since submerged organisms are protected against UV rays, *avoid digging into the medium with your loop.*
3. Place the plates under the ultraviolet lamp *with the lids removed.* Cover half of each plate with a cardboard shield as illustrated in figure 27.2. Expose the plates for the correct length of time, re-cover them with their lids, and incubate inverted at 37° C for 48 hours.

Caution: Don't look directly into the ultraviolet lamp. These rays can cause cataracts and other eye injury.

4. Record your observations on the Laboratory Report.

Organisms	Exposure Times (Student Assignments)						
Staphylococcus aureus	10 sec	20 sec	40 sec	80 sec	2½ min	5 min	10 min
Student number:							
Bacillus megaterium	1 min	2 min	4 min	8 min	15 min	30 min	60 min
Student number:							

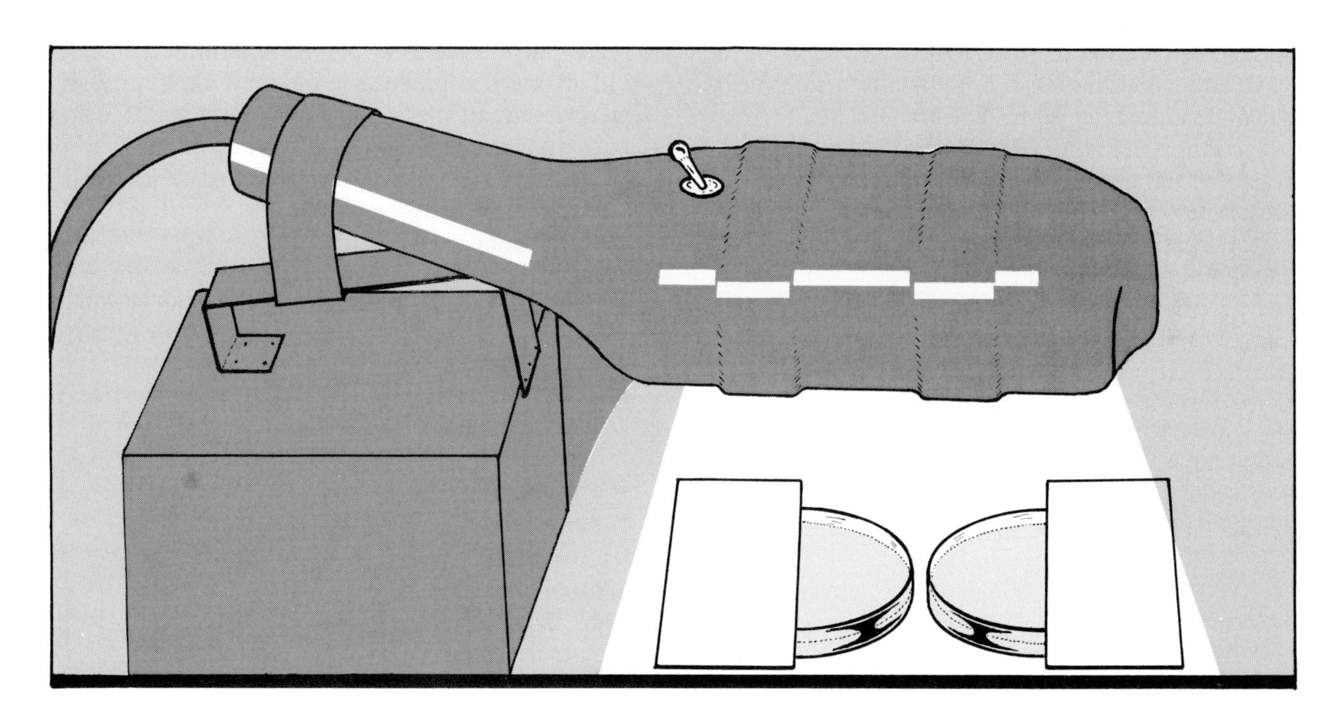

Figure 27.2 Plates are exposed to UV light with 50% covered.

28 Disinfectants: Evaluation (Use-Dilution Method)

When considering the relative effectiveness of different chemical agents against bacteria, some yardstick of comparison is necessary. Many different methods have been developed over the years since Robert Koch, in 1881, worked out the first scientific procedure by measuring the killing power of various germicides on silk threads that were impregnated with spores of *Bacillus anthracis*. Koch's method and many that followed proved unreliable for various reasons. Finally, in 1931, the United States Food and Drug Administration adopted a method which was a modification of a test developed in England in 1903 by Rideal and Walker. In 1950 the Association of Official Agricultural Chemists adopted it as the official method of testing disinfectants. This method compares the effectiveness of various agents with phenol. A value called the **phenol coefficient** is arrived at which has significant meaning within certain limitations. The restrictions are that the test should only be used for phenollike compounds that do not exert bacteriostatic effects and are not neutralized by the subculture media used. Many excellent disinfectants cannot be evaluated with this test. Disinfectants such as bichloride of mercury, iodine, metaphen, and quaternary detergents are unlike phenol in their germicidal properties and should not be evaluated in terms of phenol coefficients. Notwithstanding, however, many pharmaceutical companies have applied this test to such disinfectants with misleading results. A more suitable test for these nonphenolic disinfectants is the **use-dilution method.**

This method makes use of small glass rods on which test organisms are dried for 30 minutes. The seeded rods are then exposed to the test solutions at 20° C for 1, 5, 10, and 30 minutes, rinsed with water or neutralizing solution, and transferred to tubes of media. After incubation at 37° C for 48

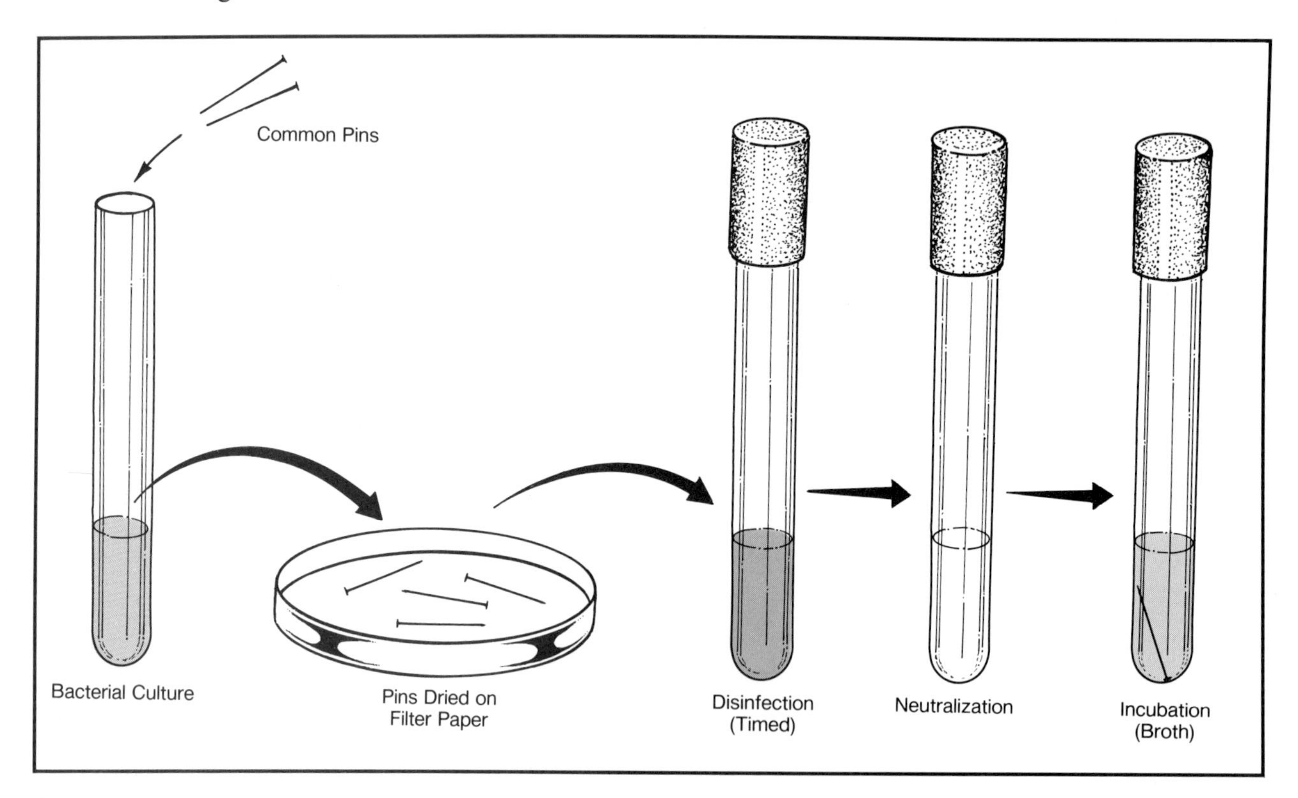

Figure 28.1 Procedure for use-dilution evaluation of a disinfectant.

hours, the tubes are examined for growth. When the results of this test are applied to practical conditions of use, they are found to be completely reliable.

In this experiment you will follow a modified procedure of the use-dilution method to compare the relative merits of three different disinfectants on two kinds of bacteria: a spore-former, *Bacillus megaterium,* and a non-spore-former, *Staphylococcus aureus.* Instead of glass rods, you will use rustproof common pins.

Since this experiment will be performed with student assignments for specific agents, a table is provided at the bottom of the page for time (minutes) assignments. Note that a blank column is provided for students to write in any substitution of agents that may be desired.

Materials:

per student:

- 2 tubes of one of the following disinfectants:
  - 1:750 Zephiran for students 1, 4, 7, 10, 13, 16, 19, 22, 25, 28
  - 5% phenol for students 2, 5, 8, 11, 14, 17, 20, 23, 26, 29
  - 8% formaldehyde for students 3, 6, 9, 12, 15, 18, 21, 24, 27, 30
- 2 tubes of sterile water (about 7 ml each)
- 2 tubes of nutrient broth (about 7 ml each)
- forceps

demonstration table:

- 1 nutrient broth culture of *Staphylococcus aureus*
- 1 physiological saline suspension of a 48-hour nutrient agar slant culture of *Bacillus megaterium*
- 2 sterile petri plates with filter paper in the bottom
- several forceps and Bunsen burner
- 2 test tubes containing 36 sterile common pins in each one (pins must be plated brass, which are rustproof)

1. Consult the materials list above to determine which disinfectant you are to use. Get two tubes of this disinfectant, two tubes of sterile water, and two tubes of nutrient broth from the table. Label one of each *B. megaterium* and the other *S. aureus.*

   While you are getting your supplies, the instructor will start the experiment by pouring the broth culture of Staphylococcus aureus into one of the tubes of pins and the saline suspension of B. megaterium into the other tube of pins. The instructor will then pour off the organisms and empty the pins onto the blotting paper in separate petri dishes. After the pins have dried, you may proceed with the next step.
2. Gently flame a pair of forceps, let cool, and transfer one pin from each petri plate to the separate tubes of disinfectant. Be sure to put them into the right tubes.
   Record the time of this transfer: __________
3. Leave the pins in the disinfectant for the length of time indicated below. Find your number under the time indicated for your disinfectant.
4. At the end of the assigned time, flame the mouths of the tubes of disinfectant and *carefully pour the disinfectant into the sink without discarding the pins.* Then pour the pins into the separate tubes of sterile water.
5. After one minute in the tubes of water, flame the mouths of the water and broth tubes, pour off the water, and shake the pins out of the emptied tubes into separate, labeled tubes of nutrient broth.

   At this point the instructor should put one pin from each of the petri plates on the demonstration table into separate tubes of nutrient broth. Label the tubes with the name of the organism and indicate that these tubes are controls.
6. Incubate all nutrient broth tubes with pins for 48 hours at 37° C. Examine them and record results on the Laboratory Report.

Disinfectant		Time in Minutes				
	Substitution	1	5	10	30	60
1:750 Zephiran		1, 16	4, 19	7, 22	10, 25	13, 28
5% phenol		2, 17	5, 20	8, 23	11, 26	14, 29
8% formaldehyde		3, 18	6, 21	9, 24	12, 27	15, 30

29 Antiseptics: Evaluation (Filter Paper Disk Method)

The term **antiseptic** has, unfortunately, been somewhat ill-defined. Originally, the term was applied to any agent that prevents *sepsis,* or putrefaction. Since sepsis is caused by growing microorganisms, it follows that an antiseptic inhibits microbial multiplication without necessarily killing them. By this definition, we can assume that antiseptics are essentially bacteriostatic agents. Part of the confusion that has resulted in the definition of this term is that the United States Food and Drug Administration rates antiseptics essentially the same as disinfectants. Only when an agent is to be used in contact with the body for a long period of time do they rate its bacteriostatic properties instead of its bactericidal properties.

If we are to compare antiseptics on the basis of their bacteriostatic properties, the **filter paper disk method** is a simple, satisfactory method to use. In this method, a disk of filter paper (½″ diameter) is impregnated with the chemical agent and placed on a seeded nutrient agar plate. The plate is incubated for 48 hours. If the substance is inhibitory, a clear zone of inhibition will surround the disk. The size of this zone is an expression of the agent's effectiveness and can be compared quantitatively against other substances.

In this exercise we will measure the relative effectiveness of three agents (phenol, formaldehyde, and iodine) against two organisms: *Staphylococcus aureus* (gram-positive) and *Pseudomonas aeruginosa* (gram-negative). A table at the bottom of the page will be used to assign each student one chemical agent to be tested against one organism. Note that space has been provided in the table for different agents to be written in as substitutes for the three agents listed. Your instructor may wish to make substitutions. Figure 29.1 illustrates the overall procedure.

Materials:

per student:
- 1 nutrient agar pour and 1 petri plate
- broth cultures of *S. aureus* and *P. aeruginosa*

demonstration table:
- petri dish containing sterile disks of filter paper (½″ diameter)
- forceps and Bunsen burner
- chemical agents in small beakers (5% phenol, 5% formaldehyde, 5% aqueous iodine)

1. Consult the chart below to determine your assignment.
2. Liquefy a nutrient agar pour in a water bath and cool to 50° C.
3. Label the bottom of a petri plate with the names of the organism and chemical agent.
4. Inoculate the agar pour with one loopful of the organism and pour into the plate.
5. After the medium has solidified in the plate, pick up a sterile disk with *lightly flamed* forceps, dip the disk *halfway* into a beaker of the chemical agent, and place the disk in the center of the medium. To secure the disk to the medium, press slightly on it with the forceps.
6. Incubate the plate at 37° C for 48 hours.
7. Measure the zone of inhibition from the edge of the disk to the edge of the growth. See illustration 5, figure 29.1.
8. Exchange plates with other members of the class so that you will have an opportunity to complete the table on the Laboratory Report.

Chemical Agent		Student Number	
	Substitution	*S. aureus*	*P. aeruginosa*
5% phenol		1, 7, 13, 19, 25	2, 8, 14, 20, 26
5% formaldehyde		3, 9, 15, 21, 27	4, 10, 16, 22, 28
5% iodine		5, 11, 17, 23, 29	6, 12, 18, 24, 30

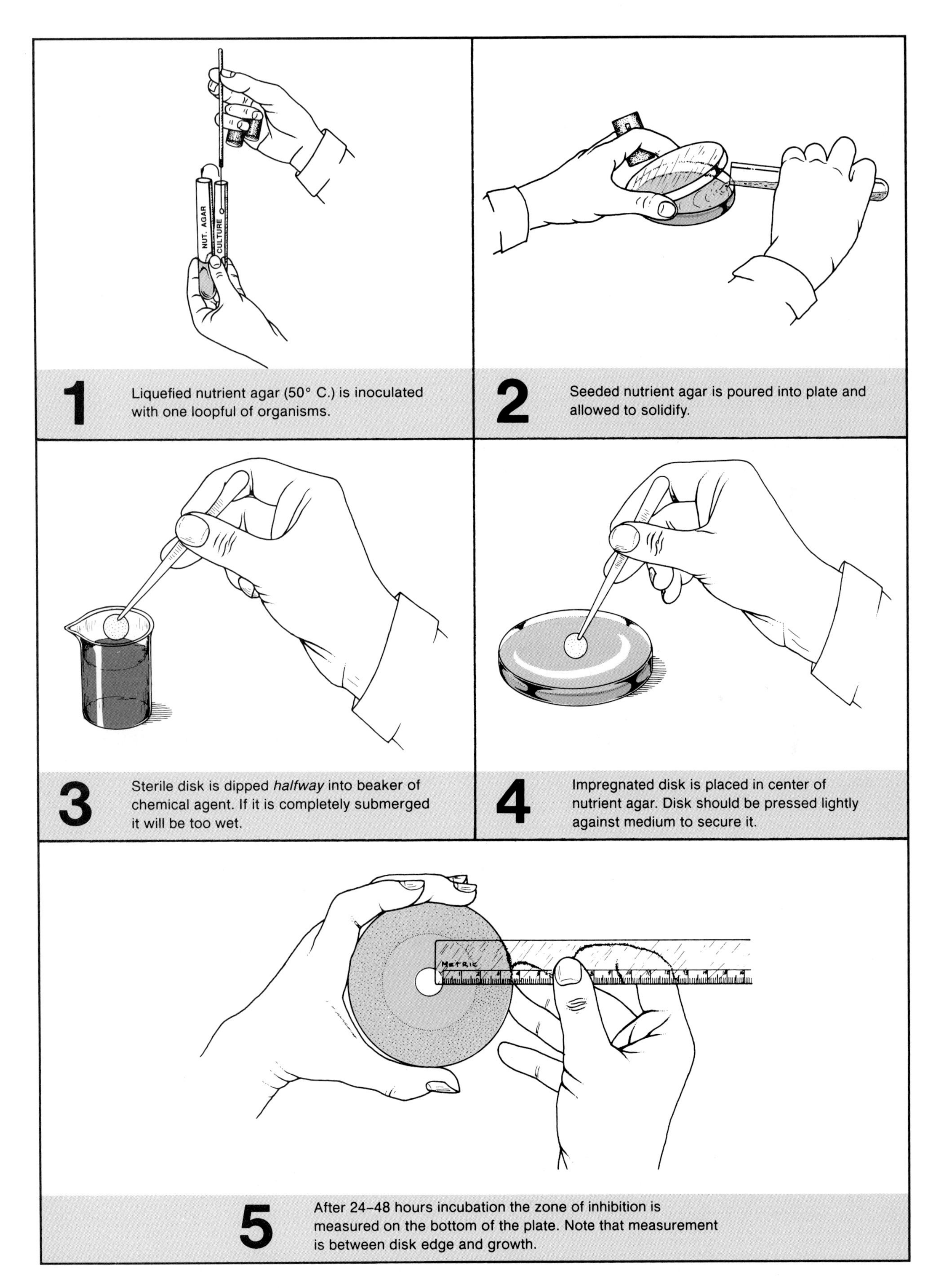

Figure 29.1 Filter paper disk method of evaluating an antiseptic.

30 Alcohol Evaluation: Its Effectiveness as a Skin Degerming Agent

As a skin disinfectant, 70% alcohol is undoubtedly the most widely used agent. The ubiquitous prepackaged alcohol swabs used by nurses and technicians are evidence that these items are indispensible. The question that often arises is: How effective is alcohol in routine use? When the skin is swabbed prior to penetration, are all, or mostly all, of the surface bacteria killed? To determine alcohol effectiveness as it might be used in routine skin disinfection, we are going to perform a very simple experiment which utilizes four thumbprints and a plate of enriched agar. Class results will be pooled to arrive at a statistical analysis.

Figure 30.1 illustrates the various steps in this test. Note that the petri dish is divided into four parts. On the left side of the plate an unwashed left thumb is first pressed down on the agar in the lower quadrant of the plate. Next the left thumb is pressed down on the upper left quadrant. With the left thumb we are trying to establish the percentage of bacteria that are removed by simple contact with the agar.

On the right side of the plate an unwashed right thumb is pressed down on the lower right quadrant of the plate. The next step is to either dip the right thumb into alcohol or to scrub it with an alcohol swab and dry it. Half of the class will use the dipping method and the other half will use alcohol swabs. Your instructor will indicate what your assignment will be. The last step is to press the dried right thumb on the upper right quadrant of the plate.

After inoculating the plate it is incubated at 37° C for 24 to 48 hours. Colony counts will establish the effectiveness of the alcohol.

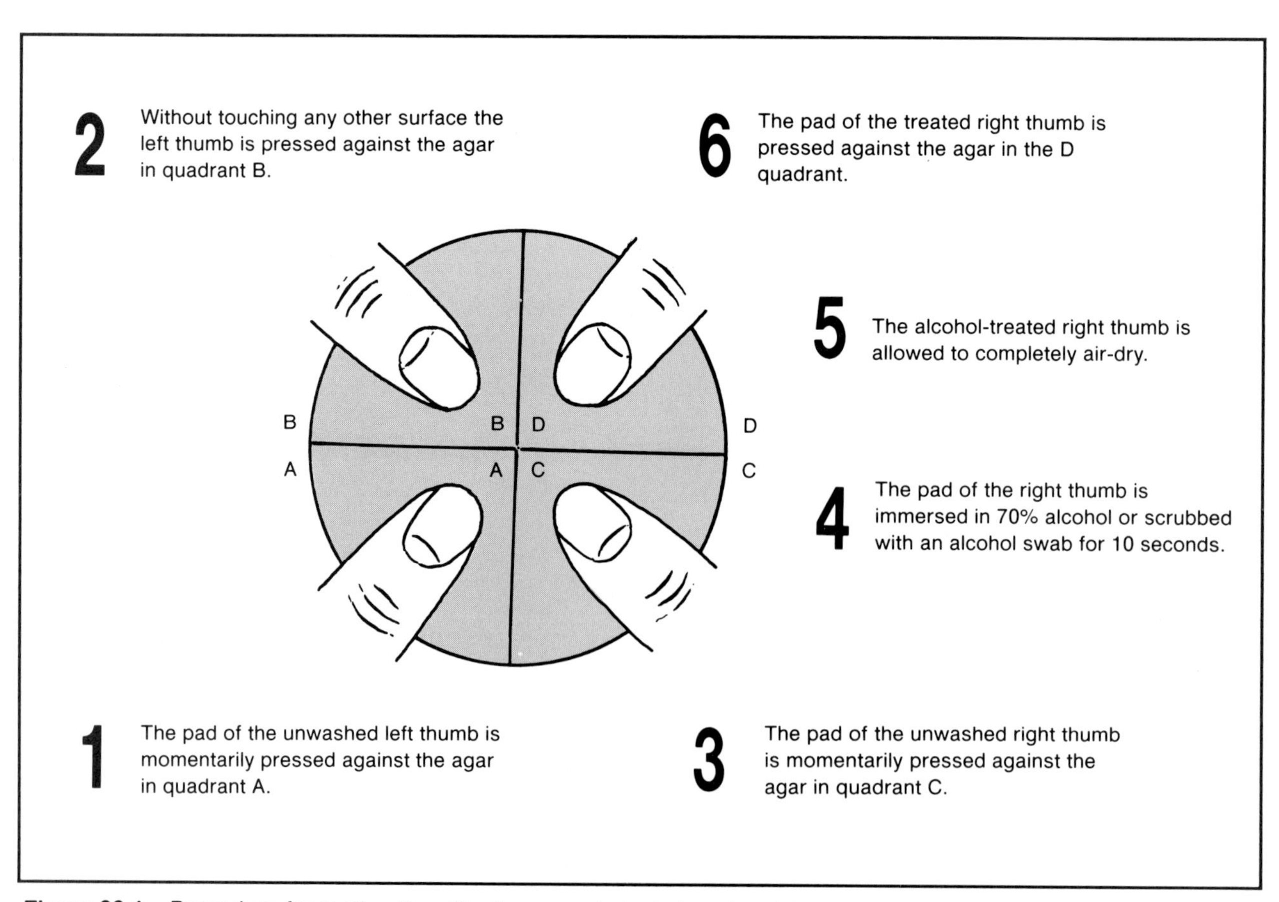

Figure 30.1 Procedure for testing the effectiveness of alcohol on the skin.

Materials:

1 petri plate of veal infusion agar
small beaker
70% ethanol
alcohol swab

1. Perform this experiment with unwashed hands.
2. With a china marking pencil mark the bottom of the petri plate with two perpendicular lines that divide it into four quadrants. Label the left quadrants **A** and **B,** and the right quadrants **C** and **D** as shown in figure 30.1. (*Keep in mind that when you turn the plates over to label them, the A and B quadrants will be on the right and C and D will be on the left.*)
3. Press the pad of your left thumb against the agar surface in the A quadrant.
4. Without touching any other surface, press the left thumb into the B quadrant.
5. Press the pad of your right thumb against the agar surface of the C quadrant.
6. Disinfect the right thumb by one of the two following methods:
   - dip the thumb into a beaker of 70% ethanol for 5 seconds, or
   - scrub the entire pad surface of the right thumb with an alcohol swab.
7. Allow the alcohol to completely evaporate from the skin.
8. Press the right thumb against the agar in the D quadrant.
9. Incubate the plate at 37° C for 24 to 48 hours.
10. Follow the instructions on the Laboratory Report for evaluating the plate.

31 Antimicrobic Sensitivity Testing (Kirby-Bauer Method)

The principal drugs used in the treatment of infectious disease fall into three categories: antibiotics, sulfonamides, and chemotherapeutics. Collectively, they may be referred to as **antimicrobics.** Once the causative organism of a specific disease has been isolated, the physician needs to know, as soon as possible, which antimicrobic will be most effective. The use of sensitivity disks in a procedure illustrated in figure 31.1 can readily provide this information.

Antimicrobic impregnated disks were first introduced in the late 1940s as penicillin came into widespread use. As the decades rolled by and a multitude of new drugs were discovered, a great deal of experimentation took place with the hope of developing a test method that would accommodate the large variety of antimicrobics with a high degree of reliability. Many problems were encountered.

The effectiveness of an antimicrobic in sensitivity testing is based on the size of the zone of inhibition. The zone of inhibition, however, varies with the diffusibility of the agent, the size of the inoculum, the type of medium, and many other factors. Only by taking all these variables into consideration could a reliable method be worked out. The **Kirby-Bauer method,** which is described here, is such a method and is the accepted procedure in use today. It is sanctioned by the U.S. FDA and the Subcommittee on Antimicrobial Susceptibility Testing of the National Committee for Clinical Laboratory Standards. Although time is insufficient here to consider all facets of the test, the basic procedure will be followed.

The recommended medium in this test is *Mueller-Hinton agar.* Its pH should be between 7.2 and 7.4, and it should be poured to a uniform thickness of 4 mm in the petri plate. This requires 60 ml in a 150-mm plate and 25 ml in a 100-mm plate. For certain fastidious microorganisms, 5% defibrinated animal blood (sheep, horse, or other) is added to the medium.

Inoculation of the surface of the medium is made with a cotton swab from a broth culture. In clinical applications, the broth turbidity has to match a defined standard. Care must be taken to express the excess broth from the swab prior to inoculation.

High potency disks are used in this test. The disks may be placed on the agar with a mechanical dispenser or sterile forceps. Regardless of how they are placed on the medium, it is desirable to press down on each disk to ensure close contact of the disk to the medium.

After 16 to 18 hours incubation, the plates are examined and the diameters of the zones are measured to the nearest millimeter. To determine whether the zone diameters are significantly large, it is necessary to consult a table in appendix A.

In this exercise, we will work with four microorganisms: *Staphylococcus aureus, Escherichia coli, Proteus vulgaris,* and *Pseudomonas aeruginosa.* Each student will inoculate one plate with one of the four organisms and place the disks on the medium by whichever method is available.

First Period
(Plate Preparation)

Materials:

1 petri plate of Mueller-Hinton agar per student
nutrient broth cultures (with swabs) of *S. aureus, E. coli, P. vulgaris,* and *P. aeruginosa*
disk dispenser (BBL or Difco)
cartridges of high-potency antimicrobic disks (BBL or Difco)
forceps

1. Select the organism that you are going to work with from the following table:

Organism	Student Number
S. aureus	1, 5, 9, 13, 17, 21, 25
E. coli	2, 6, 10, 14, 18, 22, 26
P. vulgaris	3, 7, 11, 15, 19, 23, 27
P. aeruginosa	4, 8, 12, 16, 20, 24, 28

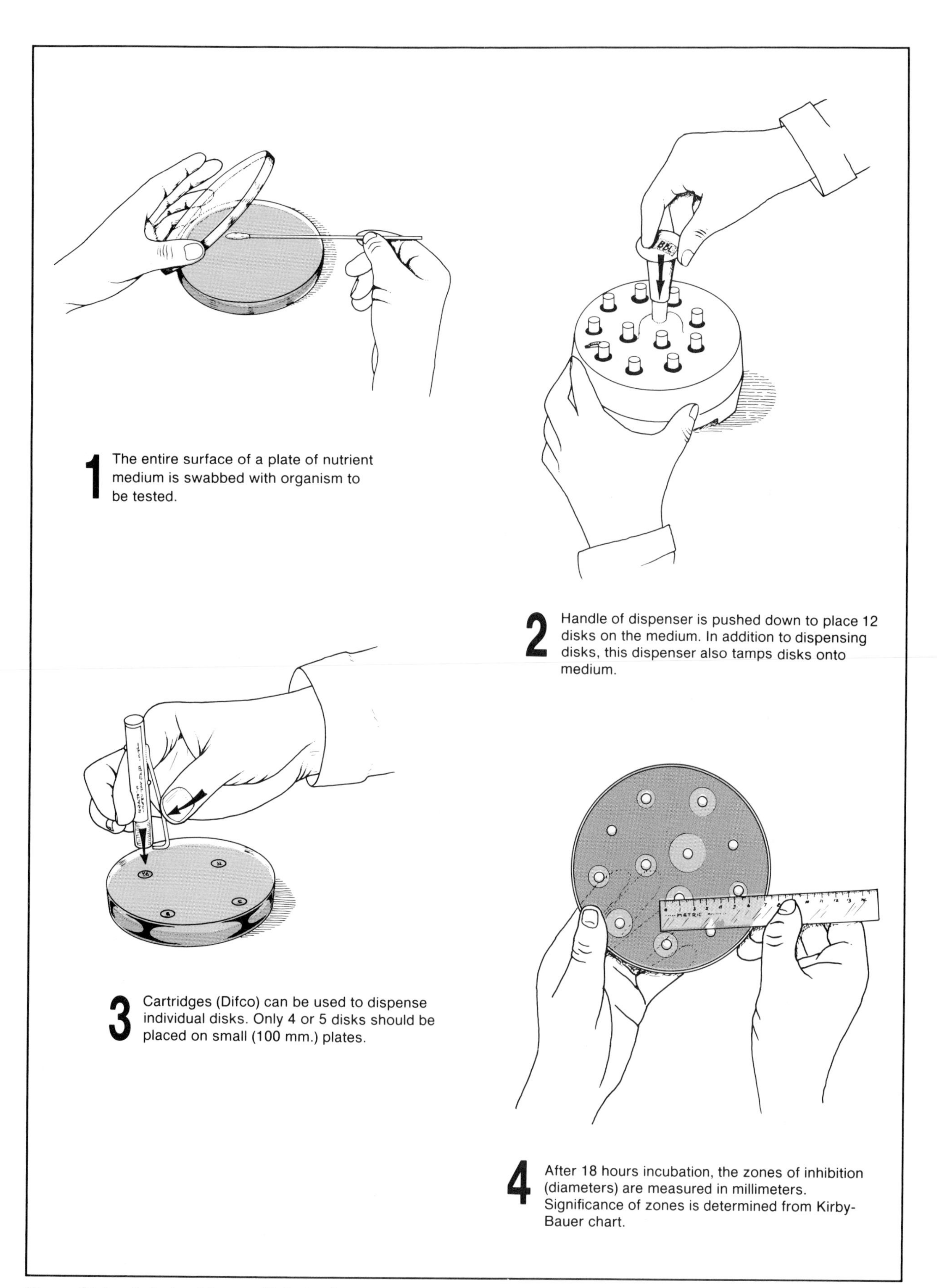

Figure 31.1 Antimicrobic sensitivity testing.

2. Label your plate with the name of your organism.
3. Inoculate the surface of the medium with the swab after expressing excess fluid from the swab by pressing and rotating the swab against inside walls of tube above fluid level. Cover the surface of the agar evenly by swabbing in three directions. A final sweep should be made of the agar rim with the swab.
4. Allow three to five minutes for the agar surface to dry before applying disks.
5. Dispense disks as follows:
   a. If automatic dispenser is used, remove lid, place dispenser over plate, and push down firmly on plunger.
   b. If forceps are used, sterilize them first by flaming before picking up the disks. Keep each disk at least 15 mm from edge of plate. Place no more than 13 on 150-mm plate, nor more than 5 on 100-mm plate. Apply light pressure to each disk on the agar with the tip of a sterile forceps or inoculating loop.
6. Invert and incubate for 16 to 18 hours at 37° C.

Second Period
(Interpretation)

After incubation, measure the zone diameters with a metric ruler to the nearest whole millimeter. The zone of complete inhibition is determined without magnification. Ignore faint growth or tiny colonies that can be detected by very close scrutiny. Large colonies growing within the clear zone might represent resistant variants or a mixed inoculum, and may require reidentification and retesting in clinical situations. Ignore the "swarming" characteristic of *Proteus,* measuring to the margin of heavy growth.

Record the zone measurements on the table in the Laboratory Report and on the chalkboard. To identify the Difco disks, consult the code chart in table 31.1. Consult the code chart in table 31.2 to identify the BBL disks. To determine the degree of sensitivity (R, I, and S) of the organism to each antimicrobic, refer to table VII in appendix A. Answer all questions on the Laboratory Report.

Table 31.1 Identification Code for Difco Antimicrobial Disks

Code	Antimicrobial	Code	Antimicrobial	Code	Antimicrobial
A	Aureomycin	GM	Gentamycin	PB	Polymyxin B
AM	Ampicillin	K	Kanamycin	RA	Rifampin
AN	Amikacin	L	Lincomycin	S	Streptomycin
B	Bacitracin	LR	Cephaloridine	SD	Sulfadiazine
C	Chloromycetin	ME	Methicillin	SM	Sulfamerazine
CB	Carbenicillin	N	Neomycin	SSS	Triple sulfa
CC	Clindamycin	NA	Nalidixic acid	ST	Sulfathiazole
CL	Colistin	NB	Novobiocin	T	Terramycin
CR	Cephalothin	NF	Nafcillin	TE	Tetracycline
CX	Cloxacillin	OA	Oxolinic acid	TM	Tobramycin
E	Erythromycin	OL	Oleandomycin	V	Viomycin
FD	Nitrofurantoin	P	Penicillin G	VA	Vancomycin

Table 31.2 Identification Code for BBL Antimicrobial Disks

AM–10	Ampicillin	E–15	Erythromycin	OA–2	Oxolinic acid
AN–10	Amikacin	FM–300	Nitrofurantoin	OL–15	Oleandomycin
B–10	Bacitracin	GM–10	Gentamycin	P–10	Penicillin G
C–30	Chloramphenicol	K–30	Kanamycin	PB–300	Polymyxin B
CB–50	Carbenicillin	L–2	Lincomycin	RA–5	Rifampin
CB–100	Carbenicillin	N–30	Neomycin	S–10	Streptomycin
CC–2	Clindamycin	NA–30	Nalidixic acid	SSS–25	Triple sulfonamides
CF–30	Cephalothin	NB–30	Novobiocin	Te–30	Tetracycline
CL–10	Colistin	NF–1	Nafcillin	Va–30	Vancomycin
DP–5	Methicillin	NN–10	Tobramycin		

32 Bacterial Mutagenicity and Carcinogenesis: The Ames Test

The fact that carcinogenic compounds induce increased rates of mutation in bacteria has led to the use of bacteria for screening chemical compounds for possible carcinogenesis. The **Ames test,** developed by Bruce Ames at the University of California, Berkeley, has been widely used for this purpose.

The conventional way to determine whether or not a chemical substance is carcinogenic is to inject the material into animals and look for the development of tumors. If tumors develop, it is presumed that the substance can cause cancer. Although this method works well, it is costly, time-consuming, and cumbersome, especially if it is applied to all the industrial chemicals that have found their way into our food and water supplies.

The Ames test serves as a screening test for the detection of carcinogenic compounds by testing the ability of chemical agents to induce bacterial mutations. Although most mutagenic agents are carcinogenic, some are not; however, the correlation between carcinogenesis and mutagenicity is high—around 83%. Once it has been determined that a specific agent is mutagenic, it can be used in animal tests to confirm its carcinogenic capability.

The standard way to test chemicals for mutagenesis has been to measure the rate of *back mutations* in strains of auxotrophic bacteria. In the Ames test a strain of *Salmonella typhimurium,* which is auxotrophic for histidine (unable to grow in the absence of histidine), is exposed to a chemical agent. After chemical exposure and incubation on histidine-deficient medium, the rate of reversion (back mutation) to prototrophy is determined by counting the number of colonies that are seen on the histidine-deficient medium.

Although testing of chemicals for mutagenesis in bacteria has been performed for a long time, two new features are included in the Ames test that make it a powerful tool. The first is that the strain of *S. typhimurium* used here lacks DNA repair enzymes, which prevents the correction of DNA injury. The second feature of the test is the incorporation of mammalian liver enzyme preparations with the chemical agent. The latter is significant because there is evidence that liver enzymes convert many noncarcinogenic chemical agents to carcinogenic ones.

There are two ways to perform the Ames test. The method illustrated in figure 32.1 is a *spot test* which is widely used for screening purposes. The other method is the *plate incorporation test* which is used for quantitative analysis of the mutagenic effectiveness of compounds. Our concern here will be with the spot test; however, since the concentration of the liver extract is very critical we will omit using it in our test. The test as performed here will work well without it.

Success in performing the spot test requires considerable attention to careful measurements and timing. It is for this reason that students will work in pairs to perform the test.

Note in figure 32.1 that 0.1 ml of *S. typhimurium* is first added to a small tube containing 2 ml of top agar that is held at 45° C. This top agar contains 0.6% agar, 0.5% NaCl, and a trace of histidine and biotin. The histidine allows the bacteria to go through several rounds of cell division, which is essential for mutagenesis to occur. Since the histidine deletion extends through the biotin gene, biotin is also needed. This early growth of cells produces a faint background lawn that is barely visible to the naked eye.

Before pouring the top agar over the glucose–minimal salts agar the tube must be vortexed at slow speed for three seconds and poured quickly to get even distribution. The addition of the bacteria, vortexing, and pouring must be accomplished in 20 seconds. Failure to move quickly enough will cause stippling of the top agar.

There are two ways that one can use to apply the chemical agent to the top agar: a filter paper disk may be used, or the chemical can be applied directly to the center of the plate without a disk.

Note the unusual way in which a filter paper disk is impregnated in figure 32.1. To get it to stand on its edge it must be put in position with sterile forceps and pressed in slightly to hold it upright. Just the right amount of the chemical agent is then added with a Pasteur pipette to the upper edge of

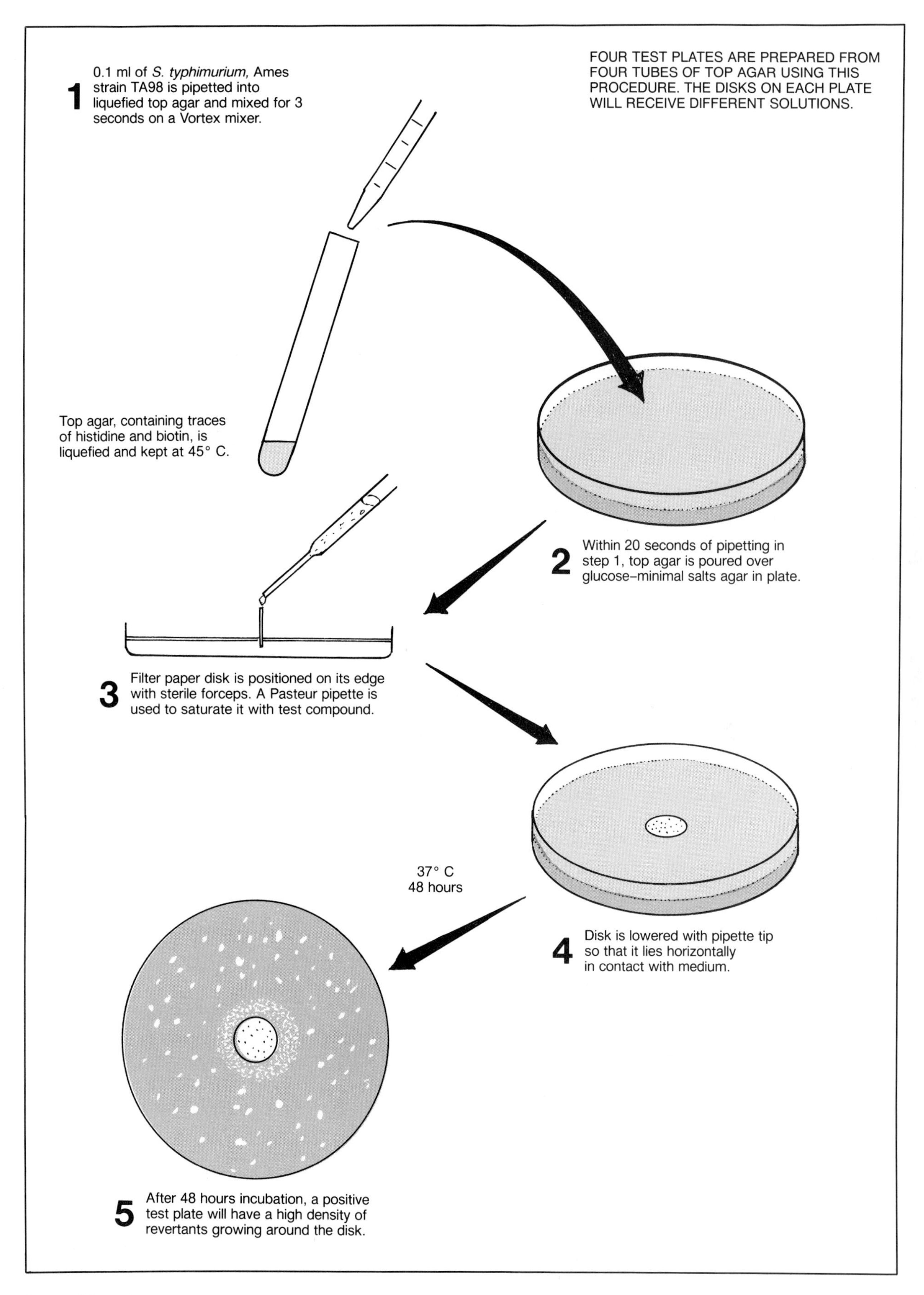

Figure 32.1 Procedure for performing a modified Ames test.

the disk to completely saturate it without making it dripping wet; then the disk is lowered onto the top agar.

Once the test reagent is deposited on the top agar, the plate is incubated at 37° C for 48 hours. If the agent is mutagenic, a halo of densely packed revertant colonies will be seen around the disk.

You will be issued an unknown chemical agent to test and you will have an opportunity to test some other substance you have brought to the laboratory. In addition to these two tests you will be inoculating positive and negative test controls; thus, each pair of students will be responsible for four plates.

Keep in mind as you perform this experiment that there is a lot more to the Ames test than revealed here. While we are using only one tester strain of *S. typhimurium,* there are several others that are used in routine testing. The additional strains are needed to accommodate different kinds of chemical compounds. While one chemical agent may be mutagenic on one tester strain, it may produce a negative result on another strain. Also, keep in mind that we are not taking advantage of using the liver extract.

First Period
(Inoculations)

Materials:

per pair of students:
- 4 plates of glucose–minimal salts agar (30 ml per plate)
- 4 tubes of top agar (2 ml per tube)
- tube of unknown possible carcinogen
- tube of sterile water
- Vortex mixer

sterile Pasteur pipettes
serological pipettes (1 ml size)
forceps
filter paper disks, sterile in petri dish
test reagents:
- 4-nitro-o-phenylenediamine solution, (4-NOPD) 10μg/ml
- substance from home for testing

culture of *S. typhimurium,* Ames strain, TA98 in trypticase soy broth

1. Working with your laboratory partner, label the bottoms of four glucose–minimal salts agar plates as follows: POSITIVE CONTROL, NEGATIVE CONTROL, UNKNOWN, and OPTIONAL.
2. Liquefy four tubes of top agar and cool to 45° C.
3. With a 1-ml serological pipette, inoculate a tube of top agar with 0.1 ml of *S. typhimurium.*
4. Thoroughly mix the organisms into the top agar by vortexing (slow speed) for three seconds, or by rolling the tube between the palms of both hands. Pour the contents onto the positive control plate of glucose–minimal salts agar. The agar plate must be at room temperature. *Work rapidly to achieve pipetting, mixing, and spreading in 20 seconds.*
5. Repeat steps 3 and 4 for each of the other three tubes of top agar.
6. With sterile forceps place a disk on its edge near the center of the positive control plate. Sterilize the forceps by dipping in alcohol and flaming.
7. With a sterile Pasteur pipette, deposit just enough 4-NOPD on the upper edge of the disk to saturate it; then, push over the disk with the pipette tip onto the agar so that it lies flat.
8. Insert a sterile disk on the negative control plate in the same manner as above. Moisten this disk with sterile water, and reposition it flat on the agar surface. Be sure to use a fresh Pasteur pipette.
9. Place a disk on the unknown plate, and, using the same procedures, infiltrate it with your unknown, and position it flat on the agar.
10. If the test substance from home is crystalline, place a few crystals directly on the top agar of the optional plate in its center. Liquid substances should be handled in same manner as above.
11. Incubate all plates for 48 hours at 37° C.

Second Period
(Evaluation)

Examine all four plates. You should have a pronounced halo of revertant colonies around the disk on the positive control plate, and no, or very few, revertants on the negative control plate. The presence of a few scattered revertants on the negative control plate are due to spontaneous back mutations, which always occur. Examine the areas beyond the halo to see if you can detect a faint lawn of bacterial growth.

CAUTION: Since much of the glassware in this experiment contains carcinogens, do not dispose of any of it in the usual manner. Your instructor will indicate how this glassware is to be handled.

Record your results on the Laboratory Report and answer all the questions.

Effectiveness of Hand Scrubbing 33

The importance of hand disinfection in preventing the spread of disease is accredited to the observations of Semmelweis at the Lying-In Hospital in Vienna in 1846 and 1847. He noted that the number of cases of puerperal fever was closely related to the practice of sanitary methods. Until he took over his assignment in this hospital, it was customary for medical students to go directly from the autopsy room to the patient's bedside and assist in deliveries without scrubbing and disinfecting their hands. When the medical students were on vacation, only the nurses, who were not permitted in the autopsy room, attended the patients. Semmelweis noted that during this time, deaths due to puerperal fever fell off markedly. As a result of his observations, he established a policy that no medical students would be allowed to examine obstetric patients or assist in deliveries until they had cleansed their hands with a solution of chloride of lime. This ruling caused the death rate from puerperal infections to drop from 12% to 1.27% in one year.

Today it is routine practice to wash hands prior to the examination of any patient and to do a complete surgical scrub prior to surgery. Scrubbing the hands involves the removal of **transient** (contaminant) and **resident** microorganisms. Depending on the condition of the skin and the numbers of bacteria present, it takes from seven to eight minutes of washing with soap and water to remove all transients; and they can be killed with relative ease using suitable antiseptics. Residents, on the other hand, are firmly entrenched and are removed slowly by washing. These organisms, which consist primarily of staphylococci of low pathogenicity, are less susceptible than the transients to the action of antiseptics.

In this exercise, an attempt will be made to evaluate the effectiveness of the length of time in removal of organisms from the hands using a surgical scrub technique. One member of the class will be selected to perform the scrub. Another student will assist by supplying the soap, brushes, and basins, as needed. During the scrub, at two-minute intervals, the hands will be scrubbed into a basin of sterile water. Bacterial counts will be made of these basins to determine the effectiveness of the previous two-minute scrub in reducing the bacterial flora of the hands. Members of the class not involved in the scrub procedure will make the inoculations from the basins for the plate counts.

Scrub Procedure

The two members of the class who are chosen to perform the surgical scrub will set up their materials near a sink for convenience. As one performs the scrub, the other will assist in reading the instructions and providing materials as needed. The basic steps, which are illustrated in figure 33.1, are also described in detail below. Before beginning the scrub, both students should read all the steps carefully.

Materials:

- 5 sterile surgical scrub brushes, individually wrapped
- 5 basins (or 2000-ml beakers) containing 1000 ml each of sterile water. These basins should be covered to prevent contamination.
- 1 dispenser of green soap
- 1 tube of hand lotion

STEP 1. To get some idea of the number of transient organisms on the hands, the scrubber will scrub all surfaces of each hand with a sterile surgical scrub brush for 30 seconds into basin A. No green soap will be used for this step. The successful performance of this step will depend on:

- spending the same amount of time on each hand (30 seconds),
- maintaining the same amount of activity on each hand, and
- scrubbing under the fingernails as well as working over their surfaces.

After completion of this 60-second scrub, notify group A that their basin is ready for the inoculations.

STEP 2. Using the *same* brush as above, begin scrubbing with green soap for two minutes, using

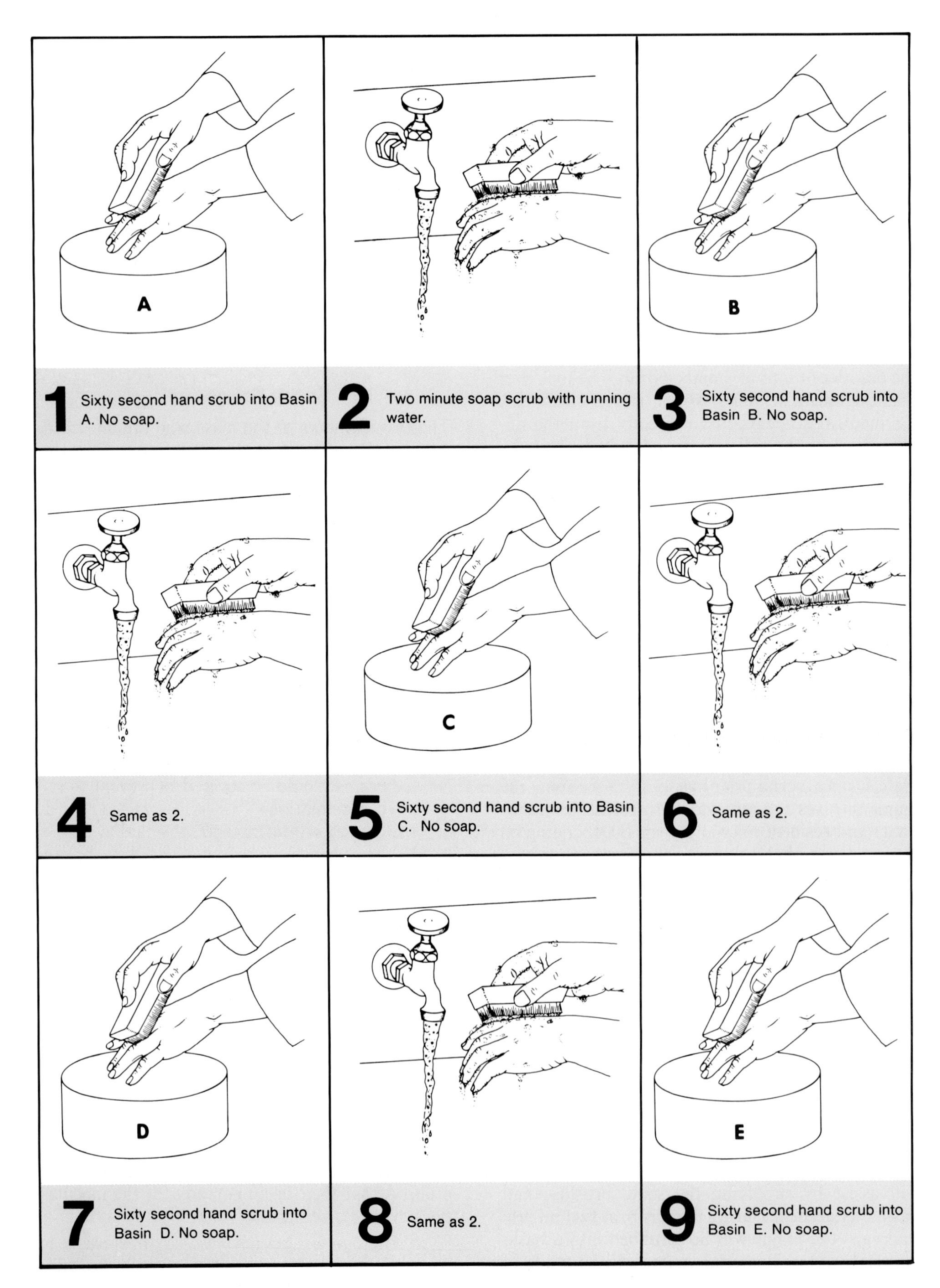

Figure 33.1 Hand scrubbing routine.

cool tap water to moisten and rinse the hands. One minute is devoted to each hand.

The assistant will make one application of green soap to each hand as it is being scrubbed.

Rinse both hands for five seconds under tap water at the completion of the scrub.

Discard this brush.

Note: This same procedure will be followed exactly in steps 4, 6, and 8 of figure 33.1.

STEP 3. With a *fresh* sterile brush, scrub the hands into basin B in a manner identical to step 1. Don't use soap. Notify group B when this basin is ready.

Note: Exactly the same procedure is used in steps 5, 7, and 9 of figure 33.1.

Remember: It is important to use a fresh sterile brush for the preparation of each of these basins.

After Scrubbing After all scrubbing has been completed, the scrubber should dry his or her hands and apply hand lotion.

Making the Pour Plates

While the scrub is being performed, the rest of the class will be divided into five groups (A, B, C, D, and E) by the instructor. Each group will make six plate inoculations from one of the five basins (A, B, C, D, or E). It is the function of these groups to determine the bacterial count per milliliter in each basin. In this way we hope to determine, in a relative way, the effectiveness of scrubbing in bringing down the total bacterial count of the skin.

Materials:

30 veal infusion agar pours—6 per group
1-ml pipettes
30 sterile petri plates—6 per group
70% alcohol
L-shaped glass stirring rod (optional)

1. Liquefy six pours of veal infusion agar and cool to 50° C. While the medium is being liquefied, label two plates each: 0.1 ml, 0.2 ml, and 0.4 ml. Also, indicate your group designation on the plate.
2. As soon as the scrubber has prepared your basin, take it to your table and make your inoculations as follows:
   a. Stir the water in the basin with a pipette or an L-shaped stirring rod for 15 seconds. If the stirring rod is used (figure 33.2), sterilize it before using by immersing it in 70% alcohol and flaming. *For consistency of results all groups should use the same method of stirring.*
   b. Deliver the proper amounts of water from the basin to the six petri plates with a sterile serological pipette. Refer to figure 33.3. If a pipette was used for stirring, it may be used for the deliveries.
   c. Pour a tube of veal infusion agar, cooled to 50° C, into each plate, rotate to get good distribution of organisms, and allow to cool.
   d. Incubate the plates at 37° C for 24 hours.
3. After the plates have been incubated, select the pair that has the best colony distribution with no fewer than 30 or more than 300 colonies. Count the colonies on the two plates and record your counts on the chart on the chalkboard.
4. After all data is on the chalkboard, complete the table and graph on the Laboratory Report.

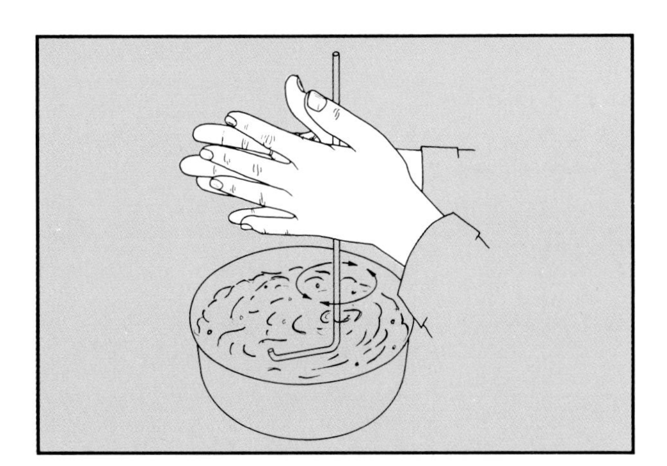

Figure 33.2 An alternate method of stirring utilizes an L-shaped glass stirring rod.

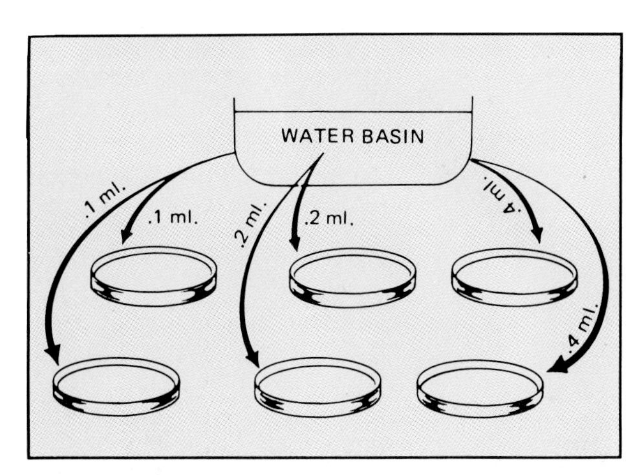

Figure 33.3 Water is distributed to six petri plates in amounts as shown.

Part 6 Identification of Unknown Bacteria

One of the most interesting experiences in introductory microbiology is to attempt to identify an unknown microorganism that has been assigned to you as a laboratory problem. The next eight exercises pertain to this phase of microbiological work. You will be given one or more cultures of bacteria to identify. The only information that might be given to you about your unknowns will pertain to their sources and habitats. All the information needed for identification will have to be acquired by you through independent study.

The first step in the identification procedure is to accumulate information that pertains to the organism's morphological, cultural, and physiological (biochemical) characteristics. This involves making different kinds of slides for cellular studies and the inoculation of various types of media to note the growth characteristics and types of enzymes produced. As this information is accumulated, it is recorded in an orderly manner on **Descriptive Charts,** which are located toward the back of the manual with the Laboratory Reports.

After sufficient information has been recorded, the next step is to consult a taxonomic key, which enables one to identify the organism. For this final step *Bergey's Manual of Systematic Bacteriology* will be used. Copies of volumes 1 and 2 of this book will be available in the laboratory, library, or both. In addition, a computer-assisted identification (CAI) program may be available, which can be used to help confirm your decision. Exercise 41 pertains to the use of *Bergey's Manual* and Shannon's CAI program.

Success in this endeavor will require meticulous techniques, intelligent interpretation, and careful recordkeeping. Your mastery of aseptic methods in the handling of cultures and the performance of inoculations will show up clearly in your results. Contamination of your cultures with unwanted organisms will yield false results, making identification hazardous speculation. If you have reason to doubt the validity of the results of a specific test, repeat it; *don't rely on chance!* As soon as you have made an observation or completed a test, record the information on the descriptive chart. Do not trust your memory—record data immediately!

34 Preparation and Care of Stock Cultures

Your unknown cultures will be used for making many different kinds of slides and inoculations. In spite of meticulous aseptic practice on your part, the chance of contamination of these cultures increases with the frequency of use. If you were to attempt to make all of your inoculations from the single tube given to you, it is very likely that somewhere along the way contamination would result.

Another problem that will arise is aging of the culture. Two or three weeks may be necessary for the performance of all tests. In this period of time, the organisms in a broth culture may die, particularly if the culture is kept very long at room temperature. To ensure against the hazards of contamination or death of your organisms, it is essential that you prepare stock cultures before any slides or routine inoculations are made.

Different types of organisms require different kinds of stock media, but for those used in this unit, nutrient agar slants will suffice. For each unknown, you will inoculate two slants. One of these will be your **reserve stock** and the other one will be your **working stock.** The reserve stock culture will not be used for making slides or routine inoculations; instead, it will be stored in the refrigerator after incubation until some time later when a transfer may be made from it to another reserve stock or working stock culture. The working stock culture will be used for making slides and routine inoculations. When it becomes too old to use or has been damaged in some way, replace it with a fresh culture that is made from the reserve stock.

Note in figure 34.1 that one slant will be incubated at 20° C and the other at 37° C. This will

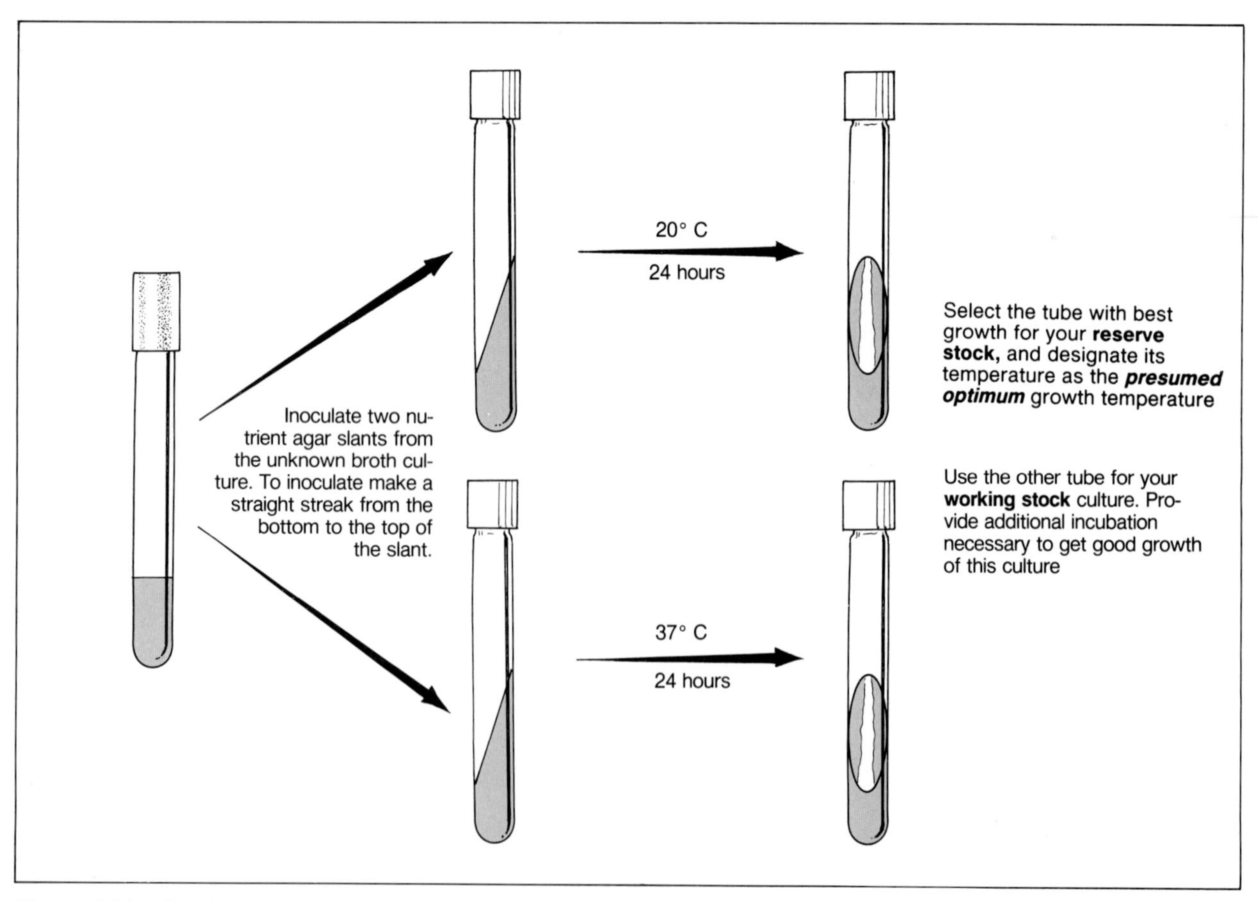

Figure 34.1 Stock culture procedure.

enable you to learn something about the optimum growth temperature of your unknown. Proceed as follows:

Materials *(for each unknown):*

2 nutrient agar slants (screw-cap type)
gummed labels

1. Label two slants with the code number of the unknown and your initials. Use gummed labels. Also, mark one tube 20° C and the other 37° C.
2. With a loop, inoculate each slant with a straight streak from bottom to top. Since these slants will be used for your cultural study, a straight streak is more useful than one that is spread over the entire surface.
3. Place the two slants in separate baskets on the demonstration table that are designated with labels for the two temperatures (20° C and 37° C).

   Although the 20° C temperature is thought of as "room temperature," it should be incubated in a biological incubator instead of leaving it out at laboratory room temperature. Laboratory temperatures are often quite variable in a 24-hour period.
4. After 24 hours, examine the slants to note the extent of growth. Examine them closely. Some organisms produce a very thin translucent growth that is difficult to detect. Determine which temperature seems to promote the best growth. Record on the Descriptive Chart the presumed optimum temperature. (Obviously, this may not be the actual optimum temperature, but for all practical purposes, it will suffice for this exercise.)

   If No Growth. If no growth is visible on either slant, there are several possible explanations. It may be that the culture you were issued was not viable. Another possibility might be that the organism grows too slowly to be visible at this time. Or, possibly, neither temperature was suitable. Think through these possibilities and decide what you should do to circumvent the problem.
5. Label the tube with the best growth **reserve stock.** Label the other tube **working stock.**
6. If both tubes have good growth, place them in the refrigerator until needed.

   If one tube has very scanty growth, refrigerate the good one (reserve stock) and incubate the other one at the more desirable temperature for another 24 hours; then refrigerate.
7. Remember these points concerning your stock cultures:

   • Most stock cultures will keep for four weeks in the refrigerator. Some fastidious pathogens will survive for only a few days. Although none of the organisms issued in this unit are of the extremely delicate type, don't wait four weeks to make a new reserve stock culture; instead, make fresh transfers every ten days.

   • Don't use your reserve stock culture for making slides or routine inoculations.

   • Don't store either of your stock cultures in your locker or desk drawer. After the initial incubation period they must be refrigerated. After two or three days at room temperature, cultures begin to deteriorate. Some die out completely.
8. Answer the questions on the Laboratory Report.

35 Morphological Study of Unknown

The first step in the identification of an unknown bacterial organism is to learn as much as possible about its morphological characteristics. One needs to know whether the organism is rod-, coccus-, or spiral-shaped; whether or not it is pleomorphic; its reaction to Gram staining; and the presence or absence of endospores, capsules, or granules. All this morphological information provides a starting point in the categorization of an unknown.

Figure 35.1 illustrates the steps that will be followed in determining morphological characteristics of your unknown. Note that fresh broth and slant cultures will be needed to make the various slides and perform motility tests. Since most of the slide techniques were covered in Part 3, you will find it necessary to refer back to that section from time to time. Note that Gram staining, motility testing, and measurements will be made from the broth culture; Gram staining and other stained slides will also be made from the agar slant. The rationale as to the choice of broth or agar slants will be explained as each technique is performed.

As soon as morphological information is acquired be sure to record your observations on the descriptive chart at the back of the manual. Proceed as follows:

Materials:

- Gram staining kit
- spore staining kit
- acid-fast staining kit
- Loeffler's methylene blue stain
- nigrosine or india ink
- tubes of nutrient broth and nutrient agar
- gummed labels for test tubes

New Inoculations

For all of these staining techniques you will need 24- to 48-hour cultures of your unknown. If your working stock slant is a fresh culture, use it. If you don't have a fresh broth culture of your unknown inoculate a tube of nutrient broth and incubate it at its estimated optimum temperature for 24 hours.

Gram Stain

Since a good gram-stained slide will provide you with more valuable information than any other slide, this is the place to start. Make gram-stained slides from both the broth and agar slant, and compare them under oil immersion.

Two questions must be answered at this time: (1) Is the organism gram-positive or is it gram-negative? and (2) Is the organism rod- or coccus-shaped? If your staining technique is correct, you should have no problem with the Gram reaction. If the organism is a long rod, the morphology question is easily settled; however, if your organism is a very short rod, you may incorrectly decide it is coccus-shaped.

Keep in mind that short rods with round ends (coccobacilli) look like cocci. If you have what seems to be a coccobacillus, examine many cells before you make a final decision. Also, keep in mind that *while rod-shaped organisms* ***frequently*** *appear as cocci under certain growth conditions, cocci* ***rarely*** *appear as rods.* (*Streptococcus mutans* is unique in forming rods under certain conditions.) Thus, it is generally safe to assume that if you have a slide on which you see both coccuslike cells and short rods, the organism is probably rod-shaped. This assumption is valid, however, only if you are not working with a contaminated culture!

Record the shape of the organism and its reaction to the stain on the Descriptive Chart.

Cell Size

Once you have a good gram-stained slide, determine the size of the organism with an ocular micrometer. Refer to Exercise 3. If the size is variable, determine the size range. Record this information on the Descriptive Chart.

Motility and Cellular Arrangement

If your organism is a nonpathogen make a hanging drop slide from the broth culture. Refer to Exercise 15. This will enable you to determine whether or

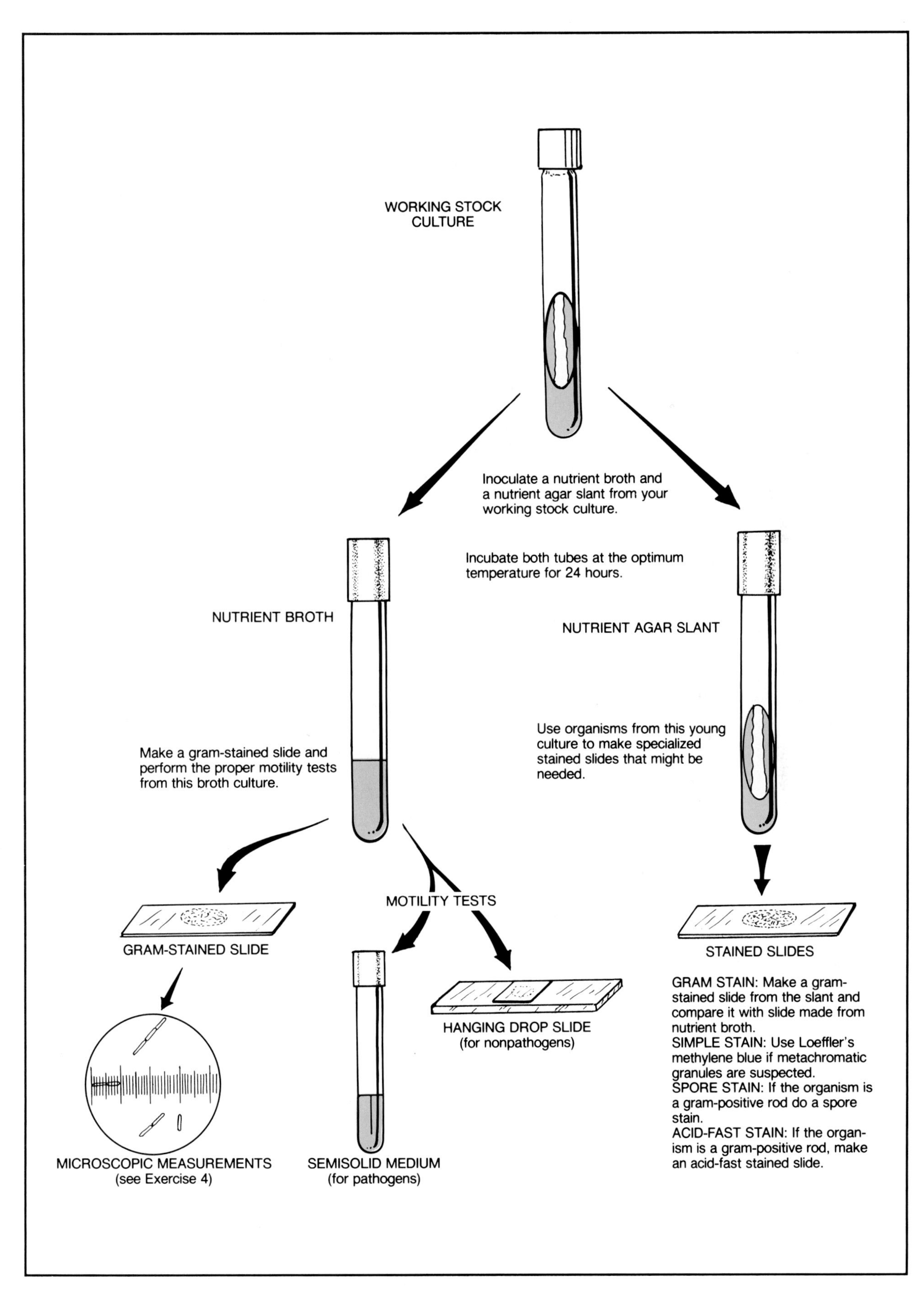

Figure 35.1 Procedure for morphological study.

not the organism is motile, and allow you to confirm the cellular arrangement. By making this slide from broth instead of the agar slant, the cells will be well dispersed in natural clumps. Note whether the cells occur singly, in pairs, masses, or chains.

If your organism happens to be a pathogen do not make a hanging drop slide; instead, stab the organism into a tube of semisolid medium to determine motility (Exercise 15). It won't be necessary to do this tube culture if the organism is a nonpathogen.

Be sure to record your observations on the Descriptive Chart.

Endospores

If your unknown is a gram-positive rod, check for endospores. *Only rarely is a coccus or gram-negative rod a spore-former.* Examination of your gram-stained slide made from the agar slant should provide a clue, since endospores show up as transparent holes in gram-stained spore-formers. Endospores can also be seen on unstained organisms if studied with phase-contrast optics.

If there seems to be evidence that the organism is a spore-former, make a slide using one of the spore-staining techniques you employed in Exercise 13.

Record on the Descriptive Chart whether the spore is *terminal, subterminal,* or in the *middle* of the rod.

Acid-Fast Staining

If your organism is a gram-positive, non-spore-forming rod, you should determine whether or not it is acid-fast. Although some bacteria require four or five days growth to exhibit acid-fastness, most species become acid-fast within two days. For best results, therefore, do not use cultures that are too old.

Another point to keep in mind is that most acid-fast bacteria do not produce cells that are 100% acid-fast. An organism is considered acid-fast if only portions of the cells exhibit this characteristic. Refer to Exercise 14 for this staining technique.

Other Structures

If the protoplast in gram-stained slides stains unevenly, you might wish to do a simple stain with Loeffler's methylene blue for evidence of metachromatic granules.

Although a capsule stain may be performed at this time, if desired, it might be better to wait until a later date when you have the organism growing on blood agar. Capsules usually are more apparent when the organisms are grown on this medium.

Laboratory Report

There is no Laboratory Report to fill out for this exercise. All information is recorded on the Descriptive Chart.

Determination of Oxygen Requirements 36

The oxygen requirements of bacteria range from the **strict** (obligate) **aerobes** that cannot exist without this gas to the **strict** (obligate) **anaerobes** that die in its presence. In between these extremes are the facultatives, indifferents, and microaerophilics. The **facultatives** are bacteria that have enzyme systems enabling them to utilize free oxygen or some alternative oxygen source such as nitrate. If oxygen is present, they tend to utilize it in preference to the alternative. The **indifferents,** however, show no preference for either condition, growing equally well in aerobic and anaerobic conditions. **Microaerophiles,** on the other hand, are organisms that require free oxygen, but only in limited amounts. Figures 36.1 and 36.2 illustrate where these types tend to grow with respect to the degree of oxygen tension in a medium.

In attempting to identify an unknown bacterium, it is important to learn into what category the organism fits. *Bergey's Manual of Systematic Bacteriology* (volume 1), lists five groups (sections 2, 4, 5, 6, and 8) of bacteria that are described in terms of oxygen requirements. If you know its oxygen needs and morphology, you have the basic information required to determine where your organism fits into the schema of bacterial classification.

In this exercise fluid thioglycollate medium will be used to determine the oxygen requirements of your unknown. Six test organisms that are representative of each type will also be used.

Fluid thioglycollate medium utilizes glucose, cystine, and sodium thioglycollate to lower its oxidation-reduction potential. The dye resazurin is included to indicate the presence of oxygen. In the presence of oxygen the dye becomes pink. Since the oxygen tension is always higher near the surface of the medium, the medium will be pink at the top and colorless in the middle and bottom. The medium also contains some agar that helps to localize the organisms and favor anaerobiosis in the bottom of the tube.

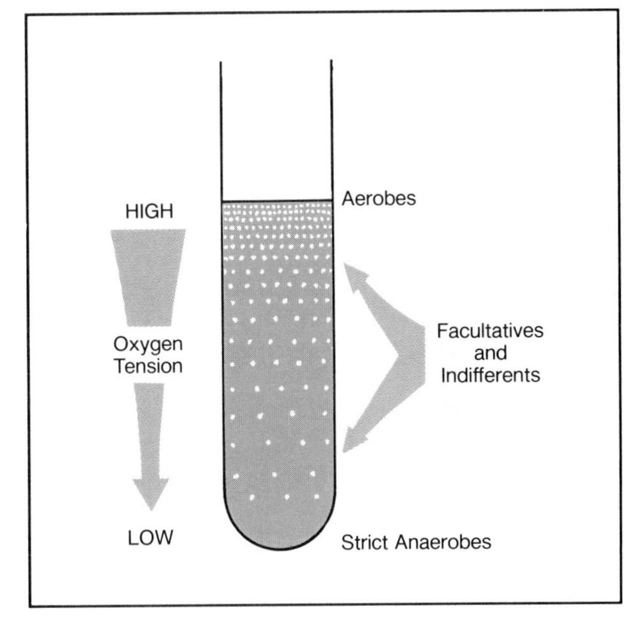

Figure 36.1 Oxygen needs.

First Period

Proceed as follows to inoculate one tube of fluid thioglycollate medium with your unknown and three tubes of the same medium with three test control organisms. To economize on media **odd-numbered students** will use *E. coli, B. subtilis,* and *C. sporogenes* for their test control inoculations. **Even-numbered students** will use *S. aureus, S. faecalis,* and *C. rubrum.*

Materials:

- 3 tubes of fluid thioglycollate medium for test organisms
- 1 tube of fluid thioglycollate medium for each unknown
- broth cultures of *Escherichia coli, Bacillus subtilis, Clostridium sporogenes, Staphylococcus aureus, Streptococcus faecalis, Clostridium rubrum*

1. After procuring the tubes of fluid thioglycollate medium from the demonstration table, examine them to note what percentage of the medium is pink. If the upper 30% is pink (oxidized resazurin), place the tubes in boiling water for a few minutes to drive off the ab-

sorbed oxygen. Cool to a safe inoculating temperature by inserting them in a beaker of cool tap water. Avoid shaking these tubes after heating to prevent reabsorption of oxygen.

2. Label three of the tubes according to the test control organisms assigned to you (odd or even, see previous page).
3. Label one tube with the number of your unknown.
4. Inoculate each tube. Shake each tube gently from side to side or rotate between the palms of hands to get good dispersion, *but do not shake so vigorously that more oxygen is taken into the medium.*
5. Incubate at 37° C for 48 hours.

Second Period

1. Carefully remove the cultures from the incubator, taking care not to unduly agitate the tubes.
2. Combine your test controls with those of your laboratory partner, and compare all six of the tubes with those in figure 36.2. Determine the oxygen requirements of each of the control organisms.
3. Determine the oxygen requirements of your unknown organism.
4. Record your conclusions on the Laboratory Report.
5. Record your observations concerning your unknown on the Descriptive Chart.

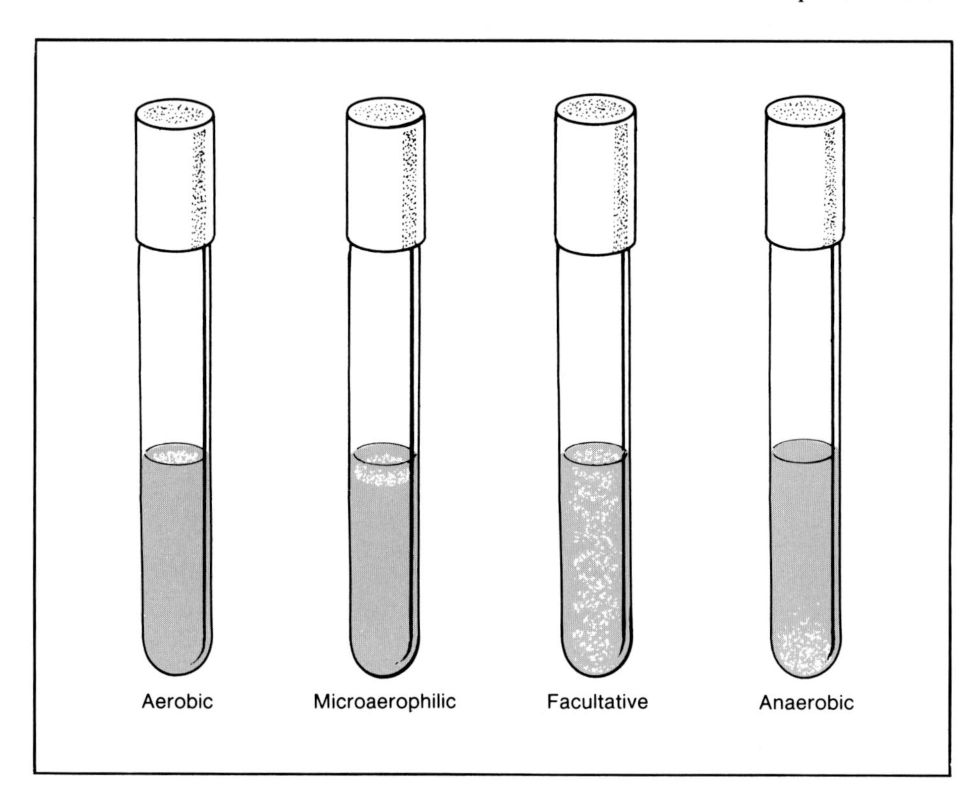

Figure 36.2 Types of bacteria according to oxygen needs.

Cultural Characteristics 37

The cultural characteristics of an organism pertain to its macroscopic appearance on different kinds of media. Descriptive terms, which are familiar to all bacteriologists and are used in *Bergey's Manual,* must be used in recording the characteristics. The most frequently used media for a cultural study are nutrient agar, nutrient broth, and nutrient gelatin. For certain types of unknowns it is desirable to inoculate a blood agar plate. This will be done later, if necessary.

Inoculations

In this exercise, one nutrient agar plate, one nutrient gelatin deep, and two nutrient broths will be inoculated. The nutrient agar slant reserve stock culture that was prepared in Exercise 34 will be used here also. The reason for inoculating two tubes of nutrient broth is to recheck the optimum growth temperature of your unknown. In Exercise 34 you incubated your nutrient agar slants at 20° C and 37° C. It may well be that the optimum growth temperature is closer to 30° C. It is to check out this intermediate temperature that an extra nutrient broth is inoculated.

Materials *(for each unknown):*

1 nutrient agar pour
1 nutrient gelatin deep
2 nutrient broths
1 petri plate
photocolorimeter and cuvettes
hand lens

1. Pour a petri plate of nutrient agar for each unknown and streak it with a method that will give good isolation of colonies. Use the original broth culture for streaking.
2. Inoculate the tubes of nutrient broth with a loop.
3. Make a stab inoculation into the gelatin deep by stabbing the inoculating needle (straight wire) directly down into the medium to the bottom of the tube and pulling it straight out. The medium must not be disturbed laterally.
4. Place all tubes except one nutrient broth into a basket and incubate for 24 hours at the temperature that seemed best in Exercise 36. Incubate the other nutrient broth separately at 30° C. Incubate plate at best temperature.

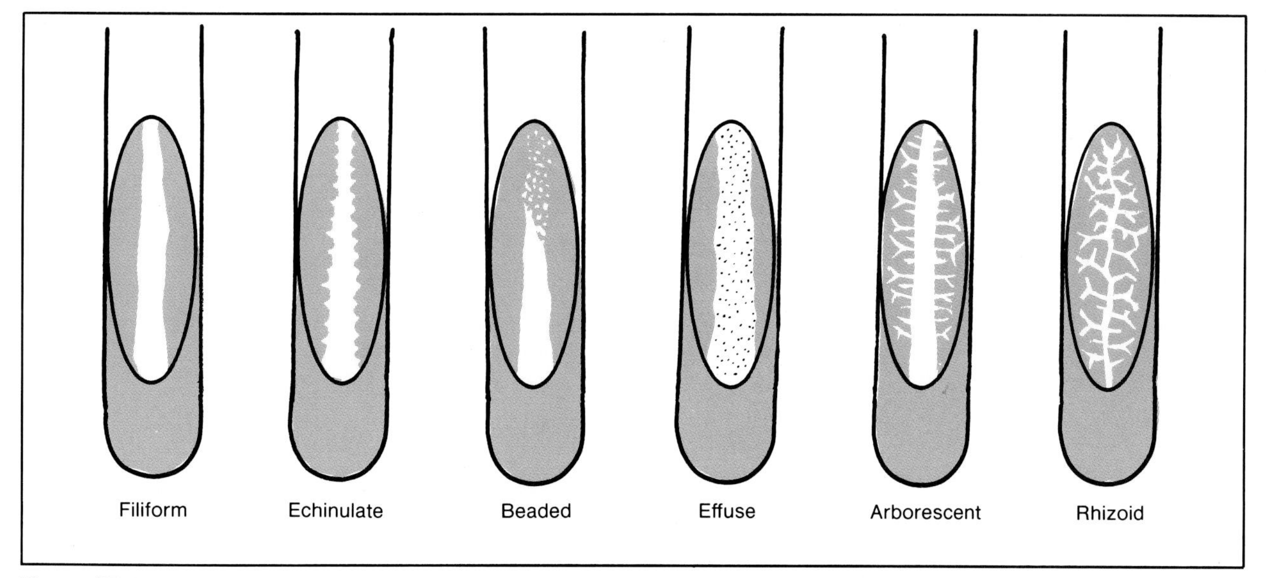

Figure 37.1 Types of bacterial growth on nutrient agar slants.

Evaluation of Cultures

After the cultures have been properly incubated, *carry them to your desk in a careful manner to avoid disturbing the growth pattern in the nutrient broth.* Also, place the tube of nutrient gelatin in ice water, which will be set up near the sink. As you examine each culture, look for the specific characteristics that are described in the following paragraphs. Record these features on the Descriptive Chart.

Nutrient Agar Slant

Procure the nutrient agar slant from the refrigerator that was set aside in Exercise 34 as a reserve stock culture. Evaluate the slant with respect to the following characteristics, recording your observations on the Descriptive Chart.

Amount of Growth The abundance of growth may be described as none, slight, moderate, and abundant.

Color Pigmentation should be looked for on the organisms and within the medium. Most organisms will lack chromogenesis, exhibiting a white growth; others are various shades of different colors. Some microorganisms produce soluble pigments that diffuse into the medium. Hold the slant up to a strong light to examine it for diffused pigmentation.

Opacity Organisms that grow prolifically on the surface of a medium will appear more opaque than those that exhibit a small amount of growth. Degrees of opacity may be expressed in terms of *opaque, transparent,* and *translucent* (partially transparent).

Form The gross appearance of different types of growth are illustrated in figure 37.1. The following descriptions of each type will help in differentiation:

- ***Filiform:*** characterized by uniform growth along the line of inoculation
- ***Echinulate:*** margins of growth exhibit toothed appearance
- ***Beaded:*** separate or semiconfluent colonies along the line of inoculation
- ***Effuse:*** growth is thin, veillike, unusually spreading
- ***Arborescent:*** branched, treelike growth
- ***Rhizoid:*** rootlike appearance

Nutrient Broth

The nature of growth on the surface, subsurface, and bottom of the tube are significant in nutrient broth cultures. Describe your cultures as thoroughly as possible on the Descriptive Chart with respect to these characteristics.

Surface Figure 37.2 illustrates different types of surface growth. A *pellicle* type of surface differs from the *membranous* type in that the latter is much thinner. A *flocculent* surface is made up of floating adherent masses of bacteria.

Subsurface Below the surface, the broth may be described as *turbid* if it is cloudy; *granular* if specific small particles can be seen; *flocculent* if small masses are floating around; and *flaky* if large particles are in suspension.

Sediment The amount of sediment in the bottom of the tube may vary from none to a great deal. To describe the type of sediment, agitate the tube,

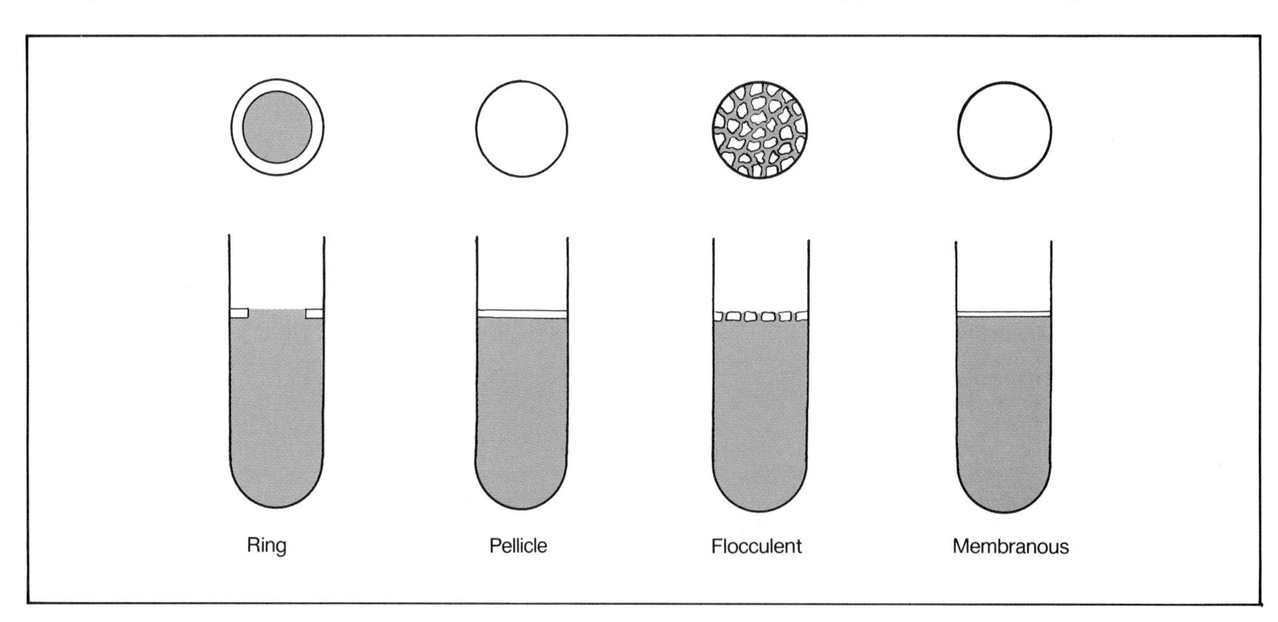

Figure 37.2 Types of surface growth in nutrient broth.

putting the material in suspension. The type of sediment can be described as *granular, flocculent, flaky,* and *viscid.* Test for viscosity by probing the bottom of the tube with a sterile inoculating loop.

Amount of Growth To determine the amount of growth, it is necessary to shake the tube to disperse the organisms. Terms such as *slight* (scanty), *moderate,* and *abundant* adequately describe the amount.

Optimum Temperature To determine which temperature produced the best growth, dispense the contents from each tube of nutrient broth into separate cuvettes and measure their percent transmittances. If the percent transmittance is less at 30° C than at the other presumed optimum temperature, revise the optimum temperature on your Descriptive Chart.

Gelatin Stab Culture

If growth occurs in nutrient gelatin, we look for two things: type of growth and the presence or absence of liquefaction. After the tubes have cooled for about five minutes in ice water, make your evaluation.

Type of Growth (no liquefaction) Not all organisms will grow on nutrient gelatin, but if growth does occur without liquefaction, it may appear as one of the forms shown on the left side of figure 37.3. From a categorization standpoint, however, the nature of growth in unliquefied nutrient gelatin is not as important as some other cultural characteristics.

Liquefaction Some organisms elaborate the enzyme *gelatinase,* which is capable of digesting gelatin. Once the medium lacks gelatin, it loses its firm characteristics. The configuration of liquefaction may appear as indicated on the right side of figure 37.3. A description of each type follows:

Crateriform: saucer-shaped liquefaction
Napiform: turniplike
Infundibuliform: funnellike or inverted cone
Saccate: elongated sac, tubular, cylindrical
Stratiform: liquefied to the walls of the tube in the upper regions

Note: The configuration of liquefaction is not as significant as the mere fact that liquefaction takes place. If your organism liquefies gelatin but you are unable to determine the exact configuration of liquefaction, do not worry about it at this time; however, be sure to record on the Descriptive Chart the presence or absence of gelatinase production.

If your organism did not liquefy the gelatin, incubate the tube of nutrient gelatin for another four or five days to see if subsequent liquefaction occurs. Some bacteria produce gelatinase only after several days.

Nutrient Agar Plate Culture

Colonies grown on plates of nutrient agar should be studied with respect to size, color, opacity, form, elevation, and margin. With a dissecting microscope or hand lens study individual colonies carefully. Refer to figure 37.4 for descriptive terminology. Record your observations on the Descriptive Chart.

Laboratory Report

There is no Laboratory Report for this exercise. Record all information on the Descriptive Chart.

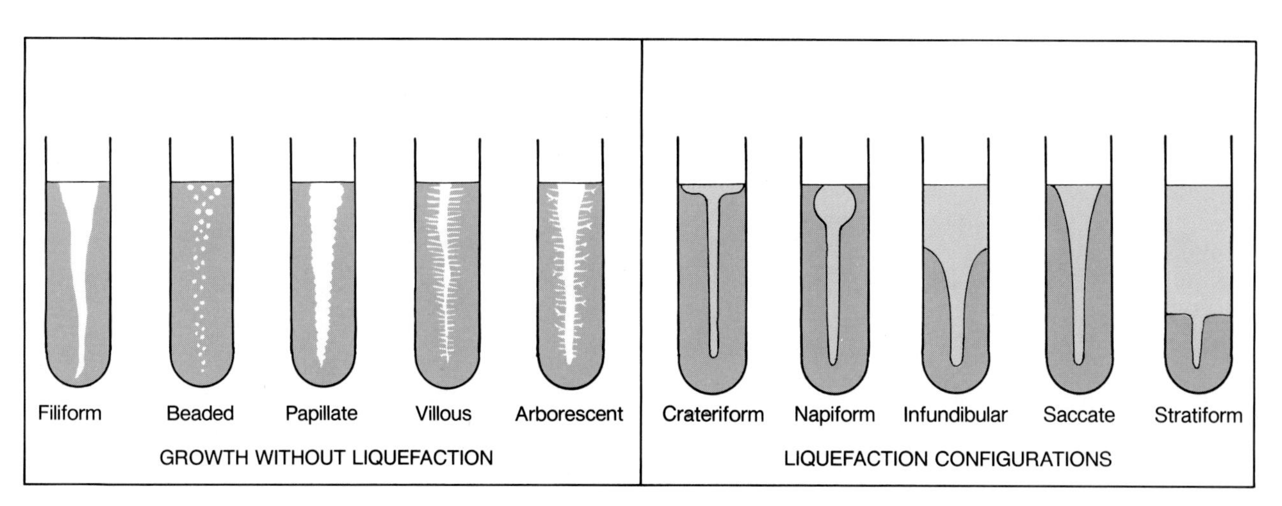

Figure 37.3 Growth in gelatin stabs.

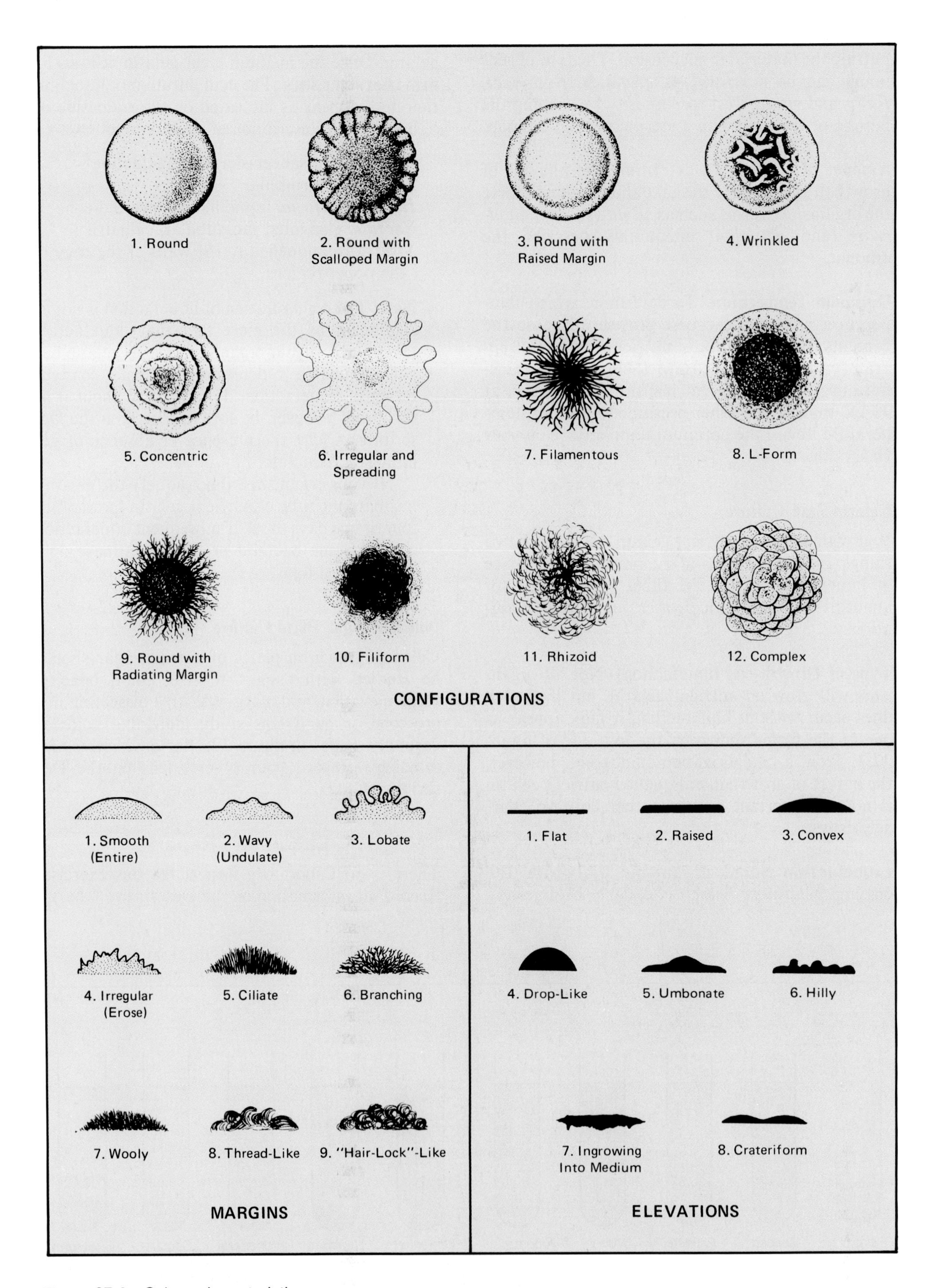

Figure 37.4 Colony characteristics.

Physiological Characteristics: Biooxidations 38

The assemblage of morphological and cultural characteristics on your Descriptive Chart during the past few laboratory periods may be leading you to believe that you already know the name of your unknown. Students at this stage often begin to draw premature conclusions. To provide you with a clearer perspective of where you are in the categorization process, refer to the separation outlines in figures 41.1 and 41.2, pages 152 and 153. Note that morphological and cultural characteristics can lead you to eleven separate groups of genera. It is very likely that one of these groups contains the genus which includes your unknown.

Although morphological and cultural characteristics are essential in getting to the genus, species determination requires a good deal more information. The physiological information that will be accumulated here and in the next two exercises will make species identification possible.

Since all physiological (biochemical) reactions in organisms are catalyzed by enzymes, and since each enzyme is produced by individual genes, we are, essentially, formulating a genetic profile of an organism as we discover what enzymes are produced. The physiological characteristics of concern in this exercise pertain to the chemistry of biooxidations. The enzymatic reactions that fall in this category pertain to respiration and fermentation.

Biooxidation reactions in bacteria pertain to the manner in which they get their energy: some are oxidative and others are fermentative. Strict aerobes that oxidize organic substances to produce the end products carbon dioxide and water are said to be *oxidative*. By utilizing organic compounds as electron donors, with oxygen as the ultimate electron (and hydrogen) acceptor, they produce CO_2 and water as end products to release energy. The ability to utilize free oxygen is accomplished by a cytochrome enzyme system. This process is also called *respiration.*

Fermentative bacteria are organisms that also utilize organic compounds for energy but lack a cytochrome system. Instead of producing only CO_2 and H_2O, they produce complex end products, such as acids, aldehydes, and alcohols, that are oxidizable and reducible. In these organisms oxygen is not the ultimate electron acceptor, and the reactions occur under anaerobic conditions. Various gases, such as carbon dioxide, hydrogen, and methane, are also produced. In fermentative bacteria the organic compounds act both as electron donors and electron acceptors.

Sugars, particularly glucose, are the compounds most widely used by fermenting organisms. Other substances such as organic acids, amino acids, purines, and pyrimidines also can be fermented by some bacteria. The end products of a particular fermentation are determined by the nature of the organism, the characteristics of the substrate, and environmental conditions such as temperature and pH.

Although fermentation and respiration represent two different types of energy-yielding biooxidations, they can both be present in the same organism, as is true of facultative anaerobes. It was pointed out in Exercise 36 that in the presence of molecular oxygen these organisms shift from fermentation to respiration. An exception, however, is seen in the lactic acid bacteria where fermentation occurs in the presence of air.

Six types of biooxidation reactions will be studied in this exercise:

1. Durham tube sugar fermentations
2. Mixed acid fermentation
3. Butanediol fermentation
4. Catalase production
5. Oxidase production
6. Nitrate reduction

The performance of these tests on your unknown will involve a considerable number of inoculations because a set of positive test controls will also be needed. Although photographs of positive test results are provided in this exercise, seeing the actual test result in a test tube is still important.

As you perform these various tests, attempt to keep in mind what groups of bacteria relate to each test. Although some tests are not very specific in pointing the way to unknown identification, others are very narrow in application.

One last comment of importance: *It is not routine practice to perform all these tests in identifying every unknown.* Although it might appear that our prime concern here is to identify an organism, our most important goal is to learn about the various types of tests for biooxidation enzymes that are available. The use of unknown bacteria to learn about them simply makes it more of a challenge. In actual practice physiological tests are used very selectively. The "shotgun approach" employed here is used to expose you to the multitude of tests that are available.

Inoculations

The following two sets of inoculations (unknowns and test controls) may be done separately or combined into one operation. The media for each set of inoculations are listed separately under each heading.

Unknown Inoculations

Figure 38.1 illustrates the procedure for inoculating seven test tubes and one petri plate with your unknown. Since your instructor may want you to inoculate some different sugar broths, blanks have been provided in the materials list for any extra broths.

If different media are distinguished from each other with differently colored tube caps, write down the identifying color after each type of medium in the materials list.

***Materials** (for each unknown):*

Durham tubes with phenol red indicator:
- 1 glucose broth
- 1 lactose broth
- 1 mannitol broth
- ____________________
- ____________________
- ____________________

2 MR-VP medium
1 nitrate broth
1 nutrient agar slant
1 petri plate of trypticase soy agar (TSA)

1. Label each tube with the number of your unknown and an identifying letter as designated in figure 38.1.
2. Label one half of the petri plate UNKNOWN and the other half *P. AERUGINOSA.*
3. Inoculate all broths and the slant with a loop. Inoculate one half of the TSA plate with your unknown, using an isolation technique.

Test Control Inoculations

Figure 38.3 on page 139 illustrates the procedure that will be used for inoculating five test tubes to be used for positive test controls. The petri plate shown on the right side is the same one that is shown in figure 38.1; thus, it will not be listed in the materials list below.

Materials:

1 glucose broth (Durham tube)
2 MR-VP medium
1 nitrate broth
1 nutrient agar slant
nutrient broth cultures of *Escherichia coli, Enterobacter aerogenes, Staphylococcus aureus,* and *Pseudomonas aeruginosa*

1. Label each tube with the code letter assigned to it as follows:

glucose broth	A[1]
MR-VP medium	D[1]
MR-VP medium	E[1]
nitrate broth	F[1]
nutrient agar slant	G[1]

2. Inoculate each of these tubes with a loopful of appropriate test organism according to figure 38.3.
3. Inoculate the other half of the TSA plate with *P. aeruginosa.*

Incubation

Except for tube E (MR-VP), all the unknown inoculations should be incubated for 24 to 48 hours at the unknown's optimum temperature. Tube E should be incubated for three to five days at the optimum temperature.

Except for tube E[1] of the test controls, incubate all the test control tubes and the TSA plate at 37° C for 24 to 48 hours. Tube E[1] should be incubated at 37° C for three to five days.

Evaluation of Tests

After 24 to 48 hours incubation, arrange all your tubes (except tubes E and E[1]) in a test-tube rack

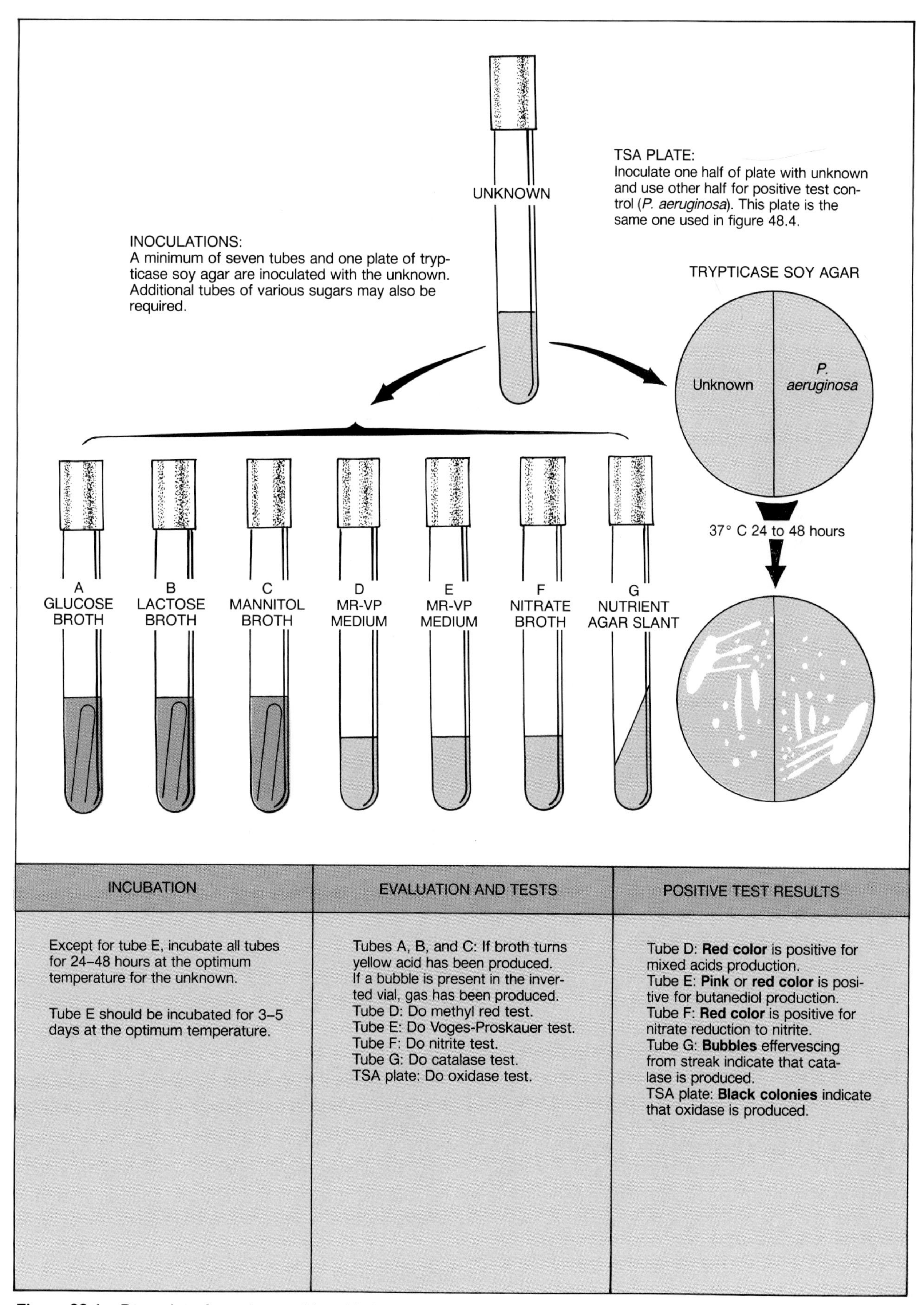

INCUBATION	EVALUATION AND TESTS	POSITIVE TEST RESULTS
Except for tube E, incubate all tubes for 24–48 hours at the optimum temperature for the unknown. Tube E should be incubated for 3–5 days at the optimum temperature.	Tubes A, B, and C: If broth turns yellow acid has been produced. If a bubble is present in the inverted vial, gas has been produced. Tube D: Do methyl red test. Tube E: Do Voges-Proskauer test. Tube F: Do nitrite test. Tube G: Do catalase test. TSA plate: Do oxidase test.	Tube D: **Red color** is positive for mixed acids production. Tube E: **Pink** or **red color** is positive for butanediol production. Tube F: **Red color** is positive for nitrate reduction to nitrite. Tube G: **Bubbles** effervescing from streak indicate that catalase is produced. TSA plate: **Black colonies** indicate that oxidase is produced.

Figure 38.1 Procedure for unknown biooxidation tests.

in alphabetical order, with the unknown tubes in one row and the test controls in another row. As you interpret the results record the information on the Descriptive Chart immediately. Don't trust your memory. Any result that is not properly recorded will have to be repeated.

Durham Tube Sugar Fermentations

When we use a bank of Durham tubes containing various sugars, we are able to determine what sugars an organism is able to ferment. If the organism does ferment a particular sugar, acid will be produced and gas *may* be produced. The presence of acid is detectable with the color change of a pH indicator in the medium. Gas production is revealed by the formation of a void in the inverted vial of the Durham tube. If it were important that we know the composition of the gas, we would have to use a Smith tube as shown in figure 38.2. For our purposes here the Durham tube is preferable.

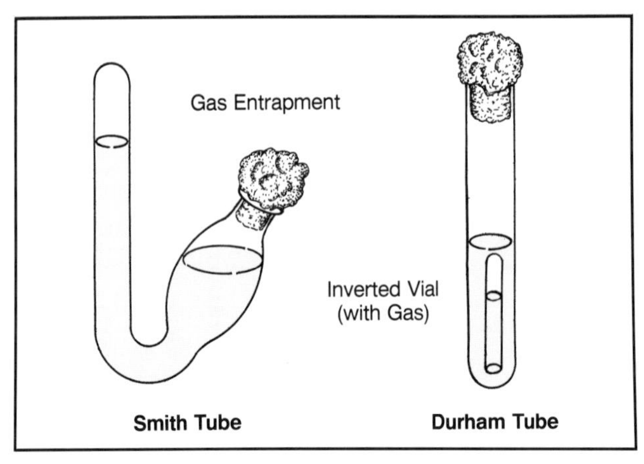

Figure 38.2 If gas analysis is necessary, the Smith type of tube is preferable to the Durham type.

Media The sugar broths used here contain 0.5% of the specific carbohydrate plus sufficient amounts of beef extract and peptone to satisfy the nitrogen and mineral needs of most bacteria. The pH indicator phenol red is included for acid detection. This indicator is red when the pH is above 7 and yellow below this point.

Although there are many sugars that one might use, glucose, lactose, and mannitol are logical ones to begin with. Your instructor may have had you include one or more additional kinds, and it is very likely that you may wish to use some others later.

Interpretation: Examine the glucose test control tube (tube A^1) which you inoculated with *E. coli.* Note that the phenol red has turned yellow, indicating acid production. Also, note that the inverted vial has a gas bubble in it. These observations tell us that *E. coli* ferments glucose to produce acid and gas. The left-hand illustration in figure 38.4 illustrates how this positive tube compares with a negative tube and an uninoculated one.

Now examine the three sugar broths (tubes A, B, and C) that were inoculated with your unknown and record your observations on the Descriptive Chart. If there is no color change, record NONE after the specific sugar. If the tube is yellow with no gas, record ACID. If the inverted vial contains gas and the tube is yellow, record ACID and GAS.

An important point to keep in mind at this time is that *a negative result is as important as a positive result.* Don't feel that you have failed in your technique if tubes are negative!

Mixed Acid Fermentation

(Methyl Red Test)

A considerable number of gram-negative intestinal bacteria can be differentiated on the basis of the end products produced when they ferment the glucose in MR-VP medium. Genera of bacteria such as Escherichia, Salmonella, Proteus, and Aeromonas ferment glucose to produce large amounts of lactic, acetic, succinic, and formic acids, plus CO_2, H_2, and ethanol. The accumulation of these acids lowers the pH of the medium to 5.0 and less. If methyl red is added to such a culture, the indicator turns red—an indication that the organism is a *mixed acid fermenter.* These organisms are generally great gas producers, too, because they produce the enzyme *formic hydrogenylase,* which splits the formic acid into equal parts of CO_2 and H_2:

$$HCOOH \xrightarrow{\text{formic hydrogenylase}} Co_2 + H_2$$

Medium MR-VP medium is essentially a glucose broth with some buffered peptone and dipotassium phosphate.

Test Procedure Perform the methyl red test first on your test control tube (D^1) and then on your unknown (tube D). Proceed as follows:

Materials:

dropping bottle of methyl red indicator

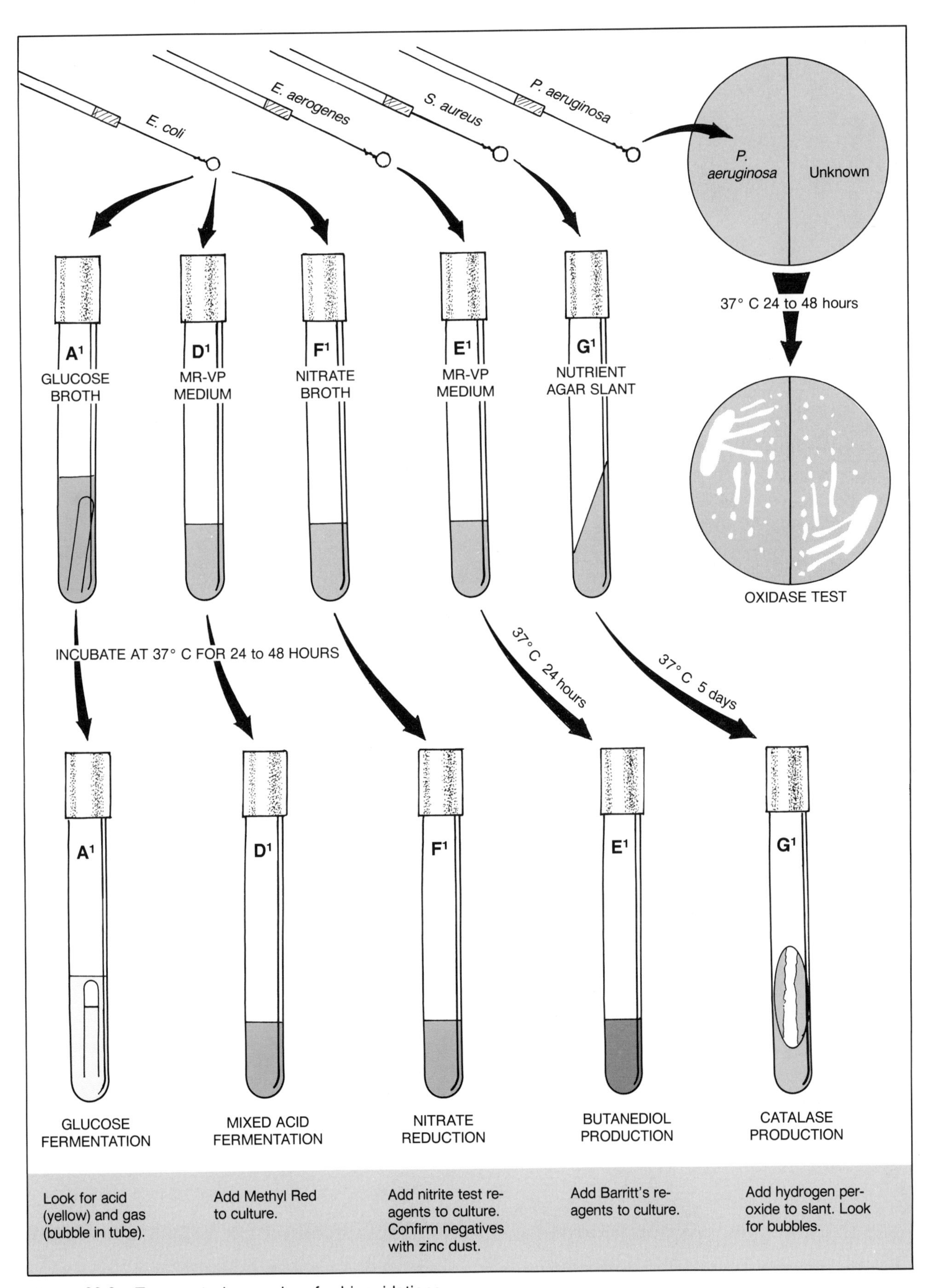

Figure 38.3 Test control procedure for biooxidations.

1. Add three or four drops of methyl red to test control tube D[1] which was inoculated with *E. coli*. The tube should become red immediately. A **reddish color,** as shown in the left hand tube of the middle illustration of figure 38.4, is a positive methyl red test.
2. Repeat the same procedure with your unknown culture (tube D) of MR-VP medium. If your unknown culture becomes yellow like the right-hand tube in figure 38.4, your unknown is negative for this test.
3. Record your results on the Descriptive Chart.

Butanediol Fermentation
(Voges-Proskauer Test)

A negative methyl red test may indicate that the organism being tested produced a lot of 2,3-butanediol and ethanol instead of acids. All species of Enterobacter and Serratia, as well as some species of Erwinia, Bacillus, and Aeromonas, do just that. The production of these nonacid end products results in less lowering of the pH in MR-VP medium, causing the methyl red test to be negative.

Unfortunately, there is no satisfactory test for 2,3-butanediol; however, acetoin (acetylmethylcarbinol), a precursor of 2,3-butanediol, is easily detected with Barritt's reagent.

Barritt's reagent consists of alpha-naphthol and KOH. When added to a three to five day culture of MR-VP medium, and allowed to stand for some time, the medium changes to pink or red in the presence of acetoin. Since acetoin and 2,3-butanediol are always simultaneously present, the test is valid. This indirect method of testing for 2,3-butanediol is called the *Voges-Proskauer test.*

Test Procedure Perform the Voges-Proskauer test on your unknown and test control tubes of MR-VP medium (tubes E and E[1]). Note that the test control tube was inoculated with *E. aerogenes.* Follow this procedure:

Materials:

Barritt's reagents
2 pipettes (1 ml)
2 empty test tubes

1. Label one empty test tube **E** (unknown) and the other **E[1]** (control).
2. Pipette 1 ml from tube E to the unknown tube and 1 ml from the tube E[1] to the control tube. Use separate pipettes for each tube.
3. Add 18 drops (about 0.5 ml) of Barritt's solution A (α-napthol) to each of these tubes that contain 1 ml of culture.
4. Add an equal amount of Barritt's solution B (KOH) to the same tubes.
5. Shake the tubes vigorously every 20 seconds until the control tube turns pink or red. Let the tubes stand for one or two hours to see if the unknown turns red. *Vigorous shaking is very important* to achieve complete aeration.

   A positive Voges-Proskauer reaction is **pink** or **red.** The left-hand tube in the right-hand illustration of figure 38.4 shows what a positive result looks like.
6. Record your results on the Descriptive Chart.

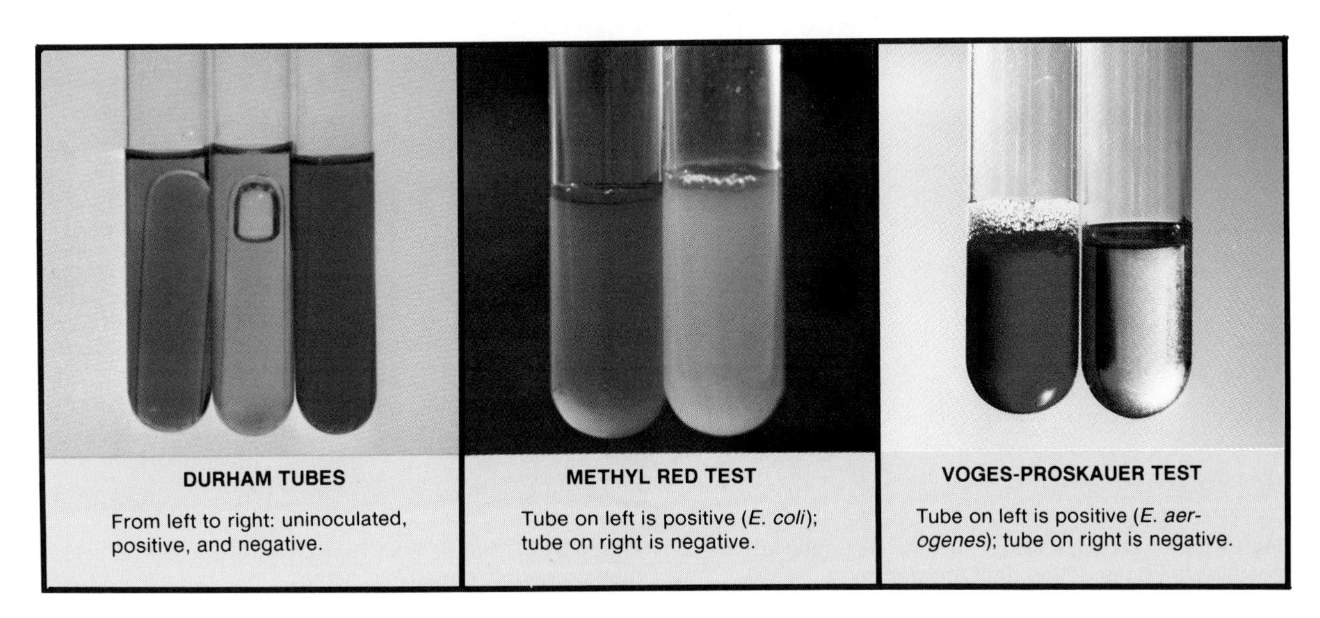

Figure 38.4 Durham tubes, mixed acid, and butanediol fermentation tests.

Catalase Production

Most aerobes and facultatives that utilize oxygen produce hydrogen peroxide, which is toxic to their own enzyme systems. This survival in the presence of this antimetabolite is possible because they produce an enzyme called *catalase,* which converts the hydrogen peroxide to water and oxygen:

$$2H_2O_2 \xrightarrow{\text{catalase}} 2H_2O + O_2$$

It has been postulated that the death of strict anaerobes in the presence of oxygen may be due to the suicidal act of H_2O_2 production in the absence of catalase production. The presence or absence of catalase production is an important means of differentiation between certain groups of bacteria.

Test Procedure To determine whether or not catalase is produced, all that is necessary is to place a few drops of 3% hydrogen peroxide on the organisms of a slant culture. If the hydrogen peroxide effervesces, the organism is catalase positive.

Materials:

3% hydrogen peroxide
test control tube G^1 with *S. aureus* growth and unknown tube G

1. While holding test control tube G^1 at an angle, allow a few drops of H_2O_2 to flow slowly down over the *S. aureus* growth on the slant. Note how bubbles emerge from the organisms.
2. Repeat the test on your unknown (tube G) and record your results on the Descriptive Chart.

Oxidase Production

The production of oxidase is one of the most significant tests we have for differentiating certain groups of bacteria. For example, all the Enterobacteriaceae are oxidase-negative and most species of *Pseudomonas* are oxidase-positive. Another important group, the *Neisseria,* are oxidase producers.

Two methods are described here for performing this test. The first method utilizes the entire TSA plate; the second method is less demanding in that only a loopful of organisms from the plate is used. Both methods are equally reliable.

Materials:

TSA plate streaked with unknown and *P. aeruginosa*
oxidase test reagents (1% solution of dimethyl-ρ-phenylenediamine hydrochloride)
Whatman No. 2 filter paper
petri dish

Entire Plate Method Onto the TSA plate that you streaked your unknown and *P. aeruginosa,* pour some of the oxidase test reagent, covering the colonies of both organisms.

Observe that the *Pseudomonas* colonies first become **pink,** then change to **maroon, dark red,** and finally **black.** Refer to figure 38.5. If your unknown follows the same color sequence, it, too, is oxidase-positive. Record your results on the Descriptive Chart.

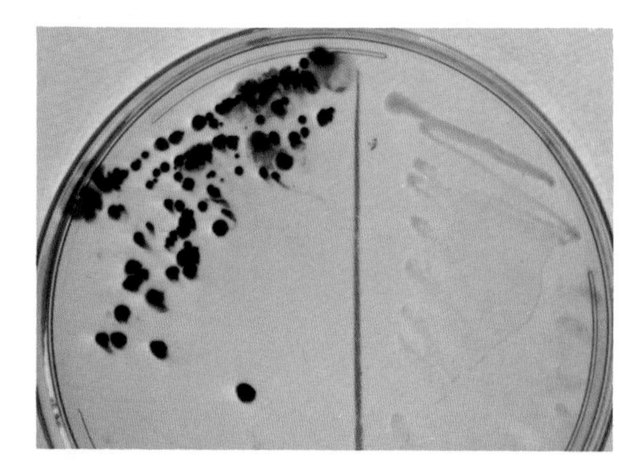

Figure 38.5 Oxidase test: The colonies on the left are positive; the ones on the right are negative.

Figure 38.6 Nitrate test: Tube on left is positive (*E. coli*); tube on right is negative.

Filter Paper Method If one does not wish to destroy a complete culture on an agar plate with this test, the following procedure will be preferred.

On a piece of Whatman No. 2 filter paper in a petri dish, place several drops of the oxidase test reagent. Remove a loopful of the organisms from one of the colonies and smear the organisms over a small area of the paper. The positive color reaction described previously will show up within 10–15 seconds. Record your results on the Descriptive Chart.

Nitrate Reduction

Many facultative bacteria are able to use the oxygen in nitrate as a hydrogen acceptor, thus converting nitrate to nitrite. This enzymatic reaction is controlled by an inducible enzyme called *nitratase.*

$$NO_3^- + 2_e^- + 2\,H^+ \xrightarrow{\text{nitratase}} NO_2^- + H_2O$$

Since the presence of free oxygen prevents nitrate reduction, actively multiplying organisms will use up the oxygen first and then utilize the nitrate. In culturing some organisms, it is desirable to use anaerobic methods to ensure nitrate reduction.

Test Procedure The nitrate broth used in this test consists of beef extract, peptone, and potassium nitrate. To test for nitrite after incubation, we use two reagents designated as A and B.

Reagent A contains sulfanilic acid and reagent B contains dimethyl-alpha-naphthylamine. In the presence of nitrite these reagents cause the culture to turn red. Negative results must be confirmed as negative with zinc dust.

Materials:

nitrate broth cultures of unknown (tube F) and test control *E. coli* (tube F^1)
nitrite test reagents (solutions A and B)
zinc dust

1. Add two or three drops of nitrite test solution A (sulfanilic acid) and an equal amount of solution B (dimethyl-α-naphthylamine) to the nitrate broth culture of *E. coli* (tube F^1). A **red color** should appear almost immediately (see figure 38.6), indicating that nitrate reduction has occurred.
   Caution: Since dimethyl-α-naphthylamine is carcinogenic, avoid skin contact.
2. Repeat this procedure with your unknown (tube F). If the red color does not develop, your unknown is negative for nitrate reduction. All negative results should be confirmed as being negative.
3. **Negative confirmation:** If your unknown appears to be negative for this test, add a pinch of zinc dust to the tube and shake it vigorously. *If the tube becomes* ***red,*** *the test is confirmed as being negative.* Zinc causes this reaction by reducing nitrate to nitrite; the newly formed nitrite reacts with the reagents to produce the red color.
4. Record your results on the Descriptive Chart.

39 Physiological Characteristics: Hydrolysis

Many bacteria produce exoenzymes, called *hydrolases,* which split complex organic compounds into smaller units. All hydrolytic enzymes accomplish this molecular splitting in the presence of water. We have already observed one example of protein hydrolysis in Exercise 37: gelatin hydrolysis by gelatinase. In this exercise we shall observe the hydrolysis of starch, casein, fat, tryptophan, and urea. Each test plays an important role in the identification of certain types of bacteria. This exercise will be performed in the same manner as the previous one with test controls being made for comparisons.

Figure 39.1 illustrates the general procedure to be used. Three agar plates and four test tubes will be inoculated. After incubation, some of the plates and tubes will have test reagents added to them; others will reveal the presence of hydrolysis by changes that have occurred during incubation. Proceed as follows:

Inoculations

If each student is working with only one unknown, students can work in pairs to share petri plates. Note in Figure 39.1 how each plate can serve for two unknowns with the test control organism streaked down the middle. If each student is working with two unknowns, the plates will not be shared. Whether or not the two tubes for test controls will be shared depends on the availability of materials.

Materials:

per *pair* of students with one unknown each, or for *one* student with two unknowns:
- 1 starch agar plate
- 1 skim milk agar plate
- 1 spirit blue agar plate
- 3 urea broths
- 3 tryptone broths

nutrient broth cultures of *B. subtilis, E. coli, S. aureus,* and *Proteus vulgaris*

1. Label and streak the three different agar plates in the manner shown in figure 39.1. Note that straight line streaks are made on each plate. Also, indicate the type of medium on each plate.

   If each student is working with only one unknown, the plates will be shared and the unknown streaks will have to be properly identified.
2. Label a tube of urea broth ***P. vulgaris*** and a tube of tryptone broth ***E. coli.*** These will be your test controls for urea and tryptophan hydrolysis. Inoculate each tube accordingly.
3. For each unknown label one tube of urea broth and one tube of tryptone broth with the code number of your unknown. Inoculate each tube with the appropriate unknown.
4. Incubate the plates and two test control tubes at 37° C. Incubate the unknown tubes of urea broth and tryptone broth at the optimum temperatures for the unknowns.

Evaluation of Tests

After 24 to 48 hours incubation of unknowns and test controls, compare your unknowns with the test controls, recording all data on the Descriptive Chart.

Starch Hydrolysis

Since many bacteria are capable of hydrolyzing starch, this test has fairly wide application. The starch molecule is a large one consisting of two constituents: amylose, a straight chain polymer of 200 to 300 glucose units, and amylopectin, a larger, branched polymer with phosphate groups. Bacteria that hydrolyze starch produce *amylases* that yield molecules of maltose, glucose, and dextrins.

Materials:

Gram's iodine
starch agar culture plate

Iodine solution (Gram's) is an indicator of starch. When iodine comes in contact with a medium containing starch, it turns blue. If starch is hydrolyzed

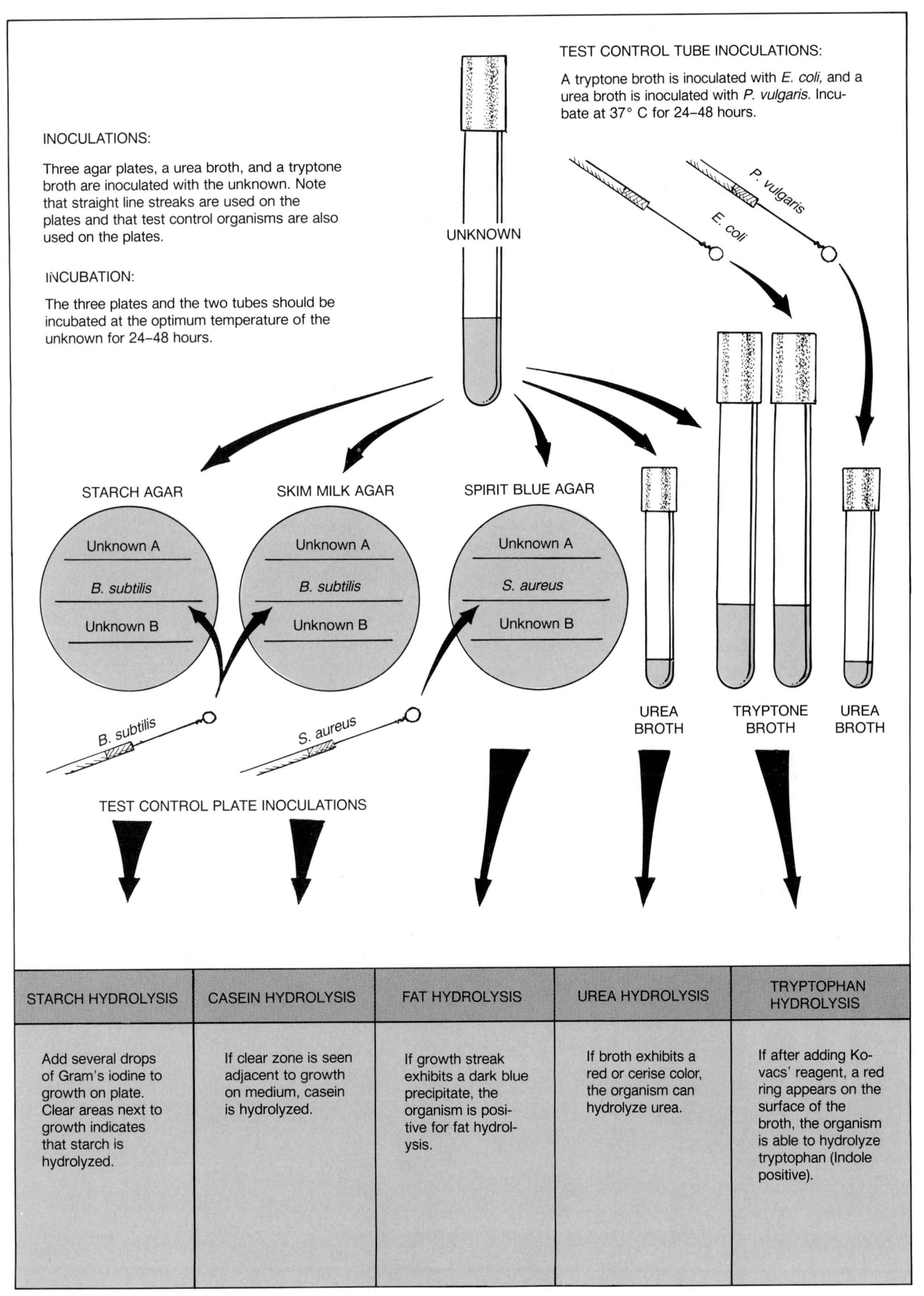

Figure 39.1 Procedure for doing hydrolysis tests on unknowns.

and starch is no longer present, the medium appears **clear** (uncolored).

By pouring a few drops of Gram's iodine over the growth on the medium, one can see clearly where starch has been hydrolyzed. If the area immediately adjacent to the growth is clear, amylase is produced.

Pour just enough iodine over each streak to wet the surface and rotate the plate gently. Compare your unknowns with the positive result seen along the growth of *B. subtilis*. The left-hand illustration in figure 39.2 illustrates what it looks like.

Casein Hydrolysis

Casein is the predominant protein in milk. Its presence causes milk to have its characteristic white appearance. Many bacteria produce the exoenzyme caseinase, which hydrolyzes casein to produce more soluble, transparent derivatives. Protein hydrolysis also is referred to as *proteolysis,* or *peptonization.*

Examine the streaks on the skim milk agar plates. Note that a **clear zone** exists adjacent to the growth of *B. subtilis*. This is evidence of casein hydrolysis. The middle illustration in figure 39.2 shows what it looks like. Compare your unknown with this positive result and record the results on the Descriptive Chart.

Fat Hydrolysis

The ability of organisms to hydrolyze fat is accomplished with the enzyme *lipase*. In this reaction the fat molecule is split to form one molecule of glycerol and three fatty acid molecules:

$$\begin{array}{lcccl} CH_2{-}O{-}\overset{O}{\overset{\|}{C}}{-}R & & CH_2OH & & RCOOH \\ | & & | & & \\ CH{-}O{-}\overset{O}{\overset{\|}{C}}{-}R' & + \; 3\,H_2O \xrightarrow{\text{lipase}} & CHOH & + & R'COOH \\ | & & | & & \\ CH_2{-}O{-}\overset{O}{\overset{\|}{C}}{-}R'' & & CH_2OH & & R''COOH \\ \text{a triglyceride} & & \text{glycerol} & & \text{3 fatty acid molecules} \end{array}$$

The glycerol and fatty acids produced in this reaction can be used by the organism to synthesize bacterial fats and other cell components. In many instances they are even oxidized to yield energy under aerobic conditions. This ability of bacteria to decompose fats plays a role in the rancidity of certain foods, such as margarine.

Spirit blue agar contains a vegetable oil which, when hydrolyzed to produce fatty acids, lowers the pH sufficiently in the medium to produce a **dark blue precipitate.**

Examine the *S. aureus* growth carefully. You should be able to see this dark blue reaction. The right-hand illustration in figure 39.2 exhibits what it should look like. Compare this positive reaction with the reaction on your unknowns. Record the results on the Descriptive Chart.

Tryptophan Hydrolysis (Indole Test)

Certain bacteria, such as *E. coli,* have the ability to split the amino acid tryptophan into indole and

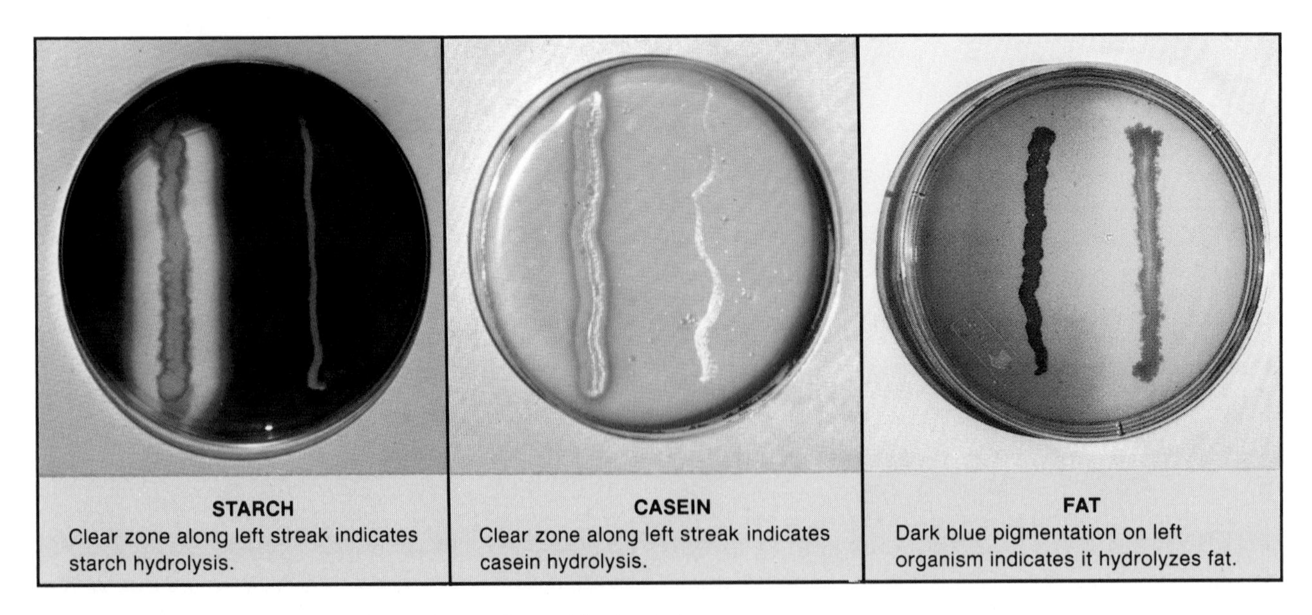

STARCH
Clear zone along left streak indicates starch hydrolysis.

CASEIN
Clear zone along left streak indicates casein hydrolysis.

FAT
Dark blue pigmentation on left organism indicates it hydrolyzes fat.

Figure 39.2 Appearances of starch, casein, and fat hydrolyses.

pyruvic acid. The enzyme that causes this hydrolysis is *tryptophanase.* Indole can be easily detected with Kovacs' reagent. This test is particularly useful in differentiating *E. coli* from some closely related enteric bacteria.

$$\text{Tryptophan} + H_2O \rightarrow \text{Indole} + \text{Pyruvic acid } (CH_3{-}C{=}O{-}COOH) + NH_3$$

TRYPTOPHAN INDOLE PYRUVIC ACID

Tryptone broth (1%) is used for this test because it contains a great deal of tryptophan. Tryptone is a peptone derived from casein by pancreatic digestion.

Materials:

Kovacs' reagent
tryptone broth cultures of unknown and *E. coli*

To test for indole add 10 or 12 drops of Kovacs' reagent to the *E. coli* culture in tryptone broth. A **red layer** should form at the top of the culture, as shown in figure 39.3. Repeat the test on your unknown and record the results on the Descriptive Chart.

Urea Hydrolysis

In the identification of pathogenic bacteria of the intestines, the differentiation of the *Proteus* group from the gram-negative pathogens is important. One characteristic peculiar to *Proteus* is its ability to produce *urease,* which splits ammonia off from the urea molecule. This characteristic is not true of the pathogens that might be confused with *Proteus.*

$$C{=}O(NH_2)_2 + H_2O \xrightarrow{\text{urease}} 2NH_3 + CO_2$$

UREA AMMONIA

Urea broth is a buffered solution of yeast extract and urea. It also contains phenol red as a pH indicator. Since urea is unstable and breaks down in the autoclave at 15 psi steam pressure, it is often sterilized by filtration. It is tubed in small amounts to hasten the visibility of the reaction.

When urease is produced by an organism in this medium, the ammonia that is released raises the pH. As the pH becomes higher, the phenol red changes from a yellow color (pH 6.8) to a **red** or **cerise color** (pH 8.1 or more).

Examine your tube of urea broth that was inoculated with *Proteus vulgaris.* Compare your unknown with this standard. Figure 39.4 reveals how positive and negative results of this test should appear. Record your result on the Descriptive Chart.

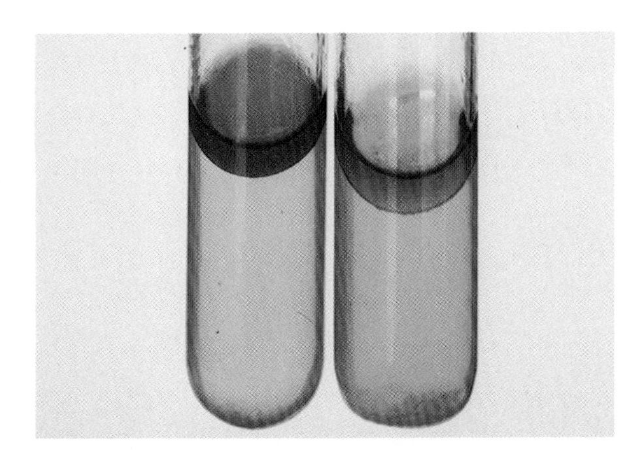

Figure 39.3 Indole test: Tube on left is positive (*E. coli*); tube on right is negative.

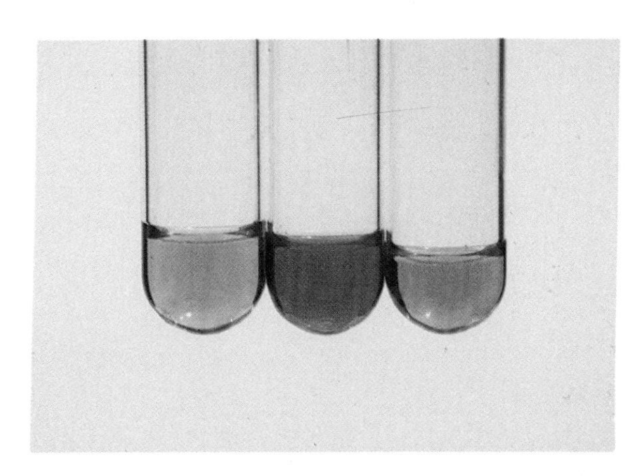

Figure 39.4 Urease test: From left to right—uninoculated, positive (*Proteus*) and negative.

40 Physiological Characteristics: Miscellaneous Tests

In addition to fermentation and hydrolysis types of physiological tests, there are a few other tests that should be performed in certain unknown characterization routines. Hydrogen sulfide production, citrate utilization, and litmus milk reactions fall into this category. This exercise includes inoculations for these three tests. An explanation of the value of the IMViC tests also is included.

Inoculations

Students will work in pairs to inoculate a set of test controls.

Materials:

for test controls, per pair of students:
- 1 Kligler's iron agar deep
- 1 Simmons citrate agar slant
- nutrient broth cultures of *Proteus vulgaris* and *Enterobacter aerogenes*

per unknown, per student:
- 1 Kligler's iron agar deep
- 1 Simmons citrate agar slant
- 1 litmus milk

1. Label one tube of Kligler's iron agar *Proteus vulgaris,* and additional tubes with your unknown numbers. Inoculate each tube by stabbing with a straight wire.
2. Label one tube of Simmons citrate agar *E. aerogenes,* and additional tubes with your unknown numbers. Use a straight wire to inoculate each slant. The slant surface is streaked first and then stabbed in its center.
3. Inoculate one tube of litmus milk with your unknown. (**Note:** A test control for this medium is not made. Figure 40.4 will take its place.)
4. Incubate unknowns at their optimum temperatures. Incubate the test controls at 37° C. for 24 to 48 hours.

Evaluation of Tests

After 24 to 48 hours incubation, examine the tubes and evaluate according to the following discussion. Record all results on the Descriptive Chart.

Hydrogen Sulfide Production

Certain bacteria, such as *Proteus vulgaris,* produce hydrogen sulfide from the amino acid cysteine. These organisms produce the enzyme *cysteine desulfurase,* which works in conjunction with the coenzyme pyridoxyl phosphate. The production of H_2S is the initial step in the deamination of cysteine as indicated in the following reactions:

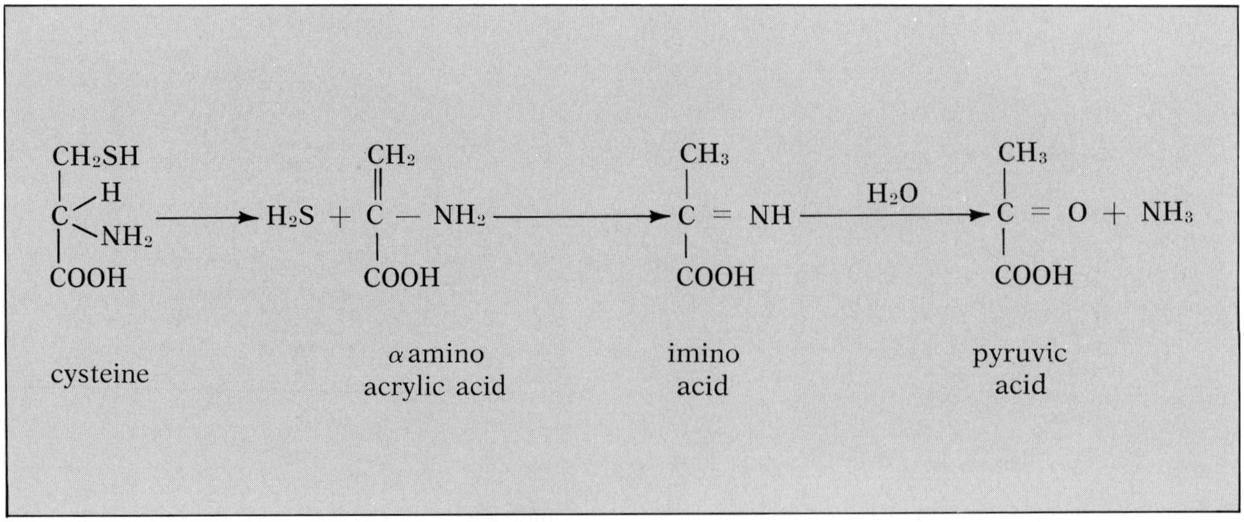

$$\underset{\text{cysteine}}{HSCH_2\text{–}CH(NH_2)\text{–}COOH} \longrightarrow H_2S + \underset{\alpha\text{ amino acrylic acid}}{CH_2{=}C(NH_2)\text{–}COOH} \longrightarrow \underset{\text{imino acid}}{CH_3\text{–}C({=}NH)\text{–}COOH} \xrightarrow{H_2O} \underset{\text{pyruvic acid}}{CH_3\text{–}C({=}O)\text{–}COOH} + NH_3$$

Figure 40.1 Deamination of cysteine with H_2S production.

Kligler's iron agar is used here to detect hydrogen sulfide production. This medium contains ferrous sulfate, which reacts with H_2S to form a **dark precipitate** of iron sulfide. It also contains glucose, lactose, and phenol red. When this medium is used in slants it is excellent for detecting glucose and lactose fermentation.

Examine the tube of this medium that was inoculated with *P. vulgaris.* It should show the typical reaction. Compare your unknown with this control tube. Your test control should appear as in figure 40.2. Record your results on the Descriptive Chart.

Citrate Utilization

The ability of some organisms, such as *E. aerogenes* and *Salmonella typhimurium,* to utilize citrate as a sole source of carbon can be a very useful differentiation characteristic in working with intestinal bacteria. Koser's citrate medium and Simmons citrate agar are two media used to detect this ability in bacteria. Both of these are synthetic media in which sodium citrate is the sole source of carbon, and nitrogen is supplied by ammonium salts instead of amino acids. Simmons citrate agar contains the indicator bromthymol blue, which changes from green to blue when growth of organisms causes alkalinity.

Examine the test control slant of this medium that was inoculated with *E. aerogenes.* Note the distinct **prussian blue color change** that has occurred. Compare this positive result with figure 40.3 and with your unknown. Record your results on the Descriptive Chart.

The IMViC Tests

In the differentiation of *E. aerogenes* and *E. coli,* as well as some other related species, four physiological tests have been grouped together into what are called the IMViC tests. The *I* stands for indole production; the *M* and *V* stand for methyl red and Voges-Proskauer tests; *i* simply facilitates pronunciation; and the *C* signifies citrate utilization. In the differentiation of the two coliforms *E. coli* and *E. aerogenes,* the test results appear as charted:

	I	M	V	C
E. coli	+	+	−	−
E. aerogenes	−	−	+	+

The significance of these tests is that when testing drinking water for the presence of the sewage indicator *E. coli,* one must be able to rule out *E. aerogenes,* which has many of the morphological and physiological characteristics of *E. coli.* Since *E. aerogenes* is not always associated with sewage, its presence in water would not necessarily indicate sewage contamination.

If you are attempting to identify a gram-negative, facultative, rod-shaped bacterial organism, group these series of tests together in this manner to see how your unknown fits this combination.

Litmus Milk Reactions

Litmus milk contains 10% powdered skim milk and a small amount of litmus as a pH indicator. When the medium is made up, its pH is adjusted to 6.8.

Figure 40.2 Hydrogen sulfide test: The tube on the left is positive (*Proteus*); the other tube is negative.

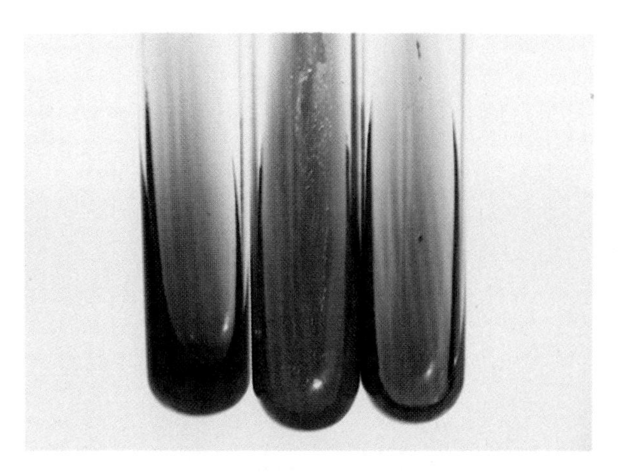

Figure 40.3 Citrate utilization: From left to right—uninoculated, positive (*E. aerogenes*), and negative.

It is an excellent growth medium for many organisms and can be very helpful in unknown characterization. In addition to revealing the presence or absence of fermentation, it can detect certain proteolytic characteristics in bacteria. A number of facultative bacteria with strong reducing powers are able to utilize litmus as an alternative electron acceptor to render it colorless. Figure 40.4 reveals the color changes that cover the spectrum of litmus milk changes. Since some of the reactions take four or five days to occur, the cultures should be incubated for at least this period of time; they should be examined every 24 hours, however. Look for the following reactions:

Acid reaction: Litmus becomes pink. This reaction is typical of fermentative bacteria.

Alkaline reaction: Litmus turns blue or purple. Many proteolytic bacteria will cause this reaction in the first 24 hours.

Litmus reduction: Culture becomes white; actively reproducing bacteria reduce the oxidation-reduction potential of the medium.

Coagulation (curd formation): Medium solidifies. Reaction is due to protein coagulation. Tilting the tube at 45° will indicate whether or not solidification has occurred.

Peptonization: Medium loses its "body" and becomes translucent. It often turns brown at this stage. This manifestation is caused by proteolytic bacteria.

Ropiness: Thick, slimy residue in the bottom of the tube. It is detected by the insertion of a sterile loop into the bottom of the tube and lifting it out slowly.

Phenylalanine Deamination

A few bacteria, such as *Proteus, Morganella,* and *Providencia,* produce the deaminase *phenylalanase,* which deaminizes the amino acid phenylalanine to produce phenylpyruvic acid. The reaction is as follows:

$$C_6H_5\text{-}CH_2CH(NH_2)COOH \xrightarrow{\text{phenylalanase}} C_6H_5\text{-}CH_2COCOOH + NH_3$$

PHENYLALANINE — PHENYLPYRUVIC ACID

No test has been set up in this series to demonstrate this test, but if it is called for in subsequent attempts to differentiate gram-negative rod-shaped unknowns, the procedure is as follows:

Heavily inoculate a slant of phenylalanine agar with the unknown. After incubating the tube for 4 hours, or 18 to 24 hours, allow four or five drops of 10% ferric chloride to run down over the slant. If phenylpyruvic acid has formed, a **green color** develops in the syneresis fluid and in the slant.

Laboratory Report

Complete Laboratory Report 38–40, which reviews all physiological tests performed in the last three exercises.

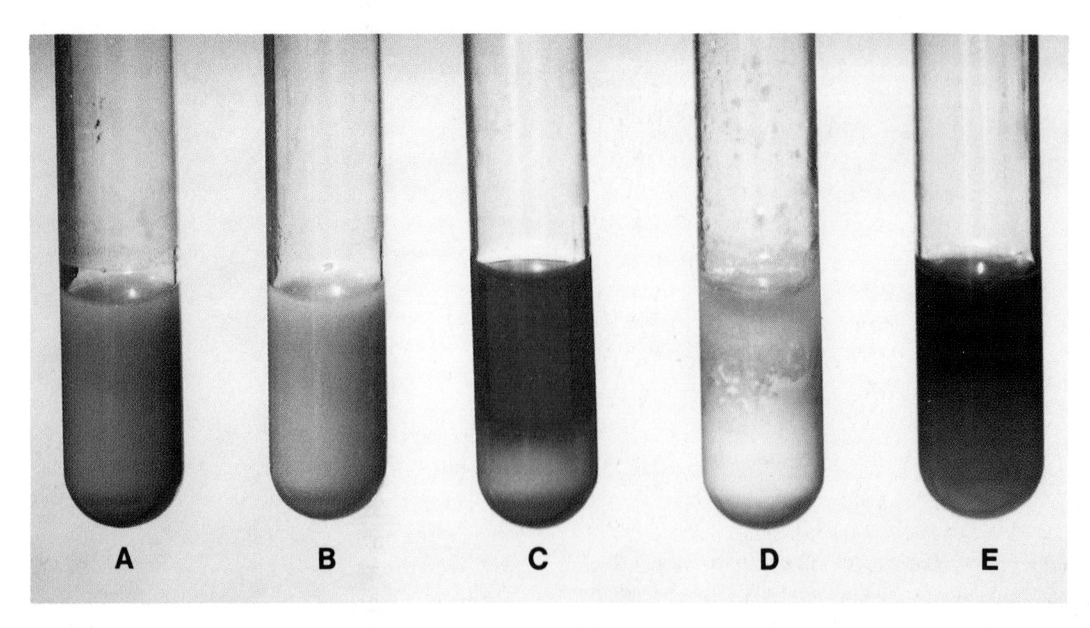

Figure 40.4 Litmus milk reactions: (A) Alkaline. (B) Acid. (C) Upper transparent portion is Peptonization; solid white portion in bottom is Coagulation and Litmus Reduction; overall redness is interpreted as Acid. (D) Coagulation and Litmus Reduction in lower half; some Peptonization (transparency) and Acid in top portion. (E) Litmus indicator is masked by production of soluble pigment (*Pseudomonas*); Some Peptonization is present, but difficult to see in photo.

41 DATA APPLICATIONS TO SYSTEMATICS: Use of *Bergey's Manual* and Computer Assistance

Once you have recorded all the data on your Descriptive Chart pertaining to morphological, cultural, and physiological characteristics of your unknown, you are ready to determine its genus and species. Determination of the genus should be relatively easy; species differentiation, however, is considerably more difficult.

The most important single source of information we have for the identification of bacteria is *Bergey's Manual of Systematic Bacteriology.* This monumental achievement, which consists of four volumes, replaced a single volume 8th edition of *Bergey's Manual of Determinative Bacteriology.* Although the more recent publication consists of four volumes, only volumes 1 and 2 will be used for the identification of the unknowns in this course.

Bergey's Manual is a worldwide collaborative effort which has an editorial board of thirteen trustees. Over two hundred specialists from nineteen different countries are listed as contributors to the first two volumes. One of the purposes of this exercise is to help you to glean the information from these two volumes that is needed to identify your unknown. Along with using *Bergey's Manual* you may also have an opportunity to check your unknown against the data base in a computer, provided that your laboratory is so equipped. Before we get into the mechanics of using *Bergey's Manual,* a few comments are in order pertaining to the problems of bacterial classification.

Classification Problems

Compared to the classification of bacteria, the classification of plants and animals has been relatively easy. In these higher forms a hierarchy of orders, families, and genera is based primarily on evolutionary evidence revealed by fossils laid down in sedimentary layers of the earth's crust. Some of the earlier editions of *Bergey's Manual* attempted to use the same hierarchical system, but had to abandon it when the 8th edition was published; without paleontological information to support the system it literally fell apart.

The present system of classification in *Bergey's Manual* uses a list of "sections" that separate the various groups. Each section is described in the vernacular so that it is easily understood (even for beginners). For example, Section 1 is entitled **The Spirochaetes.** Section 4 pertains to **gram-negative aerobic rods and cocci.** If one scans the table of contents in each volume after having completed all tests, it is possible, usually, to find a section which contains the unknown being studied.

A perusal of these sections will reveal that some sections have a semblance of hierarchy in the form of Orders, Families, and Genera. Other sections list only genera.

Thus, we see that the classification system of bacteria, as developed in *Bergey's Manual,* is not the tidy system we see in higher forms of life. The important thing is that it works.

Our dependency over the years on *Bergey's Manual* has led many to think of its classification system as the "official classification." Staley and Krieg in their Overview in volume 1 emphasize that no official classification of bacteria exists; in other words, the system offered in *Bergey's Manual* is simply a workable system, but in no sense of the word should it be designated as the official classification system.

Presumptive Identification

The first place to start in identifying your unknown is to determine what genus it fits into. If *Bergey's Manual* is available, scan the table of contents in volumes 1 and 2 to find the section which seems to describe your unknown. If these books are not immediately available you can determine the genus by referring to the separation outlines in figures 41.1 and 41.2. Note that seven groups of gram-positive bacteria are winnowed out in figure 41.1 and four groups of gram-negative bacteria in figure 41.2.

To determine which genus in the group best fits the description of your unknown, compare the genera descriptions provided below. Note that each group has a section designation to identify its position in *Bergey's Manual.*

Group I (section 13, vol. 2) Although there are only three genera listed in this group, section 13 in *Bergey's Manual* lists three additional genera, one of which is *Sporosarcina,* a coccus-shaped organism (see Group V). Most members of Group I are motile and differentiation is based primarily on oxygen needs.

Bacillus Although most of these organisms are aerobic, some are facultative anaerobes. Catalase is usually produced. For comparative characteristics of the 34 species in this genus refer to table 13.4 on pages 1122 and 1123.

Clostridium While most members of this genus are strict anaerobes, some may grow in the presence of oxygen. Catalase is not usually produced. An excellent key for presumptive species identification is provided on pages 1143–1148. Species characterization tables are also provided on pages 1149–1154.

Sporolactobacillus Microaerophilic and catalase-negative. Nitrates not reduced and indole is not formed. Spore formation occurs very infrequently (1% of cells).

Since there is only one species in this genus, one needs only to be certain that the unknown is definitely of this genus. Table 13.11 on page 1140 can be used to compare other genera that are similar to this one.

Group II (section 16, vol. 2) This group consists of Family Mycobacteriaceae, with only one genus: ***Mycobacterium.*** Fifty-four species are listed in section 16. Differentiation of species within this group depends to some extent on whether the organism is classified as a slow or fast grower. Tables on pages 1439–1442 can be used for comparing the characteristics of the various species.

Group III (section 14, vol. 2) Of the seven diverse genera listed in section 14, only three have been included here in this group.

Lactobacillus Non-spore-forming rods, varying from long and slender to coryneform (club-shaped) coccobacilli. Chain formation is common. Only rarely motile. Facultative anaerobic or microaerophilic. Catalase-negative. Nitrate usually not reduced. Gelatin not liquefied. Indole and H_2S not produced.

Listeria Regular, short rods with rounded ends; occur singly and in short chains. Aerobic and facultative anaerobic. Motile when grown at 20–25° C. Catalase-positive and oxidase-negative.

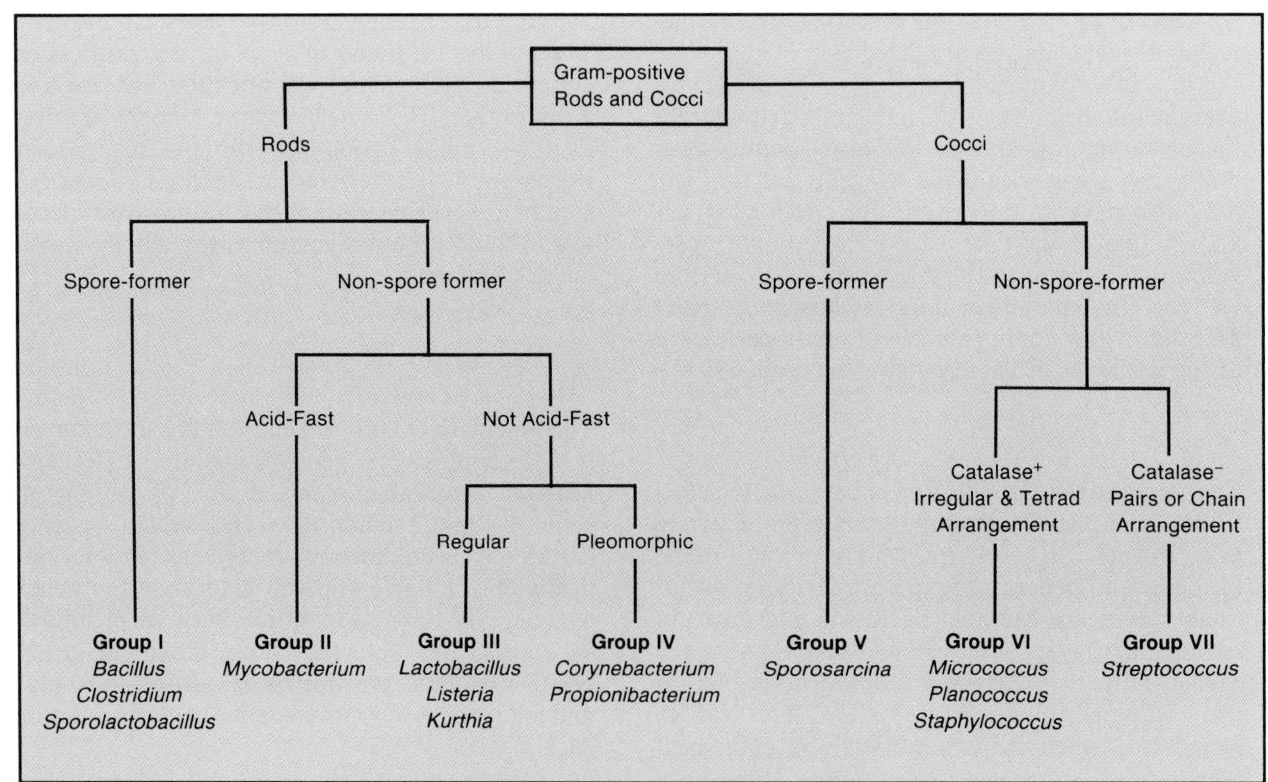

Figure 41.1 Separation outline for gram-positive rods and cocci.

Methyl red positive. Voges-Proskauer positive. Negative for citrate utilization, indole production, urea hydrolysis, gelatinase production, and casein hydrolysis. Table 14.12 on page 1241 provides information pertaining to species differentiation in this genus.

Kurthia Regular rods, 2–4 micrometers long with rounded ends; in chains in young cultures; coccoidal in older cultures. Strictly aerobic. Catalase-positive, oxidase-negative. Also negative for gelatinase production and nitrate reduction. Only two species in this genus.

Group IV (section 15, vol. 2) Although there are twenty-one genera listed in this section of *Bergey's Manual,* only two genera concern us here.

Corynebacterium Straight to slightly curved rods with tapered ends. Sometimes club-shaped. Palisade arrangements common due to snapping division of cells. Metachromatic granules formed. Facultative anaerobic. Catalase-positive. Most species produce acid from glucose and some other sugars. Often produce pellicle in broth. Table 15.3 on page 1269 provides information for species characterization.

Propionibacterium Pleomorphic rods, often diphtheroid or club-shaped with one end rounded and the other tapered or pointed. Cells may be coccoid, bifid, or even branched. Nonmotile. Some produce clumps of cells with "Chinese character" arrangements. Anaerobic to aerotolerant. Generally catalase-positive. Produce large amounts of propionic and acetic acids. All produce acid from glucose.

Group V (section 13, vol. 2) This group, which has only one genus in it, is closely related to Genus *Bacillus.*

Sporosarcina Cells are spherical or oval when single. Cells may adhere to each other when dividing to produce tetrads or packets of eight or more. Endospores formed (see photomicrographs on page 1203). Strictly aerobic. Generally motile. Only two species: *S. ureae* and *S. halophila.*

Group VI (section 12, vol. 2) This section contains two families and fifteen genera. Our concern here is with only three genera in this group. Oxygen requirements and cellular arrangement are the principal factors in differentiating the genera. Most of these genera are not closely related.

Micrococcus Spheres, occurring as singles, pairs, irregular clusters, tetrads, or cubical packets. Usually nonmotile. Strict aerobes (one species is facultative anaerobic). Catalase- and oxidase-positive. Most species produce carotenoid pigments. All species will grow in media containing 5% NaCl. For species differentiation see table 12.4 on page 1007.

Planococcus Spheres, occurring singly, in pairs, in groups of three cells, occasionally in tetrads.

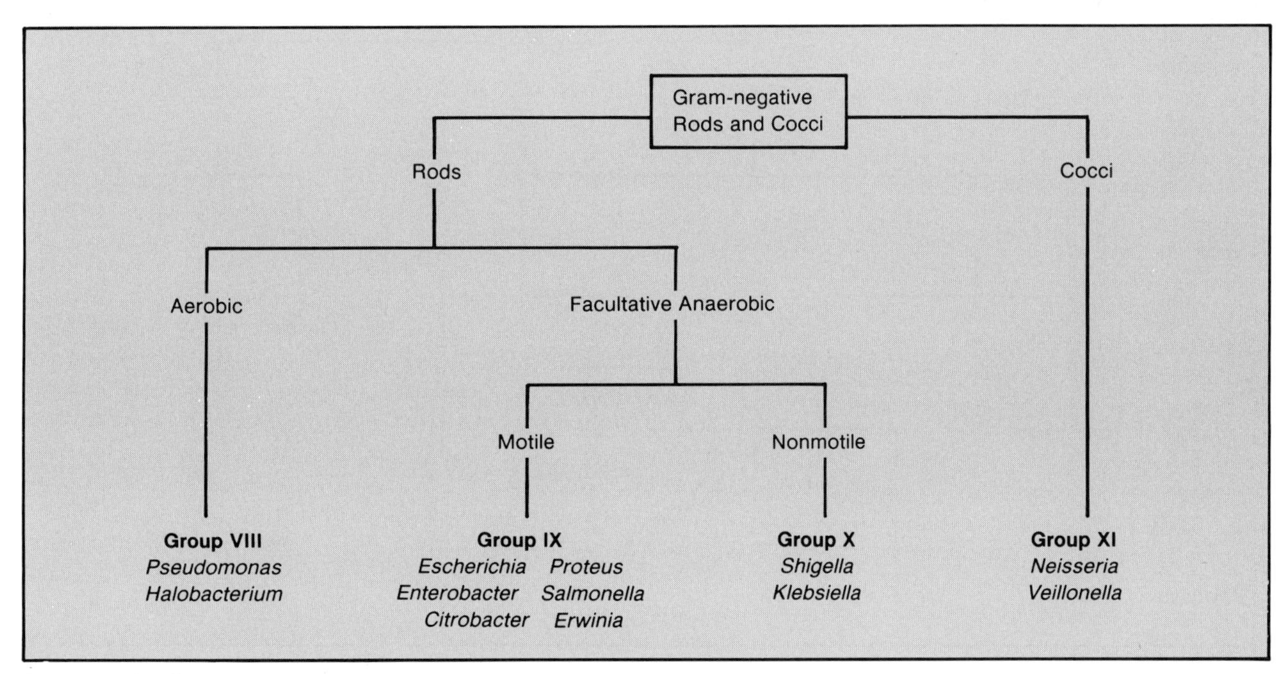

Figure 41.2 Separation outline for gram-negative rods and cocci.

Although cells are generally gram-positive, they may be gram-variable. Motility is present. Catalase- and gelatinase-positive. Carbohydrates not attacked. Do not hydrolyze starch or reduce nitrate. Refer to table 12.9 for species differentiation.

Staphylococcus Spheres, occurring as singles, pairs, and irregular clusters. Nonmotile. Facultative anaerobes. Usually catalase-positive. Most strains grow in media with 10% NaCl. Susceptible to lysis by lysostaphin. Glucose fermentation: acid, no gas. Coagulase production by some. Refer to Exercise 51 for species differentiation, or to table 12.10 on pages 1016 and 1017.

Group VII (section 12, vol. 2) Note that the single genus of this group is included in the same section of *Bergey's Manual* as the three genera above. Members of the Genus *Streptococcus* have spherical to ovoid cells that occur in pairs of chains when grown in liquid media. Some species, notably *S. mutans,* will develop short rods when grown under certain circumstances. Facultative anaerobes. Catalase-negative. Carbohydrates are fermented to produce lactic acid without gas production. Many species are commensals or parasites of man or animals. Refer to Exercise 52 for species differentiation of pathogens. Several tables in *Bergey's Manual* provide differentiation characteristics of all the streptococci.

Group VIII (section 4, vol. 1) Although there are many genera of gram-negative aerobic rod-shaped bacteria, only two genera are likely to be encountered here.

Pseudomonas Generally motile. Strict aerobes. Catalase-positive. Some species produce soluble fluorescent pigments that diffuse into the agar of a slant. Many tables are available in *Bergey's Manual* for species differentiation.

Halobacterium Cells may be rod- or disk-shaped. Cells divide by constriction. Most are strict aerobes; a few are facultative anaerobes. Catalase- and oxidase-positive. Colonies are pink, red, or red to orange. Gelatinase not produced. Most species require high NaCl concentrations in media. Cell lysis occurs in hypotonic solutions.

Groups IX and X (section 5, vol. 1) Section 5 in *Bergey's Manual* lists three families and thirty-four genera; of these thirty-four only eight genera of Family Enterobacteriaceae have been included in these two groups. If your unknown appears to fall into one of these groups, use the separation outline in figure 41.3 to determine the genus. Another useful separation outline is provided in figure 53.1 on page 202. *Keep in mind, when using these separation outlines, that there are some minor exceptions in the applications of these tests.* The diversity of species within a particular genus often presents

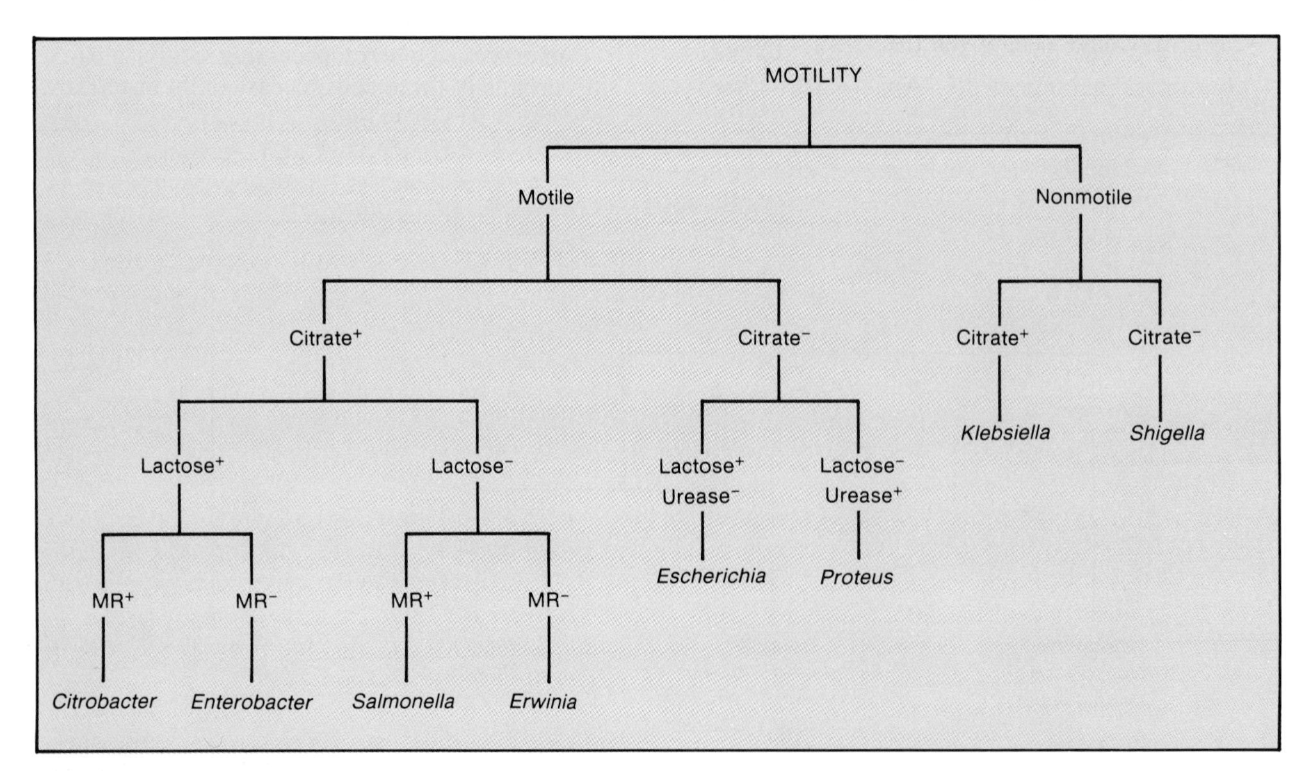

Figure 41.3 Separation outline for groups IX and X.

some problematical exceptions to the rule. Your final decision can only be made after checking the species characteristics tables for each genus in *Bergey's Manual.*

Group XI These genera are morphologically quite similar, yet physiologically they are quite different.

Neisseria (section 4, vol. 1) Cocci, occurring singly, but more often in pairs (diplococci); adjacent sides are flattened. One species (*N. elongata*) consists of short rods. Nonmotile. Except for *N. elongata,* all species are oxidase- and catalase-positive. Aerobic.

Veillonella (section 8, vol. 1) Cocci, appearing as diplococci, masses, and short chains. Diplococci have flattening at adjacent surfaces. Nonmotile. All are oxidase- and catalase-negative. Nitrate is reduced to nitrite. Anaerobic.

Problem Analysis

If you have identified your unknown by following the above procedures . . . congratulations! Not everyone succeeds at first attempt. If you are having difficulty, consider the following possibilities:

- You may have been given the wrong unknown! Although this is a remote possibility, it does happen at times. Occasionally, clerical errors are made when unknowns are put together.
- Your organism may be giving you a "false-negative" result on a test. This may be due to an incorrectly prepared medium, faulty test reagents, or improper testing technique.
- Your unknown organism may not match the description *exactly* as stated in *Bergey's Manual.* By now you are aware that the words *generally, usually,* and *sometimes* are frequently used in the book. It is entirely possible for one of these words to be inadvertently left out in Bergey's assignment of certain test results to a species. *In other words, test results, as stated in the manual, may not always apply!*
- Your culture may be contaminated. If you are not working with a pure culture, all tests are unreliable.
- You may not have performed enough tests. Check the various tables in *Bergey's Manual* to see if there is some other test that will be helpful. In addition, double-check the tables to make sure that you have read them correctly.

Confirmation of Results

There are several ways to confirm your presumptive identification. One method is to apply serological techniques if your organism is one for which typing serum is available. Another alternative is to use one of the miniature multitest systems that are described in the next section of this manual. A third method is to use a computer which has a data base programmed into it for identifying unknown bacteria. Your instructor will indicate which of these alternatives, if any, will be available.

Computer Identification

If Jim Shannon's program for the identification of bacteria with an Apple II or IBM computer is available in the laboratory, follow the instructions below for confirmation. This program contains a data base for 25 of the most common bacterial unknowns that one usually encounters in general microbiology courses. Since this program is limited to only 25 organisms, it may not include your unknown.

1. Insert the disk into the disk drive and turn on the computer. **Computer Assisted Identification of Bacterial Unknowns** will appear on the monitor screen.
2. **Press any key** and the following menu will appear on the monitor screen.
   1. List Organisms
   2. List Questions
   3. List Qualities of an Organism
   4. Input Results for an Unknown
3. **Menu Explanation:**

   List Organisms: This part of the program identifies all bacteria by both number and genus/species.

   List Questions: This portion of the program lists all the questions for which laboratory data may be entered.

List Qualities of an Organism: This part of the program displays all the questions and available data for each of the 25 bacteria programmed. This module may be used by the operator to quickly determine the characteristics of any of the programmed bacteria.

Input Results for an Unknown: Here the operator is asked to input data for the unknown. Thirty-eight data entries are available. At the end of the data input, the number of correlations are shown. For example:

Escherichia coli 22
Bacillus subtilis 10
Etc.

The computer will then print out:

The Best Match Is: . . . (Name of organism)

4. During data input, use the following key operations to accomplish the following commands:

   Return Key: Press this key to ignore a question which may not seem pertinent. This moves the operation to the next question.

   E or // : Bails out; main menu returns to the screen.

   / : Backs up one question in case of an error input.

   . : Means "Score what has been input."

Laboratory Report

There is no Laboratory Report for this exercise.

Part 7 Miniaturized Multitest Systems

Having run a multitude of tests in Exercises 35 through 40 in an attempt to identify an unknown, you undoubtedly have become aware of the tremendous amount of media, glassware, and preparation time that is involved just to set up the tests. And then, after performing all of the tests and meticulously following all the instructions, you discover that finding the specific organism in "Encyclopedia Bergey" is not exactly the simplest task you have accomplished in this course. The question must arise occasionally: "There's got to be an easier way!" Fortunately, there is: *miniaturized multitest systems.*

Miniaturized systems have the following advantages over the macromethods you have used to study the physiological characteristics of your unknown: (1) minimum media preparation, (2) simplicity of performance, (3) reliability, (4) rapid results, and (5) uniform results. These advantages have resulted in widespread acceptance of these systems by microbiologists.

Since it is not possible to describe all of the systems that are available, only four have been selected here: two by Analytab Products and two by Roche Laboratories. All four of these products are designed specifically to provide rapid identification of medically important organisms, often within five hours. Each method consists of a plastic tube or strip which contains many different media to be inoculated and incubated. To facilitate rapid identification, these systems utilize numerical coding systems that can be applied to charts or computer programs.

The four multitest systems described in this unit have been selected to provide several options. Exercises 42 and 43 pertain to the identification of gram-negative *oxidase-negative* bacteria (Enterobacteriaceae). Exercise 44 (Oxi/Ferm Tube) is used for identifying gram-negative *oxidase-positive* bacteria. Exercise 45 (Staph-Ident) is a rapid system for the differentiation of the staphylococci.

As convenient as these systems are, one must not assume that the conventional macromethods of Part 6 are becoming obsolete. Macromethods must still be used for culture studies and confirmatory tests; confirmatory tests by macromethods are often necessary when a particular test on a miniaturized system is in question. Another point to keep in mind is that all of the miniaturized multitest systems have been developed for the identification of *medically important* microorganisms. If one is trying to identify a saprophytic organism of the soil, water, or some other habitat, there is no substitute for the conventional methods.

If these systems are available to you in this laboratory, they may be used to confirm your conclusions that were drawn in Part 6 or they may be used in conjunction with some of the exercises in Part 9. Your instructor will indicate what applications will be made.

Enterobacteriaceae Identification: The API 20E System 42

The **API 20E System** is a miniaturized version of conventional tests that is used for the identification of members of the Family Enterobacteriaceae and other gram-negative bacteria. It was developed by Analytab Products, of Plainview, New York. This system utilizes a plastic strip (figure 42.1) with 20 separate compartments. Each compartment consists of a depression, or *cupule,* and a small *tube* that contains a specific dehydrated medium (see illustration 4, figure 42.2). The system has a capacity of 23 biochemical tests.

To inoculate each compartment it is necessary to first make up a saline suspension of the unknown organism; then, with the aid of a Pasteur pipette, each compartment is filled with the bacterial suspension. The cupule receives the suspension and allows it to flow into the tube of medium. The dehydrated medium is reconstituted by the saline. To provide anaerobic conditions for some of the compartments it is necessary to add sterile mineral oil to them.

After incubation for 18 to 24 hours, the reactions are recorded, test reagents are added to some compartments, and test results are tabulated. Once the test results are tabulated, a *profile number* (seven or nine digits) is computed. By finding the profile number in a code book, the *Analytical Profile Index,* one is able to determine the name of the organism. If no *Analytical Profile Index* is available, characterization can be done by using chart III in appendix D.

Although this system is intended for the identification of nonenterics as well as the Enterobacteriaceae, only the identification of the latter will be pursued in this experiment. Proceed as follows to use the API 20E system to identify your unknown enteric.

First Period

(Inoculations)

Two things will be accomplished during this period: (1) the oxidase test will be performed if it has not been previously performed, and (2) the API 20E

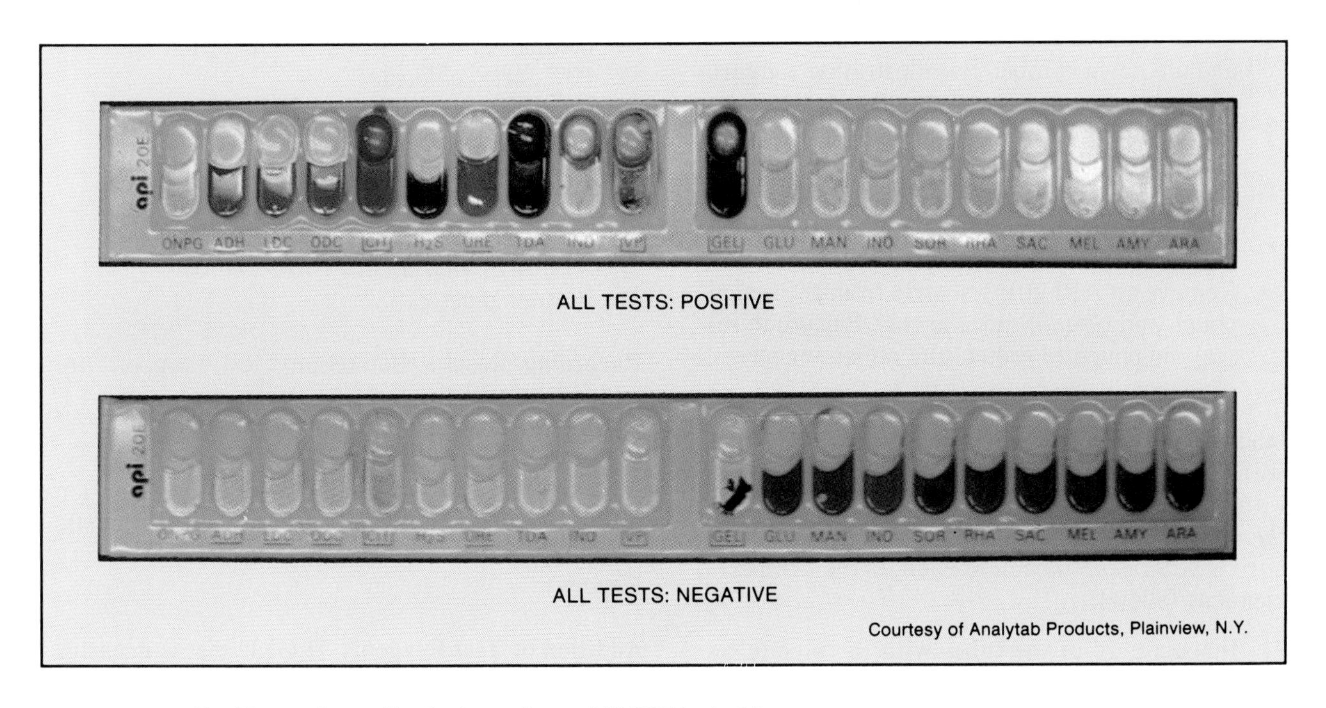

Figure 42.1 Positive and negative test results on API 20E test strips.

test strip will be inoculated. All inoculation steps are illustrated in figure 42.2.

Materials:

agar slant or plate culture of unknown
test tube of 5 ml 0.85% sterile saline
API 20E test strip
API incubation tray and cover
squeeze bottle of tap water
test tube of 5 ml sterile mineral oil
Pasteur pipettes (5 ml size)
oxidase test reagent
Whatman No. 2 filter paper
empty petri dish
Vortex mixer

Oxidase Test Since very few Enterobacteriaceae are oxidase-positive, it must be established that your unknown is oxidase-negative before proceeding with this experiment. Use the filter paper method for performing the oxidase test on your unknown if you have not already done so. Refer to page 142.

Saline Suspension Prepare a saline suspension of your unknown by transferring organisms from the center of a well-isolated colony on an agar plate (or from a slant culture) to the tube of sterile saline solution. Disperse the organisms well throughout the saline.

Tray Preparation Prepare a tray and test strip as follows:

1. Label the end strip of the tray with your name and unknown number. See illustration 2, figure 42.2.
2. To provide a moist chamber within the tray, dispense about 5 ml of tap water into the tray with a squeeze bottle. Note that the bottom of the tray has numerous depressions to accept the water.
3. Remove an API 20E test strip from the sealed pouch and place it into the tray. Be sure to re-seal the pouch to protect the remaining strips.

Inoculations Vortex the saline suspension to get uniform dispersal, and fill a sterile Pasteur pipette with the suspension. *Take care not to spill any of the organisms on the table or yourself; you may have a pathogen!* Inoculate each of the compartments as follows:

1. Inoculate all of the tubes with the pipette by depositing the suspension into the cupules as you tilt the API tray.

   Important: Slightly **underfill** ADH, LDC, ODC, H_2S, and URE. (Note that the labels for these compartments are underlined on the strip.)

   Underfilling these compartments facilitates interpretation of the results.
2. Since the media in |CIT|, |VP|, and |GEL| compartments require oxygen, **completely fill both the cupule and tube** of these compartments. (Note that the labels on these three compartments are bracketed as shown here.)
3. To provide anaerobic conditions for the ADH, LDC, ODC, H_2S, and URE compartments, dispense sterile **mineral oil** to the cupules of these compartments. Use another sterile Pasteur pipette for this step.

Incubation Place the lid on the incubation tray and incubate at 37° C for 18 to 24 hours. Refrigeration of the strip beyond 24 hours for more convenient later analysis is not recommended.

Second Period

(Evaluation of Tests)

During this period all reactions will be recorded on the Laboratory Report, test reagents will be added to four compartments, and the seven-digit profile number will be determined so that the unknown can be looked up in the *API 20E Analytical Profile Index*. Proceed as follows:

Materials:

incubation tray with API 20E test strip
10% ferric chloride
Barritt's reagents A and B
Kovacs' reagent
nitrite test reagents A and B
zinc dust or 20-mesh granular zinc
hydrogen peroxide (1.5%)
API 20E Analytical Profile Index
Pastuer pipettes

Recording Results Before any test reagents are added to any of the compartments, consult chart I, appendix D, to determine the nature of positive reactions of each test except TDA, VP, and IND. Also refer to chart II in appendix D for an explanation of the twenty symbols that are used on the plastic test strip. Record these results on the Laboratory Report.

Addition of Test Reagents If GLU test is negative (blue or blue-green), and there are less than three positive reactions before reagent addition, do not

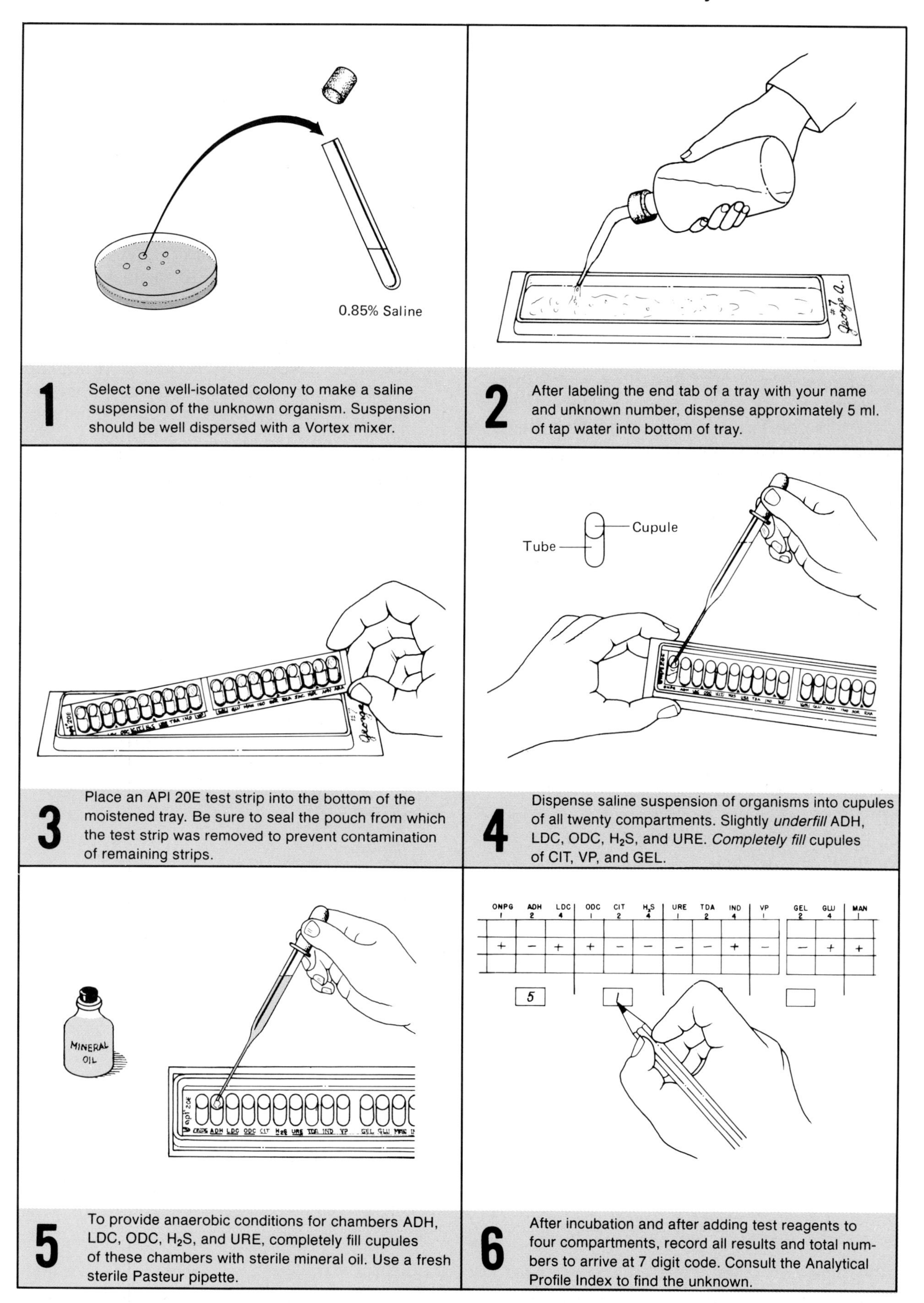

Figure 42.2 The API 20E procedure.

progress any further with this test as outlined here in this experiment. Organisms that are GLU-negative are nonenterics; for these organisms, additional incubation time is required. If you wish to follow through on an organism of this type consult your instructor for more information.

If GLU test is positive (yellow), or there are more than three positive reactions, add test reagents as follows:

1. Add **one drop** of **10% ferric chloride** to the TDA tube. A positive reaction (brown-red), if it occurs, will occur immediately. A negative reaction color is yellow.
2. Add **one drop** each of **Barritt's A** and **B** solutions to the VP tube. Read the VP tube within 10 minutes. The pale pink color which occurs immediately has no significance. A positive reaction is dark pink or red and may take 10 minutes before it shows.
3. Add **one drop** of **Kovacs'** reagent to the IND tube. Look for a positive (**red ring**) reaction within two minutes.

   After several minutes the acid in the reagent reacts with the plastic cupule to produce a color change from yellow to brownish-red, which is considered negative.
4. Examine the GLU tube closely for evidence of bubbles. Bubbles indicate the reduction of nitrate and the formation of N_2 gas. Note on the Laboratory Report that there is a place to record the presence of this gas.

   Now, add **two drops** of each **nitrite test reagent** to the GLU tube. A positive (**red**) reaction should show up within two to three minutes if nitrates are reduced.

   If this test is negative, confirm negativity with **zinc dust** or 20-mesh granular zinc. A pink-orange color after 10 minutes confirms that nitrate reduction did not occur. A yellow color results if N_2 was produced.
5. Add **one drop** of **hydrogen peroxide** to each of the MAN, INO, and SOR cupules. If catalase is produced, gas bubbles will appear within two minutes. Best results will be obtained in tubes which have no gas from fermentation.

Final Determination After all test results have been recorded and the seven-digit profile determined according to the procedures outlined on the Laboratory Report, identify your unknown by looking up the profile number in the *API 20E Analytical Profile Index.*

Clean-up When finished with the test strip be sure to place it in a container of disinfectant that has been designated for test strip disposal.

Enterobacteriaceae Identification: The Enterotube II System

43

The **Enterotube II** miniaturized multitest system was developed by Roche Diagnostics of Nutley, New Jersey, for rapid identification of Enterobacteriaceae. It incorporates twelve different conventional media and fifteen biochemical tests into a single ready-to-use tube that can be simultaneously inoculated in a moment's time with a minimum of equipment.

If you have an unknown gram-negative rod or coccobacillus that appears to be one of the Enterobacteriaceae, you may wish to try this system on it. Before applying this test, however, *make certain that your unknown is oxidase-negative,* since very few of the Enterobacteriaceae are oxidase-positive. If you have a gram-negative rod that is oxidase-positive you might try the Oxi/Ferm Tube instead, which is featured in the next exercise.

Figure 43.1 illustrates an uninoculated tube (upper) and a tube with all positive reactions (lower). Figure 43.2 outlines the entire procedure for utilizing this system.

Each of the 12 compartments of an Enterotube II contains a different agar-based medium. Compartments that require aerobic conditions have openings for access to air; those requiring anaerobic conditions have layers of paraffin wax over the media. Extending through all compartments of the entire tube is an inoculating wire. To inoculate the media, one simply picks up some organisms on the end of the wire and pulls the wire through each of the chambers in a single, rotating action.

After incubation, the indole and Voges-Proskauer tests must be performed, and the reactions in all compartments are noted. Positive reactions are given numerical values which are totaled to arrive at a five-digit code. Identification of the unknown is achieved by consulting a booklet, the *Enterotube II Interpretation Guide,* which lists these numerical codes for the Enterobacteriaceae. Proceed as follows to use an Enterotube II in the identification of your unknown.

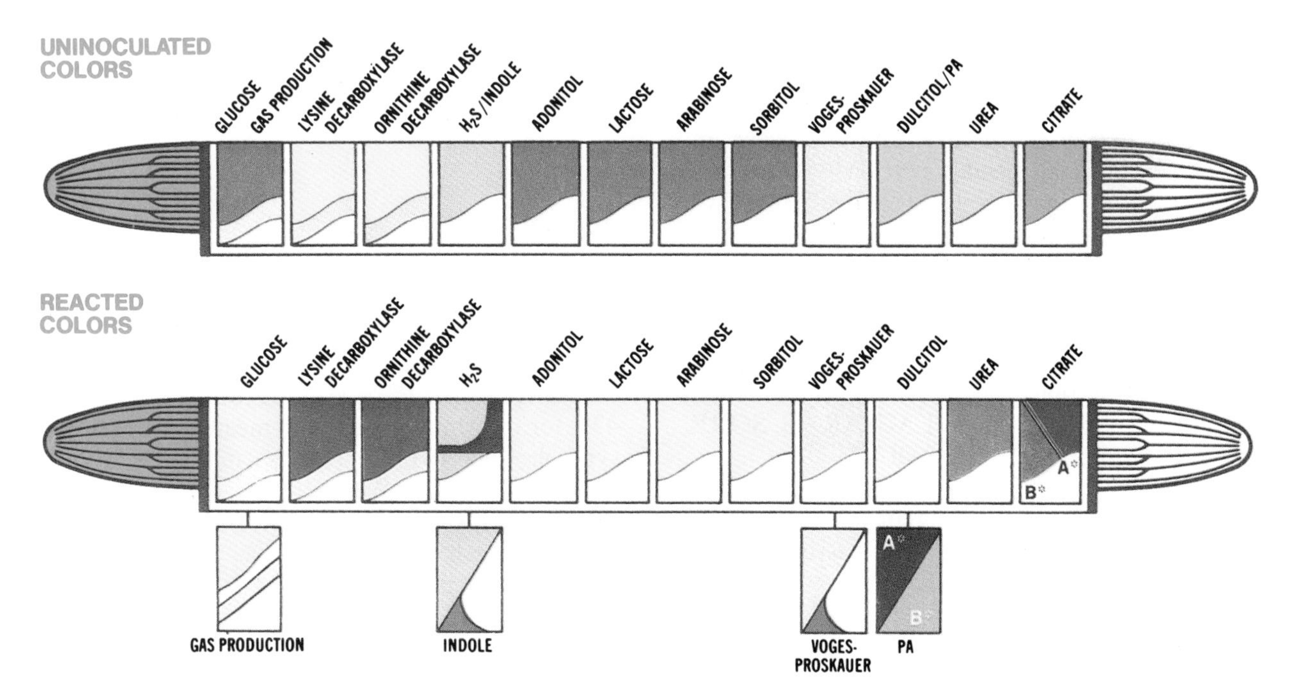

Figure 43.1 Enterotube II color differences between uninoculated and positive tests. (Courtesy of Roche Diagnostics, Nutley, N.J.)

First Period

(Inoculation and Incubation)

The Enterotube II can be used to identify Enterobacteriaceae from colonies on agar that have been inoculated from urine, blood, sputum, etc. The culture may be taken from media such as MacConkey, EMB, SS, Hektoen enteric, or trypticase soy agar.

Materials:

culture plate of unknown
1 Enterotube II

1. Write your initials or unknown number on the white paper label on the side of the tube above VP.
2. Unscrew both caps from the Enterotube II.
3. *Without heat-sterilizing* the exposed inoculating wire, insert it into a well-isolated colony.
4. Inoculate each chamber by first twisting the wire and then withdrawing it through all 12 compartments. Rotate the wire as you pull it through.
5. Reinsert the wire (without sterilizing), and with a turning motion, force it through all 12 compartments again.
6. Withdraw the wire again until the tip is in the H_2S/indole compartment. At this point, break the wire by bending. A notch in the wire makes this relatively easy.

   Leaving the wire in the last three chambers excludes oxygen from the three paraffin-covered chambers (GLU-GAS, LYS, ORN). The fermentation of glucose and the decarboxylation of lysine and ornithine occur anaerobically. The small portion of wire protruding into the H_2S/indole compartment will not interfere with these two tests.
7. Replace the caps at both ends.
8. To expose some small holes in the sides of eight of the compartments, strip off the blue plastic strip (illustration 4, figure 43.2). These holes provide aerobic conditions for the following compartments: ADON, LAC, ARAB, SORB, V-P, DUL-PA, UREA, and CIT.
9. Slide the clear band over the glucose compartment to contain any small amount of sterile wax that may escape due to excessive gas production by some bacteria.
10. Incubate at 30°–37° C for 18 to 24 hours with the Enterotube II lying on its flat surface. *When incubating several tubes together, allow space between them to allow for air circulation.*

Second Period

(Inoculation and Incubation)

The Enterotube II may be interpreted conventionally by simply comparing the results with information on chart IV, appendix D, or by using the *Enterotube II Interpretation Guide*. If the *Interpretation Guide* is available, a five-digit code number will be computed by totaling numerical values assigned to each positive test result. Since the latter method is faster than using chart IV in appendix D, use the *Interpretation Guide* if it is available.

Whether or not the *Interpretation Guide* is available, the next two steps will be followed to complete this experiment:

- Positive and negative results must be recorded on the Laboratory Report.
- Test reagents must be added to two of the compartments.

Once the above steps have been completed and a presumptive identification has been made, it may be necessary to perform some confirmatory tests. Proceed as follows to complete this experiment.

Materials:

Enterotube II, inoculated and incubated
Kovacs' reagent
10% KOH with 0.3% creatine solution
5% alpha-naphthol in absolute ethyl alcohol
syringes with needles, or disposable Pasteur pipettes
test-tube rack
Roche *Enterotube II Interpretation Guide*

Recording Results

Compare the colors of each compartment of your Enterotube II with the lower tube illustrated in figure 43.1. With a pencil, mark a small plus (+) or minus (−) near each compartment symbol on the white label on the side of the tube. Refer to table 43.1 for information as to the meaning of each test symbol. Record these results on the Laboratory Report.

Important: If at this point you discover that your unknown is GLU-negative, proceed no further with the Enterotube II; your unknown is not one of the Enterobacteriaceae. Your unknown may be *Acinebacter* sp. or *Pseudomonas maltophilia*. If an Oxi/Ferm Tube is available, try it, using the procedure outlined in the next exercise.

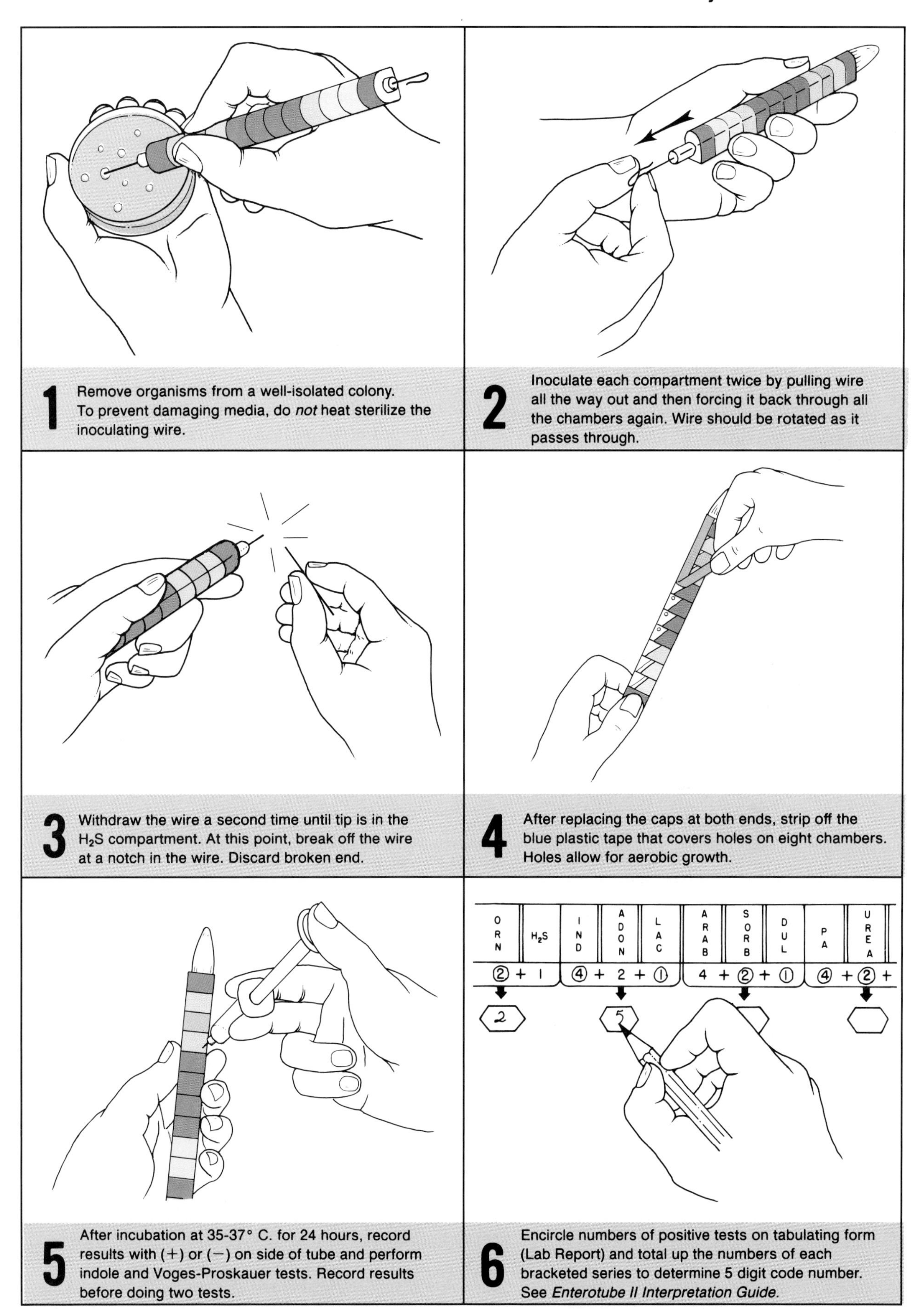

Figure 43.2 The Enterotube II procedure.

Addition of Test Reagents

Although you may have performed these two tests in Exercises 38 and 39, repeat them here.

Indole Test Place the Enterotube II into a test-tube rack, with the GLU-GAS compartment pointing downward.

Inject one or two drops of Kovacs' reagent onto the surface of the medium in the H_2S/indole compartment. This may be done with a syringe and needle through the thin mylar plastic film that covers the flat surface, or with a disposable Pasteur pipette through a small hole made in the mylar film with a hot inoculating needle.

A positive test is indicated by the development of a **red color** on the surface of the medium or mylar film within 10 seconds.

Voges-Proskauer Test With the Enterotube II in the same position as for the indole test, add two drops of potassium hydroxide containing creatine and three drops of 5% alpha-naphthol. Use syringes or disposable Pasteur pipettes.

A positive test is indicated by the development of a **red color** within 20 minutes.

Record the results of these two tests on the table on the Laboratory Report. If no coding manual is available, determine the name of your unknown from chart IV, appendix D.

Identification with Interpretation Guide

If the Roche *Interpretation Guide* for the Enterotube II is available, compute the five-digit code according to instructions on the Laboratory Report. The name of the unknown will be found opposite the code number in the booklet. Refer to pages ii and iii in the front of the booklet to note the significance of atypical test results and confirmatory tests. Do any confirmatory tests that are recommended.

Laboratory Report

Complete the Laboratory Report for this exercise.

Table 43.1 Biochemical Reactions of ENTEROTUBE II

SYMBOL	UNINOCULATED COLOR	REACTED COLOR	TYPE OF REACTION
GLU-GAS			**Glucose (GLU)** The end products of bacterial fermentation of glucose are either acid or acid and gas. The shift in pH due to the production of acid is indicated by a color change from red (alkaline) to yellow (acidic). Any degree of yellow should be interpreted as a positive reaction; orange should be considered negative. **Gas Production (GAS)** Complete separation of the wax overlay from the surface of the glucose medium occurs when gas is produced. The amount of separation between the medium and overlay will vary with the strain of bacteria.
LYS			**Lysine Decarboxylase** Bacterial decarboxylation of lysine, which results in the formation of the alkaline end product cadaverine, is indicated by a change in the color of the indicator from pale yellow (acidic) to purple (alkaline). Any degree of purple should be interpreted as a positive reaction. The medium remains yellow if decarboxylation of lysine does not occur.
ORN			**Ornithine Decarboxylase** Bacterial decarboxylation of ornithine causes the alkaline end product putrescine to be produced. The acidic (yellow) nature of the medium is converted to purple as alkalinity occurs. Any degree of purple should be interpreted as a positive reaction. The medium remains yellow if decarboxylation of ornithine does not occur.
H_2S/IND			**H_2S Production** Hydrogen sulfide, liberated by bacteria that reduce sulfur-containing compounds such as peptones and sodium thiosulfate, reacts with the iron salts in the medium to form a black precipitate of ferric sulfide usually along the line of inoculation. Some **Proteus** and **Providencia** strains may produce a diffuse brown coloration in this medium, which should not be confused with true H_2S production. **Indole Formation** The production of indole from the metabolism of tryptophan by the bacterial enzyme tryptophanase is detected by the development of a pink to red color after the addition of Kovac's reagent.

Table 43.1 Continued

SYMBOL	UNINOCULATED COLOR	REACTED COLOR	TYPE OF REACTION
ADON			**Adonitol** Bacterial fermentation of adonitol, which results in the formation of acidic end products, is indicated by a change in color of the indicator present in the medium from red (alkaline) to yellow (acidic). Any sign of yellow should be interpreted as a positive reaction; orange should be considered negative.
LAC			**Lactose** Bacterial fermentation of lactose, which results in the formation of acidic end products, is indicated by a change in color of the indicator present in the medium from red (alkaline) to yellow (acidic). Any sign of yellow should be interpreted as a positive reaction; orange should be considered negative.
ARAB			**Arabinose** Bacterial fermentation of arabinose, which results in the formation of acidic end products, is indicated by a change in color from red (alkaline) to yellow (acidic). Any sign of yellow should be interpreted as a positive reaction; orange should be considered negative.
SORB			**Sorbitol** Bacterial fermentation of sorbitol, which results in the formation of acidic end products, is indicated by a change in color from red (alkaline) to yellow (acidic). Any sign of yellow should be interpreted as a positive reaction; orange should be considered negative.
V.P.			**Voges-Proskauer** Acetylmethylcarbinol (acetoin) is an intermediate in the production of butylene glycol from glucose fermentation. The presence of acetoin is indicated by the development of a red color within 20 minutes. Most positive reactions are evident within 10 minutes.
DUL-PA		A* B*	**Dulcitol** Bacterial fermentation of dulcitol, which results in the formation of acidic end products, is indicated by a change in color of the indicator present in the medium from green (alkaline) to yellow or pale yellow (acidic). **Phenylalanine Deaminase** This test detects the formation of pyruvic acid from the deamination of phenylalanine. The pyruvic acid formed reacts with a ferric salt in the medium to produce a characteristic black to smoky gray color.
UREA			**Urea** The production of urease by some bacteria hydrolyzes urea in this medium to produce ammonia, which causes a shift in pH from yellow (acidic) to reddish-purple (alkaline). This test is strongly positive for **Proteus** in 6 hours and weakly positive for **Klebsiella** and some **Enterobacter** species in 24 hours.
CIT		A* B*	**Citrate** Organisms that are able to utilize the citrate in this medium as their sole source of carbon produce alkaline metabolites which change the color of the indicator from green (acidic) to deep blue (alkaline). Any degree of blue should be considered positive.

(Courtesy of Roche Diagnostics, Nutley, N.J.)

44 O/F Gram-Negative Rods Identification: The Oxi/Ferm Tube System

The **Oxi/Ferm Tube,** produced by Roche Diagnostics, takes care of the identification of the oxidase-positive, gram-negative bacteria that cannot be identified with the use of the Enterotube II system. The two multitest systems were developed to work together: if an unknown gram-negative rod is oxidase-negative, the Enterotube II is used; if it is oxidase-positive, the Oxi/Ferm Tube is used. If an oxidase-negative gram-negative rod turns out to be glucose-negative on the Enterotube II test, then one must also use the Oxi/Ferm Tube for identification. If you attempted to identify an unknown in the last exercise and found that your unknown is (1) oxidase-positive or (2) oxidase-negative and glucose-negative, you have been referred to this exercise.

The Oxi/Ferm Tube is intended for the identification of species in the following genera: *Alcaligenes, Achromobacter, Acinebacter, Aeromonas, Bordatella, Chromobacterium, Flavobacterium, Moraxella, Pasteurella, Pseudomonas, Plesiomonas,* and *Vibrio.*

The system incorporates eight different conventional media in a single ready-to-use tube that can be simultaneously inoculated in a moment's time with a minimum of equipment. Like the Enterotube II system, the Oxi/Ferm Tube has an inoculating wire that extends through all eight compartments of the entire tube. To inoculate the media, one simply picks up some organisms on the end of the wire and pulls the wire through each of the chambers in a single, rotating action.

After incubation, the results are recorded and Kovacs' reagent is injected into the H_2S/indole chamber to perform the indole test. Positive reactions are given numerical values which are totaled to arrive at a four-digit code. By looking up the code in a coding manual, one can quickly determine the

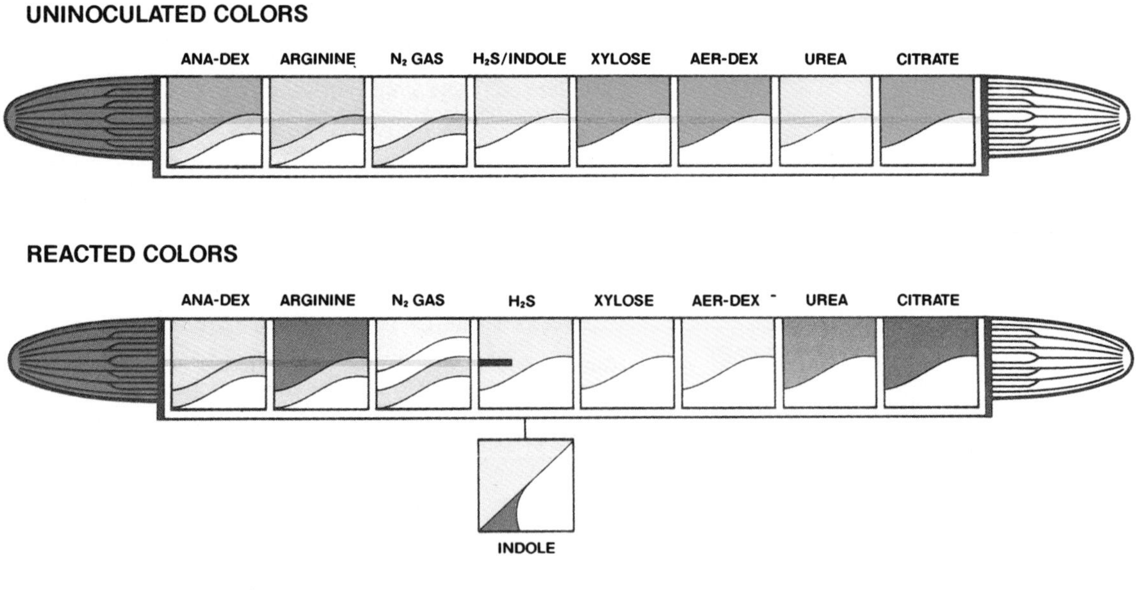

Figure 44.1 Oxi/Ferm Tube color differences between uninoculated and positive tests.

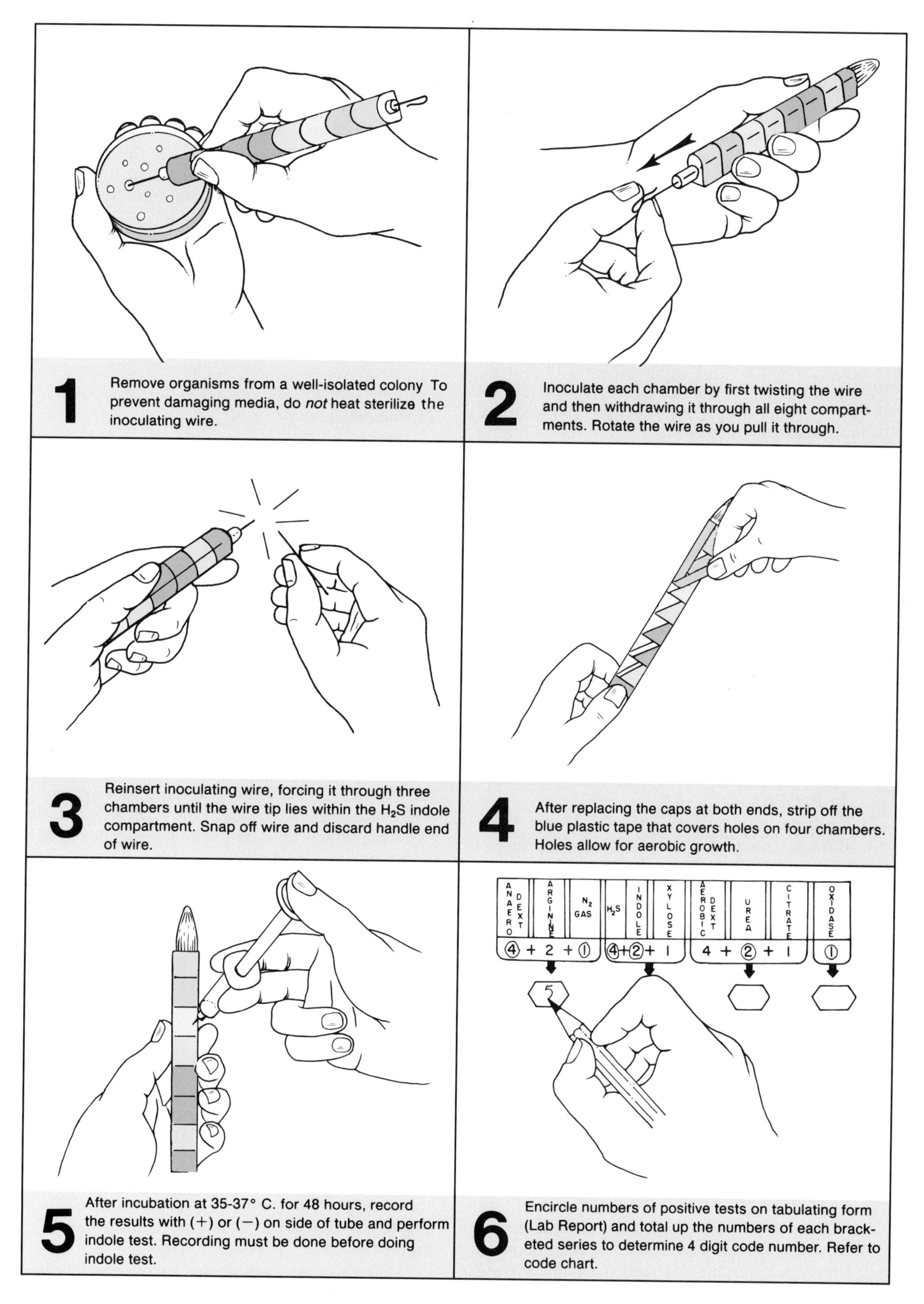

Figure 44.2 The Oxi/Ferm Tube procedure.

name of the unknown and any tests that one might need to perform to confirm the identification.

Figure 44.1 illustrates an uninoculated tube (upper), and a tube with all positive reactions (lower). Figure 44.2 outlines the entire procedure for utilizing this system.

First Period

(Inoculation and Incubation)

The Oxi/Ferm Tube must be inoculated with a large inoculum from well-isolated colonies. Culture purity, of course, is of paramount importance. If there is any doubt of purity a TSA plate should be inoculated and incubated at 35° C for 24 hours, followed by 24 hours incubation at room temperature. If no growth occurs on TSA, but growth does occur on blood agar, the organism has special growth requirements; such an organism cannot be identified with the Oxi/Ferm Tube.

Materials:

culture plate of unknown
1 Oxi/Ferm Tube
1 plate of TSA (if purity check necessary)

1. Write your initials or unknown number on the white paper label on the side of the tube above AER-DEX.
2. Unscrew both caps from the Oxi/Ferm Tube.
3. *Without heat-sterilizing* the exposed inoculating wire, insert it into a well-isolated colony.
4. Inoculate each chamber by first twisting the wire and then withdrawing it through all eight compartments. Rotate the wire as you pull it through.
5. If a purity check of the culture is necessary, streak a petri plate of TSA with the inoculating wire that has just been pulled through the tube. **Do not flame.**
6. Reinsert the inoculating wire, forcing it through three chambers until the tip of the wire is within the H_2S/indole compartment. At this point, snap off the wire at the notch and discard the handle end of the wire.
7. Replace both caps on tube.
8. To expose some small holes on the sides of four of the compartments, strip off the blue plastic tape. These holes provide aerobic conditions for the xylose, aerobic dextrose, urea, and citrate compartments.
9. Incubate at 35°–37° C for **48 hours,** with the tube lying on its flat surface.

Second Period

(Evaluation of Tests)

The Oxi/Ferm Tube may be interpreted conventionally by simply comparing the results with information on chart V, appendix D, or by using the *Computer Coding and Identification System.* The latter is a booklet provided by Roche Diagnostics. Regardless of the method used, it will be necessary to perform the indole test on the H_2S/indole compartment and record all results on the Laboratory Report.

Materials:

Oxi/Ferm Tube, inoculated and incubated
Kovacs' reagent
syringes with needles, or disposable Pasteur pipettes
Computer Coding and Identification System for the Oxi/Ferm Tube (a booklet)

Recording Results

Compare the colors of each compartment of your Oxi/Ferm Tube with the lower tube illustrated in figure 44.1. With a pencil, mark a small plus (+) or minus (−) near each compartment symbol on the white label on the side of the tube. Refer to table 44.1 for information as to the significance of each compartment label. Record these results on the Laboratory Report. All reactions must be recorded before Kovacs' reagent is added.

Indole Test

With a syringe and needle, inject two or three drops of Kovacs' indole reagent through the flat, plastic surface into the H_2S/indole compartment. Release the reagent onto the inside flat surface and allow it to drop down onto the agar.

If a Pasteur pipette is used instead of a syringe needle, it will be necessary to form a small hole in the mylar film with a hot inoculating needle to admit the tip of the Pasteur pipette.

A positive test is indicated by the development of a **red color** on the surface of the medium or mylar film within 10 seconds. Record the results of this test on the Laboratory Report.

Four-Digit Code

Follow the instructions on the Laboratory Report for determining the four-digit code. If the *Computer Coding and Identification System* booklet is available, apply the code to it to identify your unknown. If no booklet is available, use chart V, appendix D.

Laboratory Report

Complete the Laboratory Report for this exercise.

Table 44.1 Biochemical Reactions of Oxi/Ferm Tube

SYMBOL	UNINOCULATED COLOR	REACTED COLOR	TYPE OF ACTION
ANA-DEX			**Anaerobic Dextrose** Positive fermentation is shown by a change in color from green (neutral) to yellow (acid). Most oxidative fermentative gram-negative rods are negative. If the medium becomes blue after incubation, the test should be considered negative.
ARGININE			**Arginine Dihydrolase** Bacterial decarboxylation of arginine results in the formation of alkaline end products. The color change here is from yellow-green (acid) to purple (alkaline). Gray is negative.
N_2 GAS			**N_2 Gas Production** Separation of the wax overlay from the agar surface indicates that nitrogen gas has been produced. Occasionally, this gas may cause separation of the agar from the compartment wall.
H_2S INDOLE			**H_2S Production** The hydrogen sulfide produced by bacterial reduction of sulfur-bearing compounds, such as peptones and sodium thiosulfate, reacts with the iron salts in the medium to form a black precipitate of ferric sulfide. Some strains may produce a diffuse coloration in this medium; however, this should not be confused with true H_2S production. True H_2S production is characterized by black color along the inoculation channel only. This test is positive for ***Pseudomonas putrefaciens.*** **Indole Formation** The hydrolysis of tryptophan by the enzyme tryptophanase results in the production of indole. In the presence of Kovacs' reagent the indole takes on a pink to red color.
XYLOSE			**Xylose** Bacterial oxidation of this carbohydrate is evidenced by a change in color from green (neutral) to yellow (acid). If the medium becomes blue after incubation, consider the test to be negative.
AER-DEX			**Aerobic Dextrose** Bacterial oxidation of this sugar is evidenced by a change in color from green (neutral) to yellow (acid). If the medium becomes blue after incubation, consider the test to be negative.
UREA			**Urea** The presence of bacterial urease causes urea to break down with the release of ammonia (alkaline). With the elevation of pH, the phenol red in the medium changes from yellow (acid) to pink or purple (alkaline). Pale pink should be considered negative.
CITRATE			**Citrate** Organisms that can utilize sodium citrate as a sole source of carbon produce alkaline end products. The bromothymol blue indicator in this medium changes from green (neutral) to blue (alkaline) with the utilization of citrate.

(Courtesy of Roche Diagnostics, Nutley, N.J.)

45 Staphylococcus Identification: The API Staph-Ident System

The **API Staph-Ident System,** produced by Analytab Products of Plainview, New York, was developed to provide a rapid (five hour) method for identifying thirteen of the most clinically important species of staphylococci. This system consists of ten *microcupules* that contain dehydrated substrates and/or nutrient media. Except for the coagulase test, all the tests that are needed for the identification of staphylococci are included on the strip. Figure 45.1 illustrates two inoculated strips: the lower one just after inoculation and the upper one with all positive reactions. Note that the appearance of each microcupule undergoes a pronounced color change when a positive reaction occurs.

Figure 45.2 illustrates the overall procedure. The first step is to make a saline suspension of the organism from an isolated colony. A Staph-Ident strip is then placed in a tray which has a small amount of water added to it to provide humidity during incubation. Next, a sterile Pasteur pipette is used to dispense two to three drops of the bacterial suspension to each microcupule. The inoculated tray is then covered and incubated, aerobically, at 35°–37° C for five hours. After incubation a few drops of Staph-Ident reagent are added to the tenth microcupule and the results are read immediately. Finally, a four-digit profile is computed which is used to determine the species from a chart in the appendix.

As simple as this system might seem, there are a few limitations that one must keep in mind. Final species determination by a competent microbiologist must also take into consideration other factors such as the source of the specimen, the catalase reaction, colony characteristics, and antimicrobial susceptibility pattern. Very often there are confirmatory tests that must be made.

If you have been working with an unknown that appears to be one of the staphylococci, use this system to confirm your conclusions. If you have already done the coagulase test and have learned that your organism is coagulase-negative, this system will enable you to identify one of the numerous

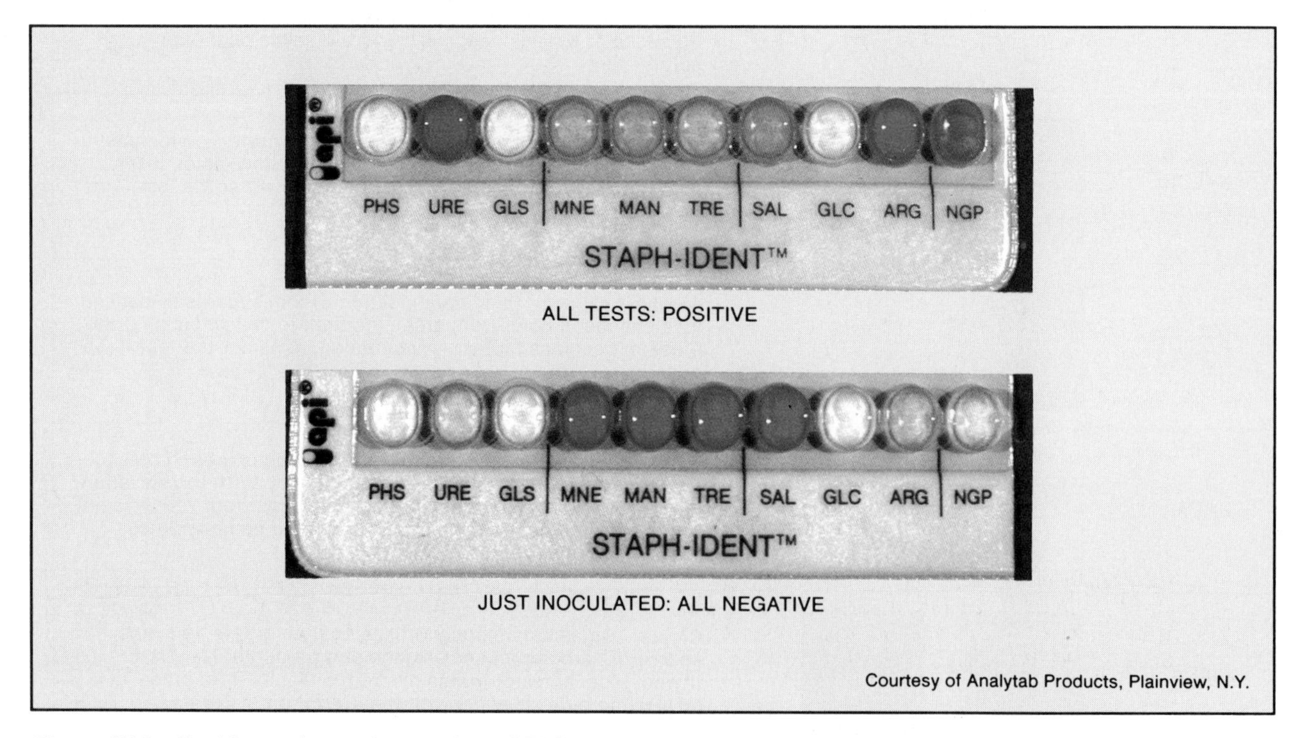

Figure 45.1 Positive and negative result on API Staph-Ident test strips.

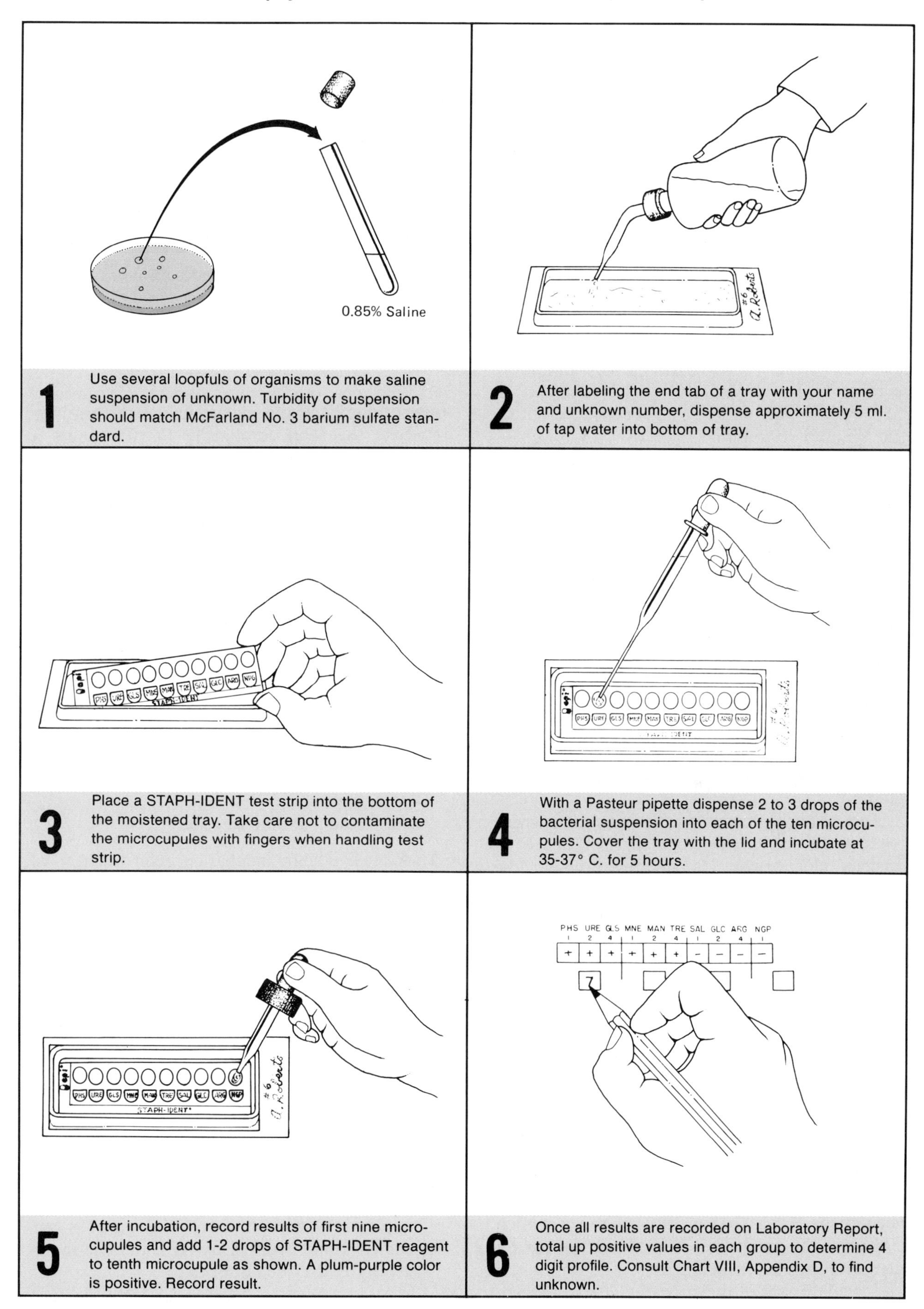

Figure 45.2 The API Staph-Ident procedure.

coagulase-negative species that are not identifiable by the procedures outlined in Exercise 72.

Coagulase Test and Strip Inoculations

Before setting up this experiment, take into consideration that it *must be completed at the end of five hours.* Holding the test strips overnight is not recommended.

Materials:

blood agar plate culture of unknown (must not have been incubated over 30 hours)
blood agar plate (if needed for purity check)
serological tube of 2 ml sterile saline
test-tube rack
API Staph-Ident test strip
API incubation tray and cover
sterile swabs (optional)
squeeze bottle of tap water
tubes containing McFarland No. 3 ($BaSO_4$) standard
sterile Pasteur pipette (5 ml size)

Coagulase Test If the coagulase test has not been performed, refer to Exercise 51, page 192, and perform the test on your unknown.

Saline Suspension Prepare a saline suspension of your unknown by transferring organisms to a tube of sterile saline from one or more colonies with a loop or sterile swab. Turbidity of suspension should match a tube of No. 3 McFarland barium sulfate standard.

Important: Do not allow the bacterial suspension to go unused for any great length of time. Suspensions older than 15 minutes become less effective.

Tray Preparation Prepare a tray and test strip as follows:

1. Label the end strip of the tray with your name and unknown number. See illustration 2 in figure 45.2.
2. Dispense about 5 ml of tap water into the bottom of the tray with a squeeze bottle. Note that the bottom of the tray has numerous depressions to accept the water.
3. Remove the API test strip from its sealed envelope and place the strip in the bottom of the tray.

Inoculations After shaking the saline suspension to disperse the organisms, fill a sterile Pasteur pipette with the suspension.

Inoculate each of the microcupules with **two to three drops** of the bacterial suspension. If a purity check is necessary, use the excess suspension to inoculate another blood agar plate.

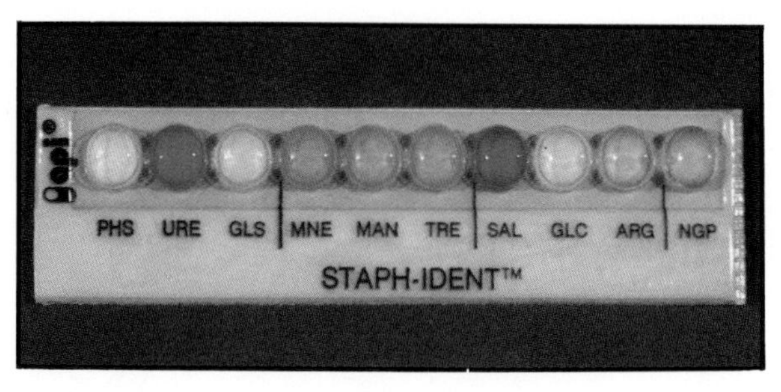

Figure 45.3 *S. aureus* test results. (Courtesy of Analytab Products, Plainview, N.Y.)

Incubation Place the plastic lid on the tray and incubate the strip aerobically for **five hours** at 35°–37° C. The blood agar plate should be incubated aerobically at the same temperature for 18 to 24 hours, or in 5–10% CO_2.

Reading and Interpreting Results

(Five Hours Later)

During this period the results will be recorded on the Laboratory Report, the profile number will be determined, and the unknown will be identified by looking up the number on the Staph-Ident Profile Register (chart VIII, appendix D).

Materials:

1 bottle of Staph-Ident reagent (room temperature)
Staph-Ident Profile Register

1. After five hours incubation, refer to chart VI, appendix D, to interpret and record the results of the first nine microcupules (PHS through ARG). Record the results on the profile determination table on the Laboratory Report. Chart VII, appendix D reveals the biochemistry involved in these tests.
2. Add **one to two drops** of Staph-Ident reagent to the NGP microcupule. Allow 30 seconds for color change to occur. A positive test results in a change of color to **plum-purple.** Record the results of this test.
3. Construct the profile number for your unknown and determine your unknown by consulting the Staph-Ident Profile Register. If no Profile Register is available, use chart VIII, appendix D. Since the Profile Register is constantly being updated by API, the one in appendix D may be out-of-date.

Disposal: Once all the information has been recorded be sure to place the entire incubation unit in a receptacle that is to be autoclaved.

Part 8 Sanitary Microbiology

Water, milk and food are excellent vehicles of disease transmission when contaminated with human pathogens. The encroachment of water supplies by human sewage has resulted in epidemics of typhoid, cholera, and bacillary dysentery on every continent. The fact that a centralized water system of a community may supply all citizenry makes it the ideal disseminator of disease when not carefully guarded.

Although milk and food are not generally supplied from a common source in a community, they excel water as transmitters of diseases from one standpoint: they support the growth of pathogens. Water lacks essential nutrients that are needed by most pathogens; so, generally speaking, they tend to die out in a relatively short period of time. On the other hand, a small number of pathogens in milk or food increase to astronomical numbers in a short period of time with ideal incubation. Many milk-borne epidemics of human diseases have been spread by contamination of milk by dirty hands of dairymen, insanitary utensils, and flies. The same thing can be said for improper handling of foods in the home, restaurants, hospitals and other institutions.

Thus, it becomes apparent that the health of a community relies heavily on governmental supervision of food, milk, and water supplies. Local and state public health departments, as well as private laboratories, are constantly performing bacteriological tests to see that our precautionary sanitary efforts have been successful. The intent is to detect the leak in the dike at the earliest moment to avert widespread disaster.

To provide uniformity of method for bacteriological examination of water and milk the American Public Health Association has published two books: *Standard Methods for the Examination of Water and Wastewater* and *Standard Methods for the Examination of Dairy Products.* Tried and tested procedures are outlined in these two books so that private and public health laboratories may have guidelines to follow. The four exercises of this unit on Sanitary Microbiology follow these accepted textbooks. The student should realize, however, that there are many acceptable procedures for bacteriological testing and that only a minimum is offered here.

46 Bacteriological Examination of Water (Multiple Tube Method)

Water that contains large numbers of bacteria may be perfectly safe to drink. The important consideration is the kinds of microorganisms that are present. Water from streams and lakes which contain multitudes of autotrophs and saprophytic heterotrophs is potable as long as pathogens for humans are lacking. The intestinal pathogens such as those that cause typhoid fever, cholera, and bacillary dysentery are of prime concern. The fact that human fecal material is carried away by water in sewage systems that often empty into rivers and lakes presents a colossal sanitary problem; thus, constant testing of municipal water supplies for the presence of fecal microorganisms is essential for the maintenance of water purity.

Routine examination of water for the presence of intestinal pathogens would be a tedious and difficult, if not impossible, task. It is much easier to demonstrate the presence of some of the nonpathogenic intestinal types such as *Escherichia coli* and *Streptococcus faecalis*. These organisms are always found in the intestines and normally are not present in soil or water; hence, when they are detected in water, it can be assumed that the water has been contaminated with fecal material. All bacteriological qualitative testing of water is based on the identification of **sewage indicators** such as these two organisms.

The series of tests depicted in figure 46.1 can be used to demonstrate the presence of coliform sewage indicators in water supplies. *A* ***coliform*** *is a facultative anaerobe that ferments lactose to produce gas and is a gram-negative, non-spore-forming rod. Escherichia coli* and *Enterobacter aerogenes* fit this description. Note that three different tests are involved: presumptive, confirmed, and completed. Each test exploits one or more of the characteristics of a coliform.

In the **presumptive test** a series of nine or twelve tubes of lactose broth are inoculated with measured amounts of water to see if the water contains any lactose-fermenting bacteria that produce gas. If, after incubation, gas is seen in any of the lactose broths, it is *presumed* that coliforms are present in the water sample. This test is also used to determine the most probable number (MPN) of coliforms present per 100 ml of water.

In the **confirmed test** plates of Levine's EMB agar or Endo agar are inoculated from positive (gas-producing) tubes to see if the organisms that are producing the gas are gram-negative (another coliform characteristic). Both of these media inhibit the growth of gram-positive bacteria and cause colonies of coliforms to be distinguishable from noncoliforms. On EMB agar, coliforms produce small colonies with dark centers (nucleated colonies). On Endo agar coliforms produce reddish colonies. The presence of coliformlike colonies *confirms* the presence of a lactose-fermenting gram-negative bacteria.

In the **completed test** our concern is to determine if the isolate from the agar plates truly fits our definition of a coliform. Our media for this test include a nutrient agar slant and a Durham tube of lactose broth. If gas is produced in the lactose tube and a slide from the agar slant reveals that we have a gram-negative non-spore-forming rod, we are certain that we have a coliform.

The completion of these three tests with positive results establishes that coliforms are present; however, there is no certainty that *E. coli* is the coliform present. The organism might be *E. aerogenes*. Of the two, *E. coli* is the better sewage indicator since *E. aerogenes* can be of nonsewage origin. To differentiate these two species, one must perform the **IMViC tests,** which were studied in Exercise 40.

In this exercise water will be tested from local ponds, streams, pools, and other sources supplied by students and instructor. Enough known positive samples will be evenly distributed throughout the laboratory so that all students will be able to see positive test results. All three tests in figure 46.1 will be performed; if time permits, the IMViC tests may also be performed.

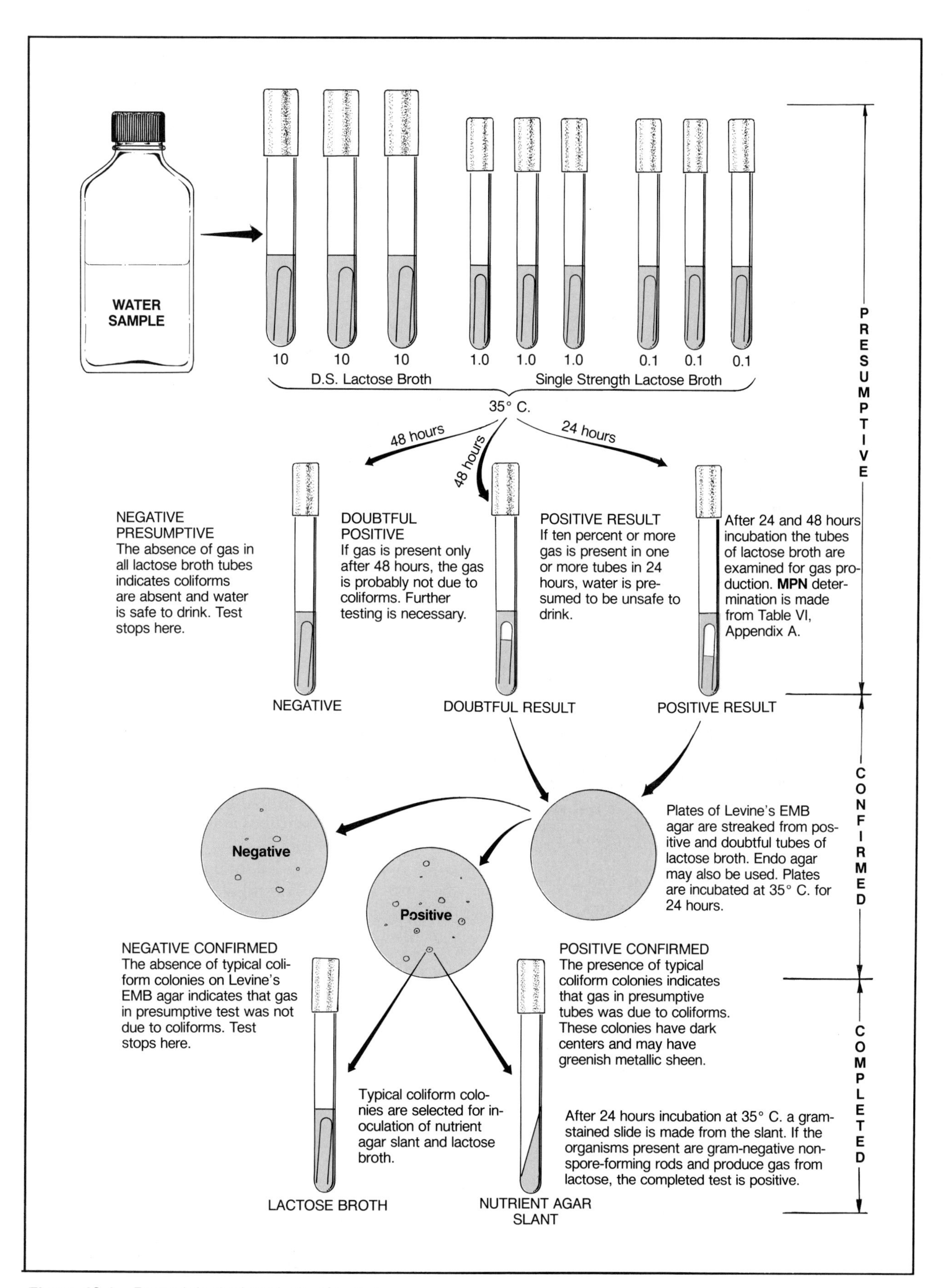

Figure 46.1 Bacteriological analysis of water.

The Presumptive Test

As stated earlier, the presumptive test is used to determine if gas-producing lactose fermenters are present in a water sample. If clear surface water is being tested, nine tubes of lactose broth will be used as shown in figure 46.1. For turbid surface water an additional three tubes of single strength lactose broth will be inoculated.

In addition to determining the presence or absence of coliforms, we can also use this series of lactose broth tubes to determine the **most probable number** (**MPN**) of coliforms present *in 100 ml of water.* A table for determining this value from the number of positive lactose tubes is provided in appendix A.

Before setting up your test, determine whether or not your water sample is clear or turbid: a separate set of instructions are provided for both types of water.

Clear Surface Water

If the water sample is relatively clear, proceed as follows:

Materials:

3 Durham tubes of DSLB
6 Durham tubes of SSLB
1 10-ml pipette
1 1-ml pipette

Note: DSLB designates double strength lactose broth. It contains twice as much lactose as SSLB (single strength lactose broth).

1. Set up three DSLB and six SSLB tubes as illustrated in figure 46.1. Label each tube according to the amount of water that is to be dispensed to it: 10 ml, 1.0 ml, and 0.1 ml, respectively.
2. Mix the bottle of water to be tested by shaking 25 times.
3. With a 10-ml pipette, transfer 10 ml of water to each of the DSLB tubes.
4. With a 1.0-ml pipette, transfer 1 ml of water to each of the middle set of tubes and 0.1 ml to each of the last three SSLB tubes.
5. Incubate the tubes at 35° C for 24 hours.
6. Examine the tubes and record the number of tubes in each set that have 10% gas or more.
7. Determine the MPN by referring to table VI, appendix A. Consider the following:

   Example: If you had gas in the first three tubes and gas only in one tube of the second series, but none in the last three tubes, your test would be read as 3–1–0. Table VI indicates that the MPN for this reading would be 43. This means that this particular sample of water would have approximately 43 organisms per 100 ml with 95% probability of there being between 7 and 210 organisms. Keep in mind that the MPN figure of 43 is a *statistical probability figure!*

8. Record this data on the Laboratory Report.

Turbid Surface Water

If your water sample appears to have considerable pollution, do as follows:

Materials:

3 Durham tubes of DSLB
9 Durham tubes of SSLB
1 10-ml pipette
2 1-ml pipettes
1 water blank (99 ml sterile water)

Note: See comment at left concerning DSLB and SSLB.

1. Set up three DSLB and nine SSLB tubes in a test-tube rack, with the DSLB tubes on the left. Label the three DSLB tubes *10 ml;* the next three SSLB tubes *1.0 ml;* the next three SSLB tubes *0.1 ml;* and the last three tubes *0.01 ml.*
2. Mix the bottle of water to be tested by shaking 25 times.
3. With a 10-ml pipette, transfer 10 ml of water to each of the DSLB tubes.
4. With a 1.0-ml pipette, transfer 1 ml to each of the next three tubes and 0.1 ml to each of the third set of tubes.
5. With the same 1.0 ml pipette, transfer 1 ml of water to the 99.0-ml blank of sterile water and shake 25 times.
6. With a *fresh* 1.0-ml pipette, transfer 1 ml of water from the blank to the remaining tubes of SSLB. This is equivalent to adding 0.01 ml of full strength water sample.
7. Incubate the tubes at 35° C for 24 hours.
8. Examine the tubes and record the number of tubes in each set that have 10% gas or more.
9. Determine the MPN by referring to table VI, appendix A. This table is set up for only nine tubes. To apply a twelve-tube reading to it, do as follows:
   a. Select the three consecutive sets of tubes that have at least one tube with no gas.
   b. If the first set of tubes (10-ml tubes) are not used, multiply the MPN by 10.

Example: Your tube reading was 3–3–3–1. What is the MPN?

The first set of tubes (10 ml) is ignored and the figures 3–3–1 are applied to the table. The MPN for this series is 460. Multiplying this by 10, the MPN is 4600.

Example: Your tube reading was 3–2–2–0. What is the MPN?

The first three numbers are applied to the table. The MPN is 210. Since the last set of tubes is ignored, 210 is the MPN.

The Confirmed Test

Once it has been established that gas-producing lactose fermenters are present in the water, it is *presumed* to be unsafe. However, gas formation may be due to noncoliform bacteria. Some of these, such as *Clostridium perfringens,* are gram-positive. To confirm the presence of gram-negative lactose fermenters, the next step is to inoculate media such as Levine's eosin methylene blue agar or Endo agar from positive presumptive tubes.

Levine's EMB agar contains methylene blue, which inhibits gram-positive bacteria. Gram-negative lactose fermenters (coliforms) which grow on this medium will produce "nucleated colonies" (dark centers). Colonies of *E. coli* and *E. aerogenes* can be differentiated on the basis of size and the presence of a greenish metallic sheen. *E. coli* colonies are small and have this metallic sheen, whereas *E. aerogenes* colonies usually lack the sheen and are larger; differentiation in this manner is not completely reliable, however. It should be understood that *E. coli* is the more reliable sewage indicator since it is not normally present in soil, while *E. aerogenes* has been isolated from grains and soil.

Endo agar contains a fuchsin sulfite indicator, which makes identification of lactose fermenters relatively easy. Coliform colonies and the surrounding medium appear red on Endo agar. Nonfermenters of lactose, on the other hand, are colorless and do not affect the color of the medium. In addition to these two media, there are several other media that can be used for the confirmed test. Brilliant green bile lactose broth, Eijkman's medium, and EC medium are a few others that can be used.

To demonstrate the confirmation of a positive presumptive in this exercise, we will use Levine's EMB agar and Endo agar.

Materials:

1 petri plate of Levine's EMB agar (odd-numbered students)
1 petri plate of Endo agar (even-numbered students)

1. Select one positive lactose broth tube from the presumptive test and streak a plate of medium according to your assignment. Use a streak method which will produce good isolation of colonies. If all of your tubes were negative, borrow a positive tube from another student.
2. Incubate the plate for 24 hours at 35° C.
3. Look for typical coliform colonies on both kinds of media. Record your results on the Laboratory Report. *If no coliform colonies are present, the water is considered safe to drink.* **Note:** In actual practice, confirmation of all presumptive tubes would be necessary to ensure accuracy of results.

The Completed Test

A final check of the colonies which appear on the confirmatory media is made by inoculating a nutrient agar slant and a tube of lactose broth. After incubation for 24 hours at 35° C, the lactose broth is examined for gas production. A gram-stained slide is made from the slant, and the slide is examined under oil. If the organism proves to be a gram-negative, non-spore-forming rod that ferments lactose, we know that coliforms were present in the initial water sample. If time permits, complete these last tests and record the results on the Laboratory Report.

The IMViC Tests

Review the discussion of the IMViC tests on page 149. The significance of these tests should be much more apparent at this time. Your instructor will indicate whether or not these tests should also be performed if you have a positive completed test.

47 Bacteriological Examination of Water (Membrane Filter Method)

In addition to the multiple tube test, a method utilizing the membrane filter has been recognized by the United States Public Health Service as a reliable method for the detection of coliforms in water. These filter disks are 150 micrometers thick, have pores of 0.45 micrometer diameter, and have 80% area perforation. The precision of manufacture is such that bacteria larger than 0.47 micrometer cannot pass through. Eighty percent area perforation facilitates rapid filtration.

To test a sample of water, the water is passed through one of these filters. All bacteria present in the sample will be retained directly on the filter's surface. The membrane filter is then placed on an absorbent pad saturated with liquid nutrient medium and incubated for 22 to 24 hours. The organisms on the filter disk will form colonies that can be counted under the microscope. If a differential medium, such as *m* Endo MF broth is used, coliforms will exhibit a characteristic golden metallic sheen.

The advantages of this method over the multiple tube test are (1) higher degree of reproducibility of results; (2) greater sensitivity since larger volumes of water can be used; and (3) shorter time (one-fourth) for getting results.

Materials:

vacuum pump or water faucet aspirators
membrane filter assemblies (sterile)
side-arm flask, 1000 ml size, and rubber hose
sterile graduates (100 ml or 250 ml size)
sterile, plastic petri dishes, 50 mm dia. (Millipore # PD10 047 00)
sterile membrane filter disks (Millipore #HAWG 047 AO)
sterile absorbent disks (packed with filters)
sterile water
5-ml pipettes
bottles of *m* Endo MF broth (50 ml)*
water samples

*See appendix C for special preparation method.

1. Prepare a small plastic petri dish as follows:
   a. With a flamed forceps, transfer a sterile absorbent pad to a sterile plastic petri dish.
   b. Using a 5-ml pipette, transfer 2.0 ml of *m* Endo MF broth to the absorbent pad.
2. Assemble a membrane filtering unit as follows:
   a. *Aseptically* insert the filter holder base into the neck of a one-liter side-arm flask.
   b. With a flamed forceps, place a sterile membrane filter disk, grid side up, on the filter holder base.
   c. Place the filter funnel on top of the membrane filter disk and secure it to the base with the clamp.
3. Attach the rubber hose to a vacuum source (pump or water aspirator) and pour the appropriate amount of water into the funnel.

   The amount of water used will depend on water quality. No less than 50 ml should be used. Waters with few bacteria and low turbidity permit samples of 200 ml or more. Your instructor will advise you as to the amount of water that you should use. Use a sterile graduate for measuring the water.
4. Rinse the inner sides of the funnel with 20 ml of sterile water.
5. Disconnect the vacuum source, remove the funnel, and carefully transfer the filter disk with sterile forceps to the petri dish of *m* Endo MF broth. *Keep grid side up.*
6. Incubate at 35° C for 22 to 24 hours. *Don't invert.*
7. After incubation, remove the filter from the dish and dry for one hour on absorbent paper.
8. Count the colonies on the disk with low-power magnification, using reflected light. Ignore all colonies that lack the golden metallic sheen. If desired, the disk may be held flat by mounting between two 2″ × 3″ microscope slides after drying. Record your count on the Laboratory Report.

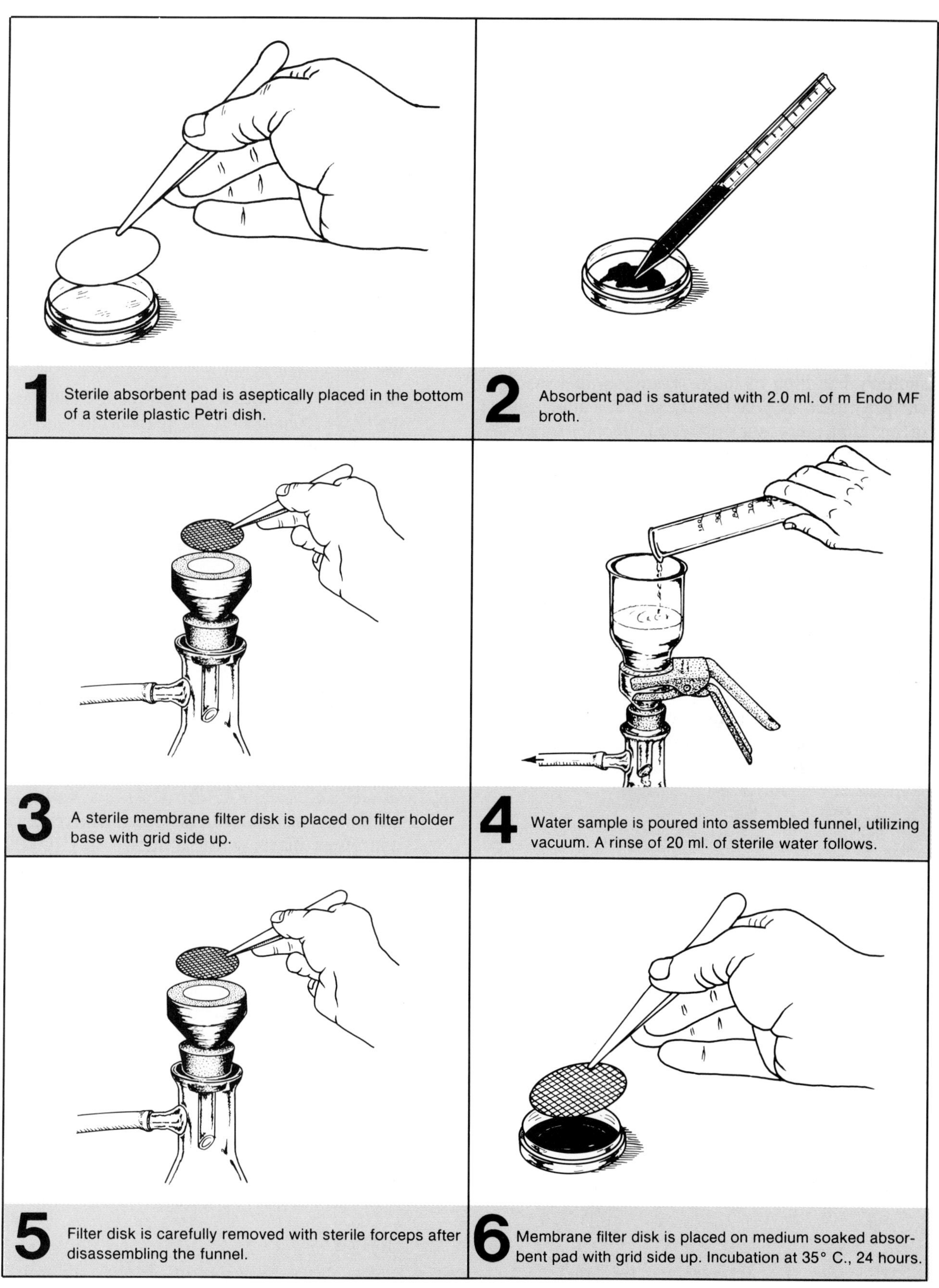

Figure 47.1 Membrane filter routine.

48 Standard Plate Count of Milk

The bacterial count in milk is the most reliable indication we have of its sanitary quality. It is for this reason that the American Public Health Association recognizes the standard plate count as the official method in its Milk Ordinance and Code. Although human pathogens may not be present in a high count, it may indicate a diseased udder, unsanitary handling of milk, or unfavorable storage temperatures. In general, therefore, a high count means that there is a greater likelihood of disease transmission. On the other hand, it is necessary to avoid the wrong interpretation of low plate counts, since it is possible to have pathogens such as the brucellosis and tuberculosis organisms when counts are within acceptable numbers. Routine examination and testing of animals act as safeguards against this latter situation.

In this exercise, standard plate counts will be made of two samples of milk: a supposedly good sample and one of known poor quality. *Odd-numbered students will work with the high-quality milk and even-numbered students will test the poor-quality sample.* A modification of the procedures in Exercise 19 will be used.

High-Quality Milk

Materials:

milk sample
1 sterile water blank (99 ml)
4 sterile petri plates
1.1-ml dilution pipettes
1 bottle TGEA (50 ml)
Quebec colony counter
mechanical hand counter

1. Following the procedures used in Exercise 19, pour four plates with dilutions of 1:1, 1:10, 1:100, and 1:1000. Before starting the dilution procedures, shake the milk sample 25 times in the customary manner.
2. Incubate the plates at 35° C for 24 hours and count the colonies on the plate which has between 30 and 300 colonies.
3. Record your results on the Laboratory Report.

Poor-Quality Milk

Materials:

milk sample
3 sterile water blanks (99 ml)
4 sterile petri plates
1.1-ml dilution pipettes
1 bottle TGEA (50 ml)
Quebec colony counter
mechanical hand counter

1. Following the procedures used in Exercise 19, pour four plates with dilutions of 1:10,000, 1:100,000, 1:1,000,000, and 1:10,000,000. Before starting the dilutions, shake the milk sample 25 times in the customary manner.
2. Incubate the plates at 35° C for 24 hours and count the colonies on the plate which has between 30 and 300 colonies. Record your results on the Laboratory Report.

Bacterial Counts of Foods 49

The standard plate count, as well as the multiple tube test, can be used on foods much in the same manner that they are used on milk and water to determine total counts and the presence of coliforms. To get the organisms in suspension, however, a food blender is necessary. In this exercise, samples of ground meat, dried fruit, and frozen food will be tested for numbers of bacteria. The instructor will indicate the specific kinds of foods to be tested and make individual assignments. Figure 49.1 illustrates the general procedure.

Materials:

per student:
- 3 petri plates
- 1 bottle (36 ml) of Eugonagar
- 1 99-ml sterile water blank
- 2 1.1-ml dilution pipettes

per class:
- food blender
- sterile blender jars (one for each type of food)
- sterile weighing paper
- 180-ml sterile water blanks (one for each type of food)
- samples of ground meat, dried fruit, and frozen vegetables, thawed two hours

1. Using aseptic techniques, weigh out on sterile weighing paper 20 grams of the food to be tested.
2. Add the food and 180 ml of sterile water to a sterile mechanical blender jar. Blend the mixture for five minutes. This suspension will provide a 1:10 dilution.
3. With a 1.1-ml dilution pipette dispense from the blender 0.1 ml to plate I and 1.0 ml to the water blank. See figure 49.1.
4. Shake the water blank 25 times, and with a fresh pipette dispense 0.1 ml to plate III and 1.0 ml to plate II.
5. Pour agar (50° C) into the three plates, and incubate at 35° C for 24 hours.
6. Count the colonies on the best plate and record the results on the Laboratory Report.

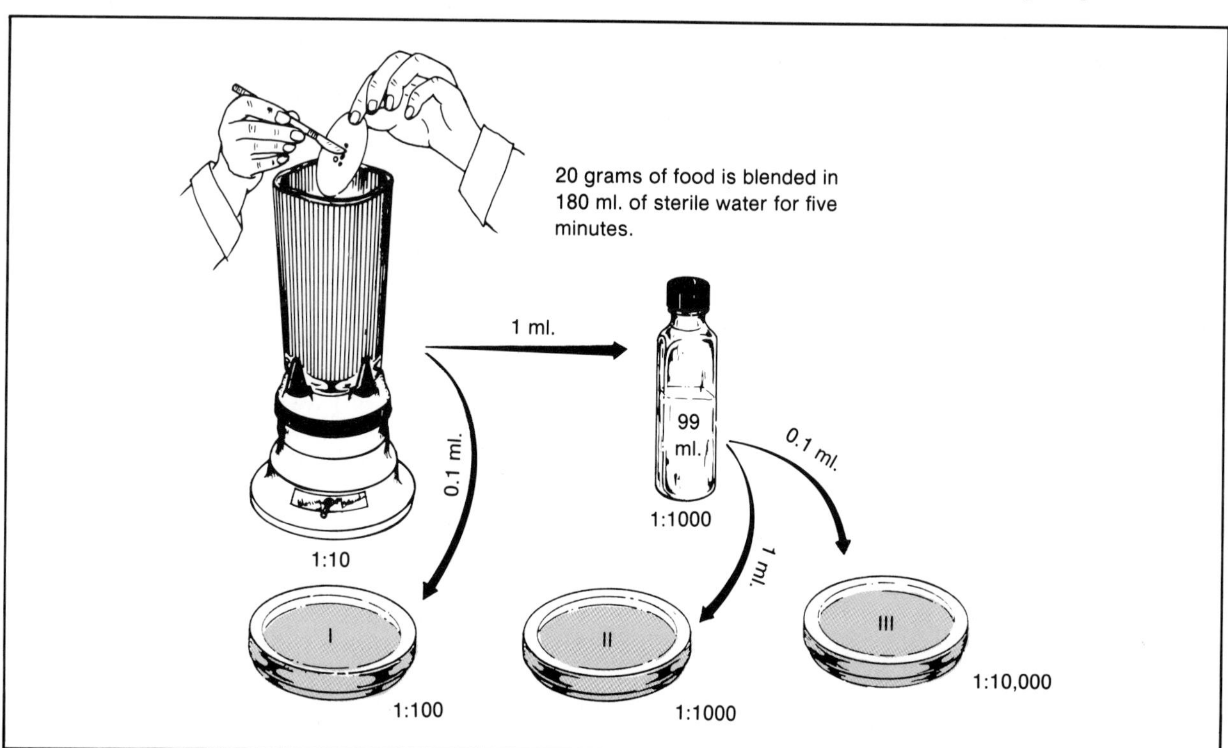

Figure 49.1 Dilution procedure for bacterial counts of food.

Part 9 Medical Microbiology and Immunology

Although many of the exercises up to this point in the manual pertain in some way to medical microbiology, they also have applications that are nonmedical. The exercises of this unit, however, are primarily medical or dental in nature.

Medical (clinical) microbiology is primarily concerned with the isolation and identification of pathogenic organisms. Naturally, the techniques for studying each type of organism are different. A complete coverage of this field of microbiology is very extensive, encompassing the Mycobacteriaceae, Brucellaceae, Enterobacteriaceae, Corynebacteriaceae, Micrococcaceae, *ad infinitum.* It is not possible to explore all of these groups in such a short period of time; however, this course would be incomplete if it did not include some of the routine procedures that are used in the identification of some of the more common pathogens.

Exercise 50 in this unit differs from the other ten exercises in that it pertains to the spread of disease (epidemiology) rather than to specific microorganisms. Its primary function is to provide an understanding of some of the tools used by public health epidemiologists to determine the sources of infection in the transmission cycle.

Since the most frequently encountered pathogenic bacteria are the gram-positive pyogenic cocci and intestinal organisms, Exercises 51, 52, and 53 have been devoted to the study of these bacteria. The exercise that provides the greatest amount of depth is Exercise 52 (the streptococci). To provide assistance in the identification of streptococci, it has been necessary to provide supplementary information in appendix E. Exercises 55 through 60 pertain to manifestations of antibody-antigen reactions (serology).

A Synthetic Epidemic 50

A disease caused by microorganisms that enter the body and multiply in the tissues is said to be an **infectious disease.** Infectious diseases that are transmissable to other persons are considered to be **communicable.** The transfer of communicable infectious agents between individuals can be accomplished by direct contact, such as in handshaking and kissing, or they can be spread indirectly through food, water, objects, etc.

Epidemiology is the study of how, when, where, what, and who are involved in the spread and distribution of diseases in human populations. An epidemiologist is, in a sense, a medical detective who searches out the sources of infection so that the transmission cycle can be broken.

Whether or not an epidemic actually exists is determined by the epidemiologist by comparing the number of new cases with previous records. If the number of newly reported cases in a given period of time in a specific area are excessive, an **epidemic** is considered in progress. If the disease spreads to two or more continents, a **pandemic** is occurring.

In this experiment we will have an opportunity to approximate, in several ways, the work of the epidemiologist. Each member of the class will take part in the spread of a "synthetic infection." For obvious safety reasons, a white powder will be used instead of a viable infectious microorganism. The mode of transmission will be by hand contact. The white powder has many of the characteristics of a microorganism in that it clings to the hands, is not evident to the naked eye once it is on the skin, and can be readily passed on to others through direct contact.

Each student will be given a numbered container of white powder. Only one member in the class will be given a powder that is to be considered the infectious agent. The other members will be issued a transmissible agent that is considered noninfectious. After each student has spread the powder on his or her hands, all members of the class will engage in two rounds of handshaking, directed by the instructor. A record of the handshaking contacts will be recorded on the chalkboard. After each round of handshaking, the hands will be rubbed on blotting paper so that a chemical test can be applied to it to determine the presence or absence of the infectious agent.

From data compiled in this experiment, an attempt will be made to determine two things: (1) the original source of the infection and (2) who the carriers are. The type of data analysis used in this experiment is not unlike the procedure that an epidemiologist would employ. Proceed as follows:

Materials:

1 numbered container of white powder
1 piece of white blotting paper
spray bottles of "developer" solution

Preliminaries After assembling your materials, write your name and unknown number at the top of your sheet of blotting paper. In addition, draw a line down the middle, top to bottom, and label the left side ROUND 1 and the right side ROUND 2.

Wash and dry your hands thoroughly. Moisten the right hand with water. Prepare it with the agent by thoroughly coating it with the white powder, especially on the palm surface. This step is similar to the contamination that would occur to one's hand if it were sneezed into during a cold.

IMPORTANT: Once the hand has been prepared, do not rest it on the tabletop or allow it to touch any other object.

Round 1 On the cue of the instructor, we will begin the first round of handshaking. Your instructor will inform you when it is your turn to shake hands with someone. You may shake with anyone, but it is best not to shake your neighbor's hand. ***Be sure to use only your treated hand, and avoid extracurricular glad-handing.***

Note: In each round of handshaking you will be selected by the instructor *only once* for handshaking; however, due to the randomness of selection by the handshakers, it is possible that you may be selected as the "shakee" several times.

Handprint 1 After every member of the class has shaken someone's hand, we need to assess just who might have picked up the "microbe." To accomplish this, wipe your fingers and palm of the contaminated hand on the left side of your blotting paper. Press fairly hard, but don't tear the surface.

IMPORTANT: Don't allow your hand to touch any other object. A second round of handshaking follows.

Round 2 Proceed to shake hands, randomly. Avoid contact with any other object.

Handprint 2 Once the second handshaking episode is finished, rub the fingers and palm of the contaminated hand on the right side of the blotting paper. **Caution:** Keep your contaminated hand off the left side of the blotting paper.

Chemical Identification To determine who has been infected we will now spray the developer solution on the handprints of both rounds.

One at a time, each student, with the help of the instructor, will spray their blotting papers with iodine solution. Color interpretation is as follows:

Blue: positive for infectious agent
Brown or yellow: negative

Tabulation of Results The instructor or a student recorder will tabulate the results on the chalkboard, using a table similar to the one on the Laboratory Report. Once all results have been recorded, you may proceed to determine the originator of the epidemic. The easiest way to determine this is to put together a flow chart of shaking and identify those persons that test positive. You will be working backward with the kind of information an epidemiologist has to work with (contacts and infections). Eventually, a pattern will emerge which shows which person started the epidemic.

Laboratory Report

Record all information on the Laboratory Report and answer the questions.

The Staphylococci: Isolation and Identification 51

Often in conjunction with streptococci, the staphylococci cause abscesses, boils, carbuncles, osteomyelitis, and fatal septicemias. Collectively, the staphylococci and streptococci are referred to as the pyogenic (pus-forming) gram-positive cocci. Originally isolated from pus in wounds, the staphylococci were subsequently demonstrated to be normal inhabitants of the nasal membranes, hair follicles, the skin, and the perineum of healthy individuals. The fact that 90% of hospital personnel are carriers of staphylococci portends serious epidemiological problems, especially since most strains are penicillin-resistant.

The **staphylococci** are gram-positive spherical bacteria that divide in more than one plane to form irregular clusters of cells. They are listed in section 12, volume 2, of *Bergey's Manual of Systematic Bacteriology.* The Genus *Staphylococcus* is grouped with three other genera in Family Micrococcaceae:

SECTION 12 GRAM-POSITIVE COCCI
- Family I Micrococcaceae
  - Genus I *Micrococcus*
  - Genus II *Stomatococcus*
  - Genus III *Planococcus*
  - Genus IV *Staphylococcus*
- Family II Deinococcaceae
  - Genus I *Deinococcus*
  - Genus *Streptococcus*

Although the staphylococci make up a coherent phylogenetic group, they have very little in common with the streptococci except for their basic similarities of being gram-positive, non-spore-forming cocci. Note that *Bergey's Manual* lists the streptococci as a separate genus from the staphylococci, independent of any family grouping.

Of the nineteen species of staphylococci listed in *Bergey's Manual,* the most important ones are *S. aureus, S. epidermidis,* and *S. saprophyticus.* The single most significant characteristic that separates these species is the ability or inability of these organisms to coagulate plasma: only *S. aureus* has this ability; the other two are coagulase-negative.

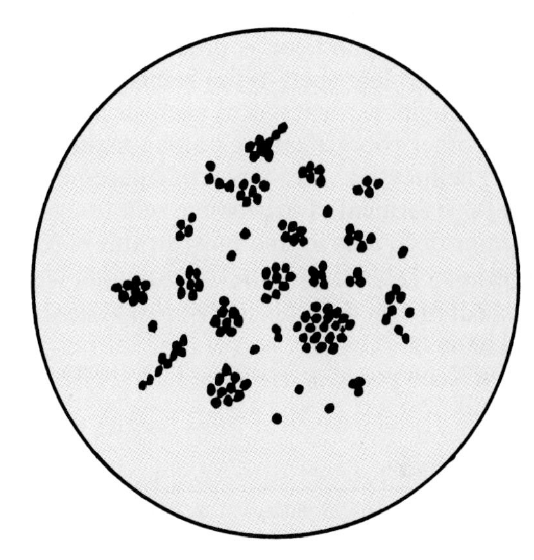

Figure 51.1 Staphylococci.

Although *S. aureus* has, historically, been considered to be the only significant pathogen of the three, the others do cause infections. Some cerebrospinal fluid infections (3), prosthetic joint infections (4), and vascular graft infections (2) have been shown to be due to coagulase-negative staphylococci. Numbers in parentheses designate references at the end of this exercise.

Our concern in this exercise will pertain exclusively to the differentiation of only three species of staphylococci. If other species are encountered, the student may wish to use the API Staph-Ident miniaturized test strip system (Exercise 45).

In this experiment we will attempt to isolate staphylococci from (1) the nose, (2) a fomite, and (3) an "unknown-control." The unknown-control will be a mixture containing staphylococci, streptococci, and some other contaminants. If the nasal membranes and fomite prove to be negative, the unknown-control will yield positive results, providing all inoculations and tests are performed correctly.

Since *S. aureus* is by far the most significant pathogen of this group, most of our concern here will be with this organism. It is for this reason that the characteristics of only this pathogen will be outlined below.

Staphylococcus aureus cells are 0.8 to 1.0 μm diameter, and may occur singly, in pairs, or as clusters. Colonies of *S. aureus* on trypticase soy agar or blood agar are opaque, 1 to 3 mm in diameter, and are yellow, orange, or white in color. They are salt tolerant, growing well on media containing 10% sodium chloride. Virtually all strains are *coagulase-positive.* Mannitol is fermented aerobically to produce acid. Alpha toxin is produced that causes a wide zone of clear (beta-type) hemolysis on blood agar; in rabbits it causes local necrosis and death.

The other two genera lack alpha toxin (do not exhibit hemolysis), and are coagulase-negative. Mannitol is fermented to produce acid (no gas) by all strains of *S. aureus* and most strains of *S. saprophyticus.* Table 51.1 lists the principal characteristics that differentiate these three species of staphylococcus.

Table 51.1 Differentiation of Three Species of Staphylococci

	S. aureus	*S. epidermidis*	*S. saprophyticus*
Alpha toxin	+	−	−
Mannitol (acid only)	+	−	(+)
Coagulase	+	−	−
Biotin for growth	−	+	NS
Novobiocin	S	S	R

Note: NS = Not significant; S = sensitive; R = resistant; (+) = mostly positive.

Enrichment and Isolation of Specimen

(First and Second Periods)

To determine the incidence of carriers in our classroom, as well as the incidence of the organism on common fomites, we will follow the procedure illustrated in figure 51.2. Results of class findings will be tabulated on the chalkboard so that all members of the class can record data required on the Laboratory Report. The physiological characteristics that we will look for in our isolates will be

1. beta-type hemolysis (alpha toxin),
2. mannitol fermentation, and
3. coagulase production.

Note in figure 51.2 that swabs that have been applied to nasal membranes and fomites will be placed in tubes of enrichment medium containing 10% NaCl (*m*-staphylococcus broth). Since your unknown-control will lack a swab, initial inoculations from this culture will have to be done with a loop.

After 4 to 24 hours incubation, mannitol salt agar and staphylococcus medium 110 will be inoculated from the enrichment broths and incubated for 24 to 36 hours. Proceed as follows:

First Period

Materials:

1 tube containing numbered unknown-control
3 tubes of *m*-staphylococcus broth
2 sterile cotton swabs

1. Label the three tubes of *m*-staphylococcus broth "nose," "fomite," and number of your unknown-control.
2. Inoculate the appropriate tube of *m*-staphylococcus broth with one or two loopfuls of your unknown-control.
3. After moistening one of the swabs by immersing partially into the "nose" tube of broth, swab the nasal membrane just inside your nostril. A small amount of moisture on the swab will enhance the pickup of organisms. Place this swab into the "nose" tube.
4. Swab the surface of a fomite with the other swab that has been similarly moistened and deposit this swab in the "fomite" tube.

   The fomite you select may be a coin, drinking glass, telephone mouthpiece, or any other item that you might think of.
5. Incubate these tubes of broth for 4 to 24 hours at 37° C.

Second Period

Materials:

3 *m*-staphylococcus broths from previous period
2 petri plates of mannitol salt agar (MSA)
2 petri plates of staphylococcus medium 110 (SM110)

1. Label the bottoms of the MSA and SM110 as shown in figure 51.2. Note that to minimize the number of plates required, it will be necessary to make half-plate inoculations for the nose and fomite. The unknown-control will be inoculated on separate plates.
2. Quadrant streak MSA and SM110 plates with the unknown-control.

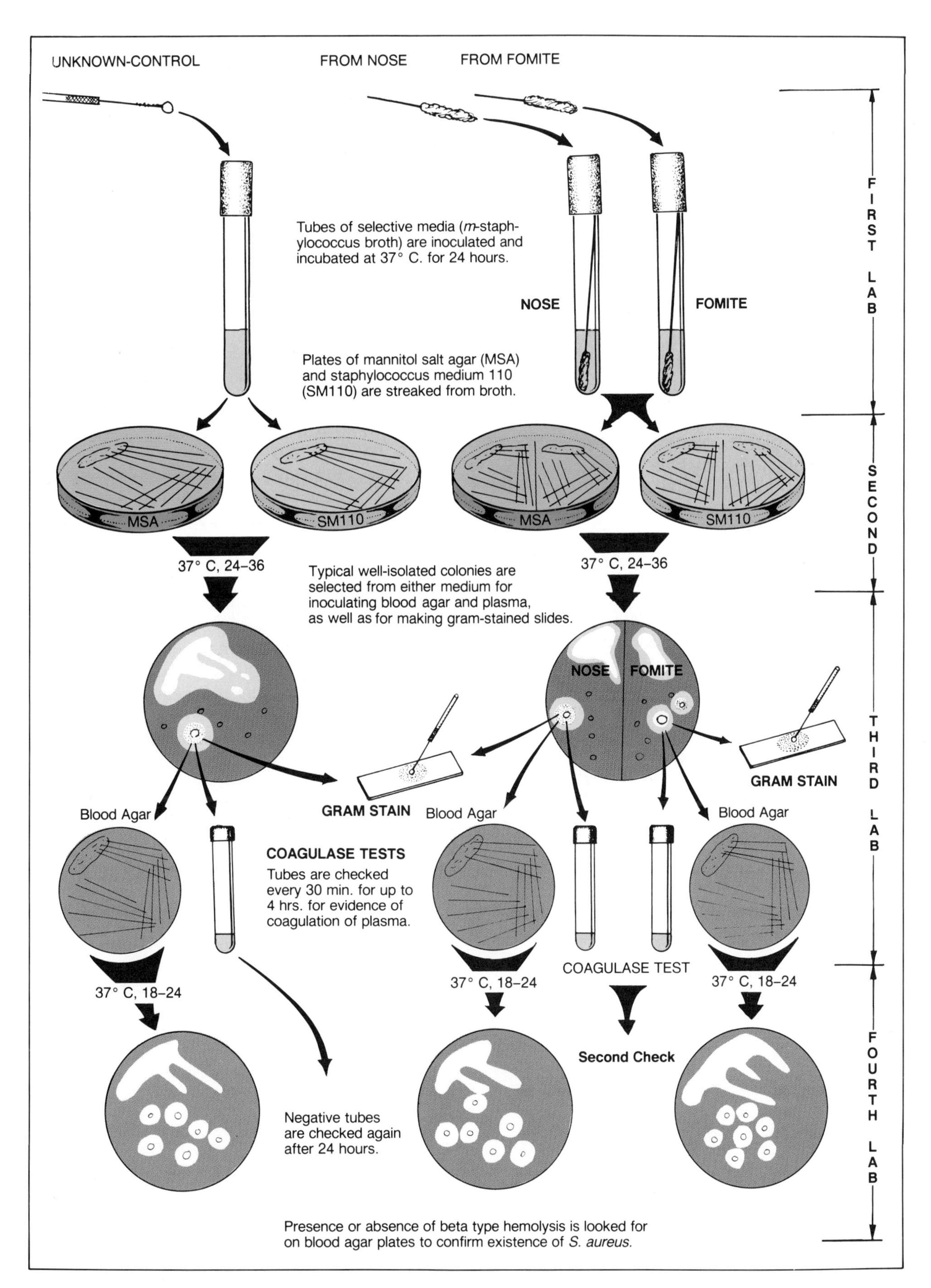

Figure 51.2 Procedure for presumptive identification of staphylococci.

3. Inoculate a portion of the nose side of each plate with the swab from the nose tube; then, with a sterile loop, streak out the organisms on the remainder of the agar on that half of each plate. The swabbed areas will provide massive growth; the streaked-out areas should yield good colony isolation.
4. Repeat step 3 to inoculate the other half of each agar plate with the swab from the fomite tube.
5. Incubate the plates aerobically at 37° C for 24 to 36 hours.

Final Tests and Evaluation
(Third and Fourth Periods)

In these last two laboratory periods, the inoculated agar plates will be evaluated to determine if any of the colonies exhibit evidence of mannitol fermentation. Colonies that appear to be *S. aureus* will be used to (1) inoculate a blood agar plate, (2) make gram-stained slides, and (3) perform coagulase tests.

Third Period

Mannitol salt agar contains beef extract, peptone, 7.5% sodium chloride, mannitol, and phenol red indicator. Staphylococcus medium 110 also contains mannitol and sodium chloride, but differs from MSA in that it is somewhat richer in protein content and contains some lactose. One advantage of SM110 over MSA is that it is more suitable for pigment production by strains of *S. aureus*. If *S. aureus* is present, the mannitol in the MSA is fermented, causing the phenol red to change color from red to yellow. The sodium chloride in both media makes them selective for *S. aureus*. Proceed as follows to evaluate your plates and make the necessary inoculations.

Materials:

Gram staining kit
MSA and SM110 plates from previous period
2 blood agar plates
1 to 3 serological tubes containing 0.5 ml of 1:4 saline dilution of rabbit or human plasma (one tube for each isolate)

Examine the mannitol salt agar plates. Has the phenol red in the medium surrounding any of the colonies turned yellow? If this color change exists around any of the colonies, it can be *presumed* that you have isolated a strain of *S. aureus*. Record your results on the Laboratory Report and chalkboard.

Examine the plates of staphylococcus medium 110. Are any pigmented colonies present? Record your results on the Laboratory Report.

If you have presumptive evidence that any of your plates have cultures of *S. aureus*, confirm your conclusions by (1) inoculating a blood agar plate from suspected colonies, (2) doing a coagulase test for each isolation, and (3) making a gram-stained slide of each isolate. Proceed as follows:

Blood Agar Plates Label the bottoms of the blood agar plates with your unknown-control number, "nose," and "fomite," as required. Select well-isolated colonies and streak out the organisms on the blood agar. Use the half-plate method of streaking if organisms are present on both the nose and fomite sides of the plates. Use a whole plate for your unknown-control.

Incubate the plates, inverted, at 37° C for 18 to 24 hours. Plates left in the incubator longer than 24 hours will degenerate, so be sure not to over-incubate.

Coagulase Test The fact that 97% of the strains of *S. aureus* have proven to be coagulase-positive and that the other two species are *always* coagulase-negative makes the coagulase test an excellent definitive test for confirming the identification of *S. aureus*.

The coagulase test consists of adding a heavy inoculum of the test organism to a small tube containing 0.5 ml of 1:4 saline dilution of rabbit or human plasma. The tube is then placed in a 37° C water bath and examined every 30 minutes up to four hours for coagulation. If no solidification is observed at the end of this period, the incubation is continued for a total of 24 hours. Any degree of clotting during this time, from a loose clot suspended in plasma to a solid, immovable clot, is considered to be a positive result. Some gram-negative rods, such as *Pseudomonas,* often cause false-positives. The mechanism of clotting in such organisms is not due to coagulase. *Because of the possibility of false-positives, this test should only be performed on gram-positive staphylococcus-like organisms.*

Although it is preferable to do Gram staining prior to the coagulase test, it is more practical here to do the coagulase test first to make it possible to observe the results of the test for a longer period of time in the laboratory. While the inoculated plasma tubes are incubating in the water bath, you can be doing the staining and microscopic work. Proceed

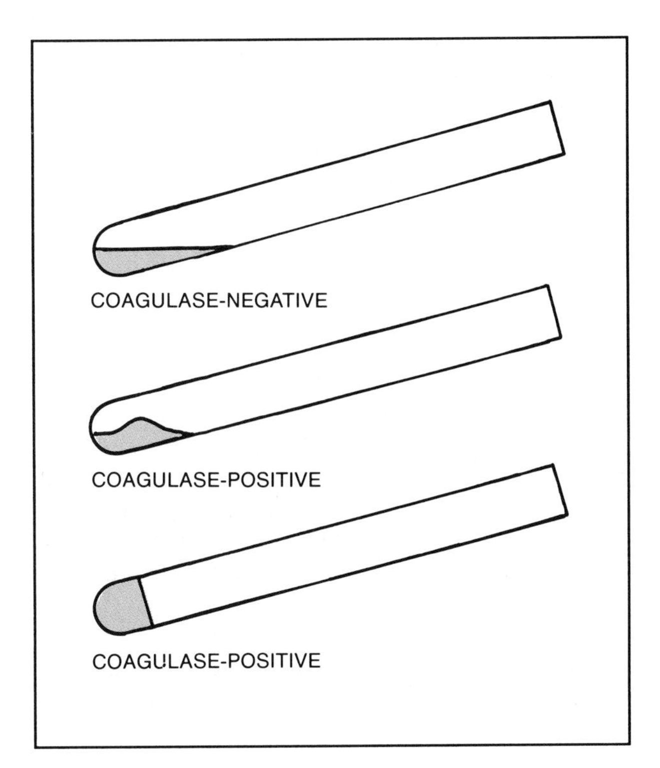

Figure 51.3 Coagulase test results: one negative and two positive tests.

as follows to do coagulase tests of suspected *S. aureus* colonies:

1. Label the plasma tubes "nose," "fomite," or "unknown."
2. With a wire loop, inoculate the appropriate tube of plasma with organisms from one or more colonies on SM110 or MSA. Use several loopfuls. Success is more rapid with a heavy inoculation. If positive colonies are present on both nose and fomite sides, inoculate a separate tube for each side.
3. Place the tubes in a 37° C water bath.
4. Check for solidification of the plasma every 30 minutes for the remainder of the period. Note in figure 51.3 that solidification may be complete, as in the lower tube, or show up as a semisolid ball, as seen in the middle tube.

   Any cultures that are negative at the end of the period will be checked by your instructor at 24 hours and placed in the refrigerator for your evaluation at the next laboratory period.

   Record your results on the Laboratory Report.

Gram-Stained Slides Select organisms for Gram staining from the same colonies that were used for the blood agar plate and coagulase tests. Examine the slides under the microscope and draw them in the appropriate circles on the Laboratory Report.

Final Confirmation
(Fourth Period)

Materials:

coagulase tubes from previous period
blood agar plates from previous period

Examine any coagulase tubes that were carried over from the last laboratory period. Record your results on the Laboratory Report.

Examine the colonies on your blood agar plates. Look for clear (beta-type) hemolysis around the colonies. Record your results on the Laboratory Report.

Laboratory Report

After recording your results on the chalkboard with the other students, complete the Laboratory Report for this exercise.

Literature Cited

1. Kloos, W. E., and K. H. Schleifer. 1975. Simplified scheme for routine identification of human *Staphylococcus* species. *J. Clin. Microbiol.* 1:82–88.
2. Liekweg, W. G., Jr., and L. T. Greenfield. 1977. Vascular prosthetic infection: Collected experience and results of treatment. *Surgery* 81:355–400.
3. Schoenbaum, S. C., P. Gardner, and J. Shillito. 1975. Infections in cerebrospinal shunts: Epidemiology, clinical manifestations, and therapy. *J. Infect. Dis.* 131:543–52.
4. Wilson, P. D., Jr., E. A. Salvati, P. Aglietti, and L. J. Kutner. 1973. The problem of infection in endoprosthetic surgery of the hip joint. *Clin. Orthop. Relat. Res.* 96:213–21.

52 The Streptococci: Isolation and Identification

The streptococci differ from the staphylococci in that they are arranged primarily in chains rather than clusters. In addition to causing many mixed infections with staphylococci, the streptococci can also separately cause diseases such as pneumonia, meningitis, endocarditis, pharyngitis, erysipelas, and glomerularnephritis.

Several species of streptococci are normal inhabitants of the pharynx. They can also be isolated from surfaces of the teeth, the saliva, skin, colon, rectum, and vagina.

The streptococci of greatest medical significance are *S. pyogenes, S. agalactiae,* and *S. pneumoniae.* Of lesser importance are *S. faecalis, S. faecium,* and *S. bovis.* Appendix E describes in greater detail the characteristics and significance of these and other streptococcal species.

In this exercise we will attempt to do two things: (1) isolate and identify prevalent streptococci of the pharynx and (2) become familiar with the principal characteristics that are used for differentiating the various groups of medically important streptococci.

To accomplish these two goals we will first use selective and enrichment media to isolate pharyngeal streptococci. To identify these isolates we will employ several physiological tests that are used for differentiation.

Each student will also be given two or three unknown streptococci to be identified along with the pharyngeal isolates. These unknowns will not be issued until physiological test media are to be inoculated. By performing tests on several streptococci, you will have an opportunity to analyze multiple test data and determine the identity of several organisms.

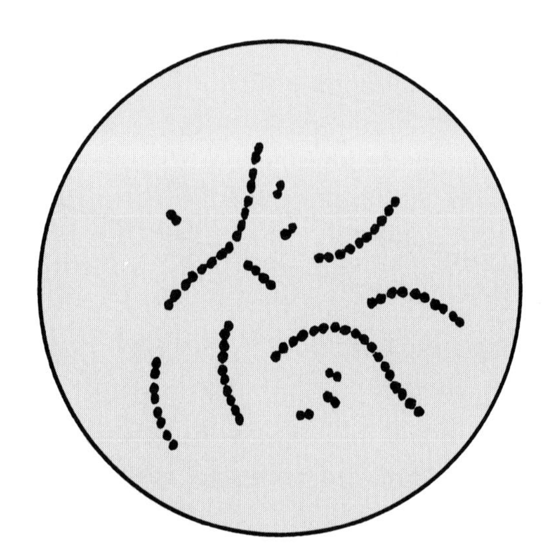

Figure 52.1 Streptococci.

Pharyngeal Isolates

Figure 52.2 illustrates the general procedure for isolating streptococci from the pharynx. Note that initial differentiation of streptococci is made on blood agar. Proceed as follows:

First Period

When working with patients who have obvious clinical symptoms of upper respiratory infection, the procedure described in figure 52.2 for the first period is the method of choice. Note that a tube of TSB is used for enrichment and a tube of TSBCV is used for selectivity. TSBCV is trypticase soy broth with 1 μg/ml of crystal violet added to inhibit staphylococci. Although this method is recommended by the Center for Disease Control, Atlanta, your instructor may wish to omit this step and inoculate directly from the pharynx to blood agar. *Your instructor will indicate which method will be used.*

Note in figure 52.2 that the organisms are removed from the posterior wall of the pharynx while the tongue is depressed. This is a procedure best performed by laboratory partners on each other.

Materials:

- 2 sterile cotton swabs
- 1 tongue depressor
- 1 tube trypticase soy broth (TSB)
- 1 tube trypticase soy broth with crystal violet (TSBCV)

1. Label one tube of each kind of broth with the initials of the subject.
2. With the subject's head tilted back and the tongue held down with the tongue depressor, rub the back surface of the pharynx up and

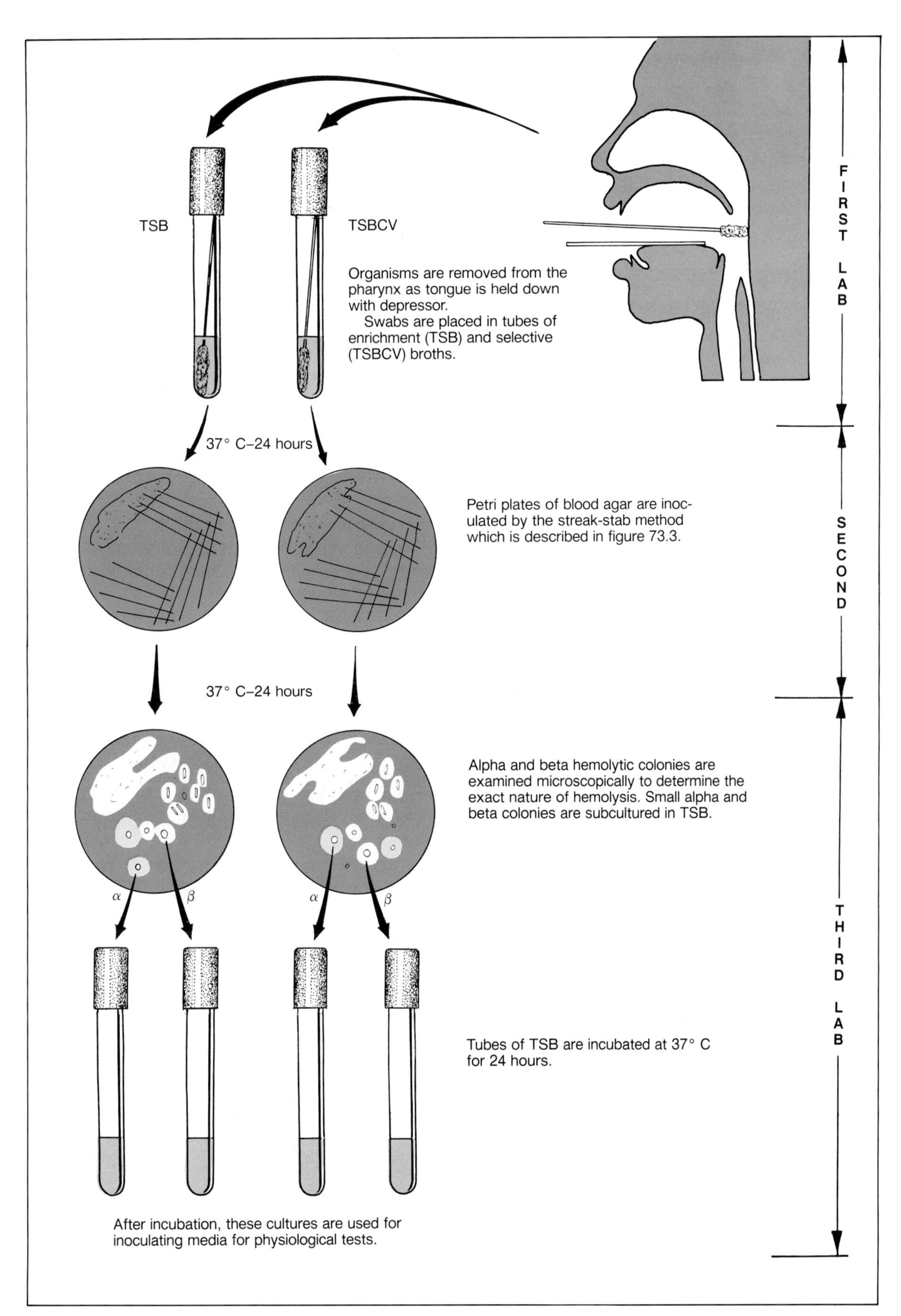

Figure 52.2 Isolation of streptococci from pharynx.

down with the swab. Also look for white patches in the tonsillar area. Avoid touching the cheeks and tongue.
3. Place the swab in one of the tubes of broth.
4. Repeat the procedure with the second swab for the other tube.

Second Period

Streptococcal hemolysis is most accurately determined when the colony develops deep in the medium below the surface. To achieve this type of anaerobic growth, it is necessary to inoculate a liquid agar medium to which blood is added, and to make a pour plate.

Since making pour plates for subsurface colonies is a rather tedious procedure as compared to streak plates, bacteriologists have discovered that a **streak-stab blood agar plate** can yield as accurate results, if interpreted properly. It is this technique that will be used here.

Materials:

tubes of TSB and TSBCV from previous period
2 blood agar plates

1. Label the bottoms of your blood agar plates with your name and type of medium (TSB or TSBCV) from which inoculations will be made.
2. Streak-stab each of the plates from the appropriate broth culture, using the technique described in figure 52.3. The essential steps are as follows:
   - Roll the swab over an area approximating one-fifth of the surface. The entire surface of the swab should contact the agar.
   - With a wire loop, streak out three areas as shown in figure 52.3 to thin out the organisms.
   - Stab the loop into the agar to the bottom of the plate at an angle perpendicular to the surface to make a clean cut without ragged edges.
   - Be sure to make one set of stabs in an unstreaked area so that streptococcal hemolysis will be easier to interpret with a microscope.
3. Incubate the plates aerobically at 37° C for 24 hours. Don't incubate longer than 24 hours.

Third Period

During this period, two things must be accomplished: first, the types of hemolysis must be correctly determined and, secondly, well-isolated colonies must be selected for making subcultures. Remember, if the subcultures are not pure cultures, subsequent procedures will fail.

Note in figure 52.2 that alpha and beta hemolytic colonies will be selected for subculturing.

Materials:

blood agar plates from previous period
tubes of TSB (one for each different type of colony)
dissecting microscope

1. Examine both of the blood agar plates, looking for both alpha and beta hemolysis. Do any of the stabs appear to exhibit hemolysis?

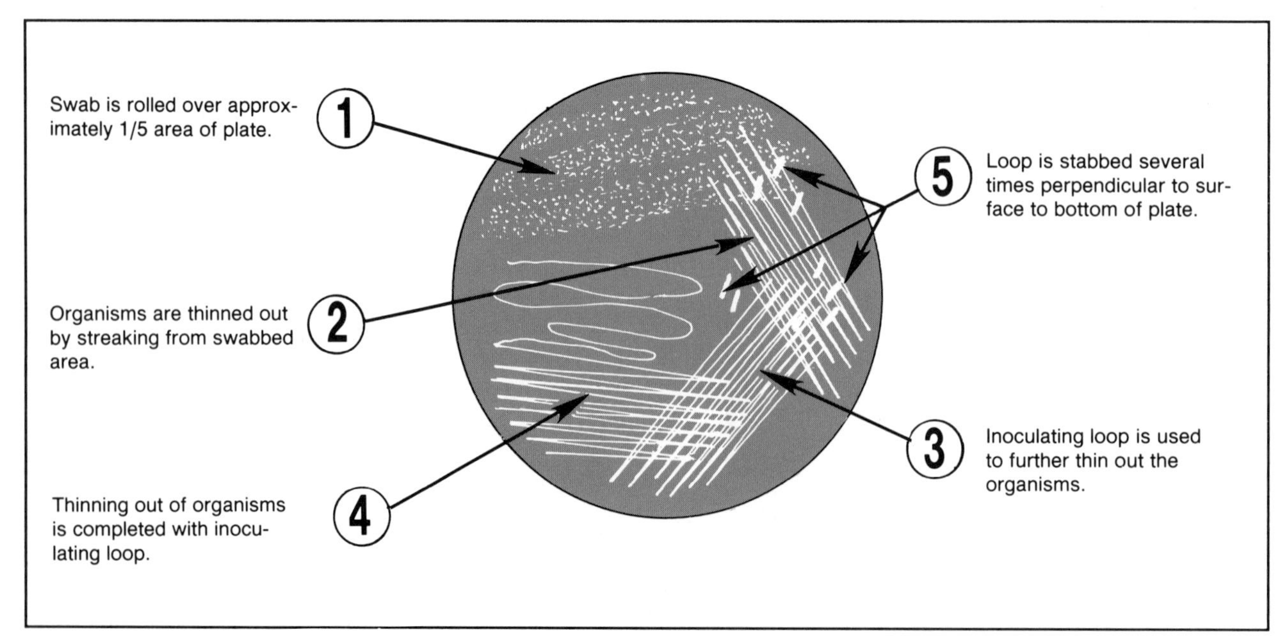

Figure 52.3 Streak-stab procedure for blood agar inoculations.

2. Examine the hemolytic zones near the stabs under 60× magnification with a dissecting microscope.

   Consult figure 52.4 to analyze the type of hemolysis, keeping in mind that this illustration is *highly diagrammatic.* The stabs on your plate will not look anything like the stabs on the right side of the illustration; the important thing to note is the presence or absence of RBCs in the hemolytic zones.

   Only those stabs that are completely free of red blood cells in the hemolytic area are considered to be **beta hemolytic.**

   If some red blood cells are seen dispersed throughout the hemolytic zone, the organism is said to be **alpha hemolytic.**

   A few organisms, such as viridans streptococci, will have RBCs near growth in the stab; these organisms are classified as being **alpha-prime hemolytic.**
3. Record your observations on the Laboratory Report.
4. Select well-isolated colonies that exhibit hemolysis (alpha, beta, or both) for inoculating tubes of TSB. Be sure to label the tubes "alpha" or "beta." Whether or not the organism is alpha or beta is crucial in identification.

   Since the chances of isolating beta hemolytic streptococci from the pharynx are usually quite slim, notify your instructor if you think you have isolated one.
5. Incubate the tubes at 37° C for 24 hours.
6. **Important:** At some time prior to the next laboratory session, review the material in appendix E that pertains to this exercise.

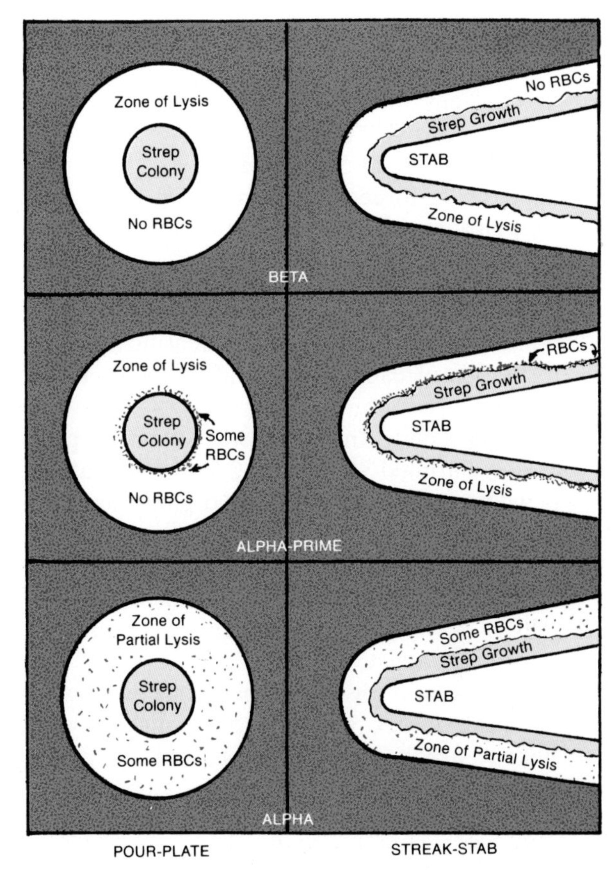

Figure 52.4 Comparison of hemolysis types as seen on pour-plates and streak-stab plates.

Inoculations for Physiological Tests

Presumptive identification of the various groups of streptococci is based on seven or eight physiological tests. Table 52.1 on page 199 reveals how they perform on these tests. Note that groups A, B, and C are all beta hemolytic; a few enterococci are also beta hemolytic. The remainder are all alpha hemolytic or nonhemolytic.

Since each of the physiological tests is specific for differentiating only two or three groups, it is not desirable to do all the tests on all unknowns. For economy and preciseness, only those tests that are recommended in the next column for each hemolytic type should be performed.

Before any inoculations are made, however, it is desirable to do a purity check on the TSB pharyngeal subcultures. To accomplish this it will be necessary to make a gram-stained slide of each of the TSB cultures.

In this laboratory period you will be issued two or three unknowns to be tested along with your pharyngeal isolates. The only information provided you about each unknown will be its hemolytic type.

Gram-Stained Slides

Make gram-stained slides of your pharyngeal isolates and examine them under oil immersion. Do they *appear* to be pure cultures? Draw the organisms in the appropriate circles on the Laboratory Report.

Beta Hemolytic Type Inoculations

Use the following media for each pharyngeal isolate or unknown that is beta hemolytic:

Materials:

1 blood agar plate
1 sodium hippurate broth
1 bacitracin differential disk
1 SXT sensitivity disk
1 broth culture of *S. aureus*
dispenser or forceps for transferring disks
TSB cultures of pharyngeal isolates
unknown broth cultures

1. Label a blood agar plate and a tube of sodium hippurate broth with proper identification information of each isolate and unknown to be tested.
2. For each isolate and unknown, follow the procedure outlined in figure 52.5 to inoculate a blood agar plate with the test organism and *S. aureus.*

   Note that a streak of the unknown is brought down perpendicular to the *S. aureus* streak, keeping the two organisms about one centimeter apart.
3. With forceps or dispenser, place one bacitracin differential disk and one SXT disk on the heavily streaked area at points shown in figure 52.5.
4. Inoculate the tubes of sodium hippurate broth for each isolate.
5. Incubate the blood agar plates at 37° C, aerobically, for 24 hours. The hippurate broth tubes should be incubated at 35° C, aerobically, for 24 hours. It may be necessary to reincubate the hippurate broths if tests prove to be negative or weakly positive.

Alpha Hemolytic Type Inoculations

Use the following media for inoculating each unknown or pharyngeal isolate that is alpha hemolytic.

Materials:

1 blood agar plate (for up to 4 unknowns)
1 6.5% sodium chloride broth
1 trypticase soy broth (TSB)
1 bile esculin (BE) slant
1 optochin (Taxo P) disk
candle jar setup

1. Mark the bottom of a blood agar plate to divide it into halves, thirds, or quarters, depending on the number of alpha hemolytic organisms to be tested. Label each space with the code number of each test organism.

   Completely streak over each area with the appropriate test organism, and place one optochin (Taxo P) disk in the center of each area. Press the disks slightly to enhance attachment.
2. Inoculate one tube each of TSB, BE, and 6.5% NaCl broth with each test organism.
3. Incubate all media at 35°–37° C as follows:
   Blood agar with Taxo P disks: 24 hours
   6.5% NaCl broths: 24, 48, and 72 hours
   Bile esculin slants: 48 hours
   Trypticase soy broths: 24 hours

   While the blood agar plates should be incubated in a candle jar or CO_2 incubator, the remaining cultures can be incubated aerobically.

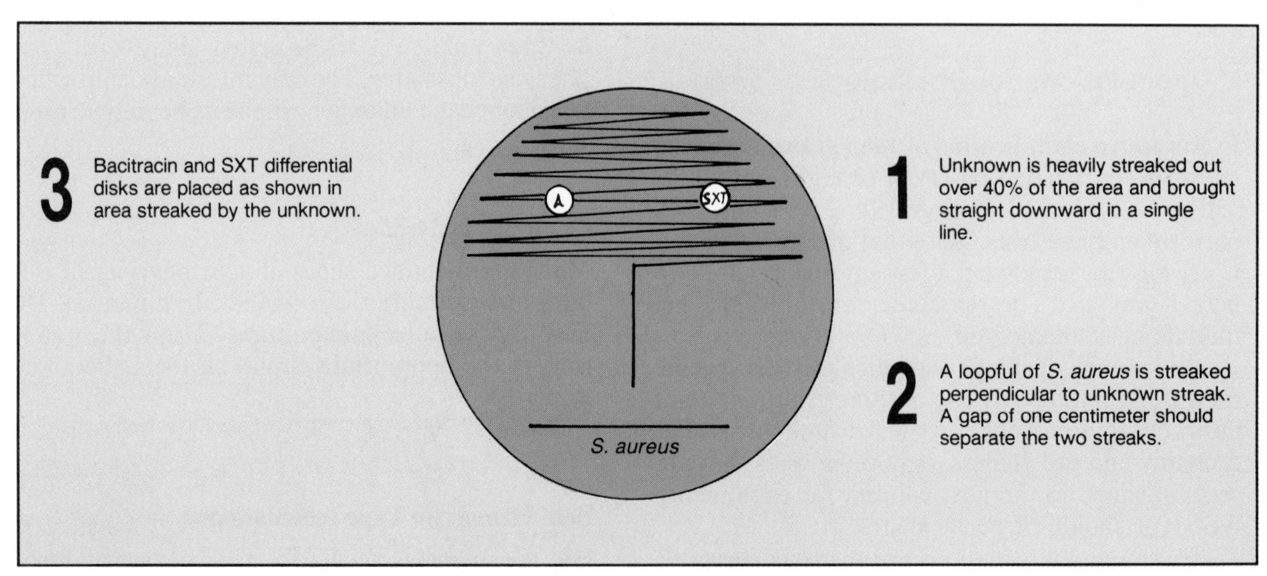

Figure 52.5 Blood agar inoculation technique for the CAMP, bacitracin, and SXT tests.

Evaluation of Physiological Tests

Once all the inoculated media have been incubated for 24 hours, you are ready to examine the plates and tubes and add test reagents to some of the cultures. Some of the tests will also have to be checked at 48 and 72 hours.

After you have assembled all the plates and tubes from the last period, examine the blood agar plates first that were double-streaked with the unknowns and *S. aureus*. **Note that the first three tests can be read from these plates.**

CAMP Reaction

If you have an unknown that produces an enlarged arrowhead-shaped hemolytic zone at the juncture where the unknown meets the *S. aureus* streak, as in figure 52.6, the organism is *S. agalactiae*. This phenomenon is due to what is called the *CAMP factor*. The only problem that can arise from this test is that if the plate is incubated anaerobically, a positive CAMP reaction can occur on *S. pyogenes*.

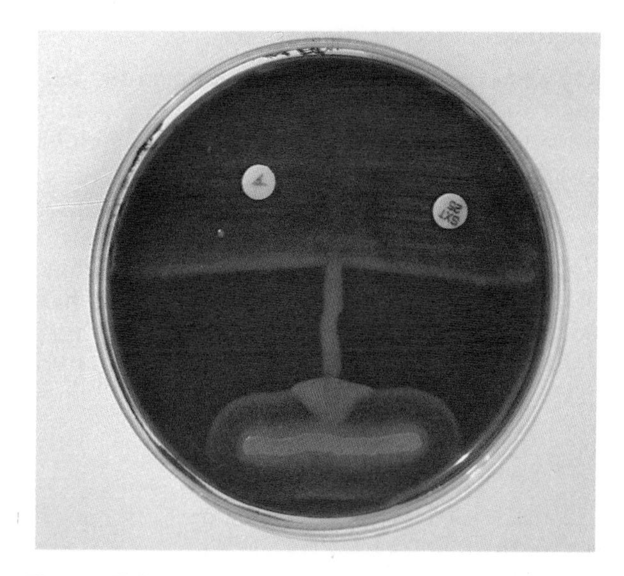

Figure 52.6 A positive CAMP test as revealed for *S. agalactiae.*

Record the CAMP reactions for each organism on the Laboratory Report.

Table 52.1 Physiological Tests for Streptococcal Differentiation

GROUP	Type of Hemolysis	Bacitracin Susceptibility	CAMP Reaction or Hippurate Hydrolysis	SXT	Bile Esculin Hydrolysis	Tolerance to 6.5% NaCl	Optochin Susceptibility	Bile Solubility	
Group A *S. pyogenes*	beta	+	−	R	−	−	−	−	
Group B *S. agalactiae*	beta	−*	+	R	−	±	−	−	
Group C *S. equi* *S. equisimilis* *S. zooepidemicus*	beta	−*	−	S	−	−	−	−	
Group D (enterococci) *S. faecalis* *S. faecium* etc.	alpha beta none	−	−	R	+	+	−	−	
Group D (nonenterococci) *S. bovis* etc.	alpha none	−	−	R/S	+	−	−	−	
Viridans *S. mitis* *S. salivarius* *S. mutans* etc.	alpha none	−*	−*	S	−	−	−	−	
Pneumococci *S. pneumoniae*	alpha	±	−		−	−	+	+	

*Exceptions occur occasionally

Bacitracin Susceptibility

Any size zone of inhibition seen around the bacitracin disks should be considered to be a positive test result. Record your results on the Laboratory Report.

This test has two limitations: (1) the disks must be of the *differential* type, not sensitivity type, and (2) the test should not be applied to alpha hemolytic streptococci. Reasons: sensitivity disks have too high a concentration of the antibiotic, and many alpha hemolytic streptococci are sensitive to these disks.

SXT Sensitivity

The disks used in this test contain 1.25 mg trimethoprim and 27.75 mg of sulfamethoxazole (SXT). The purpose of this test is to distinguish groups A and B from other beta hemolytic streptococci. Note in table 52.1 that both groups A and B are uniformly resistant to SXT.

If a beta hemolytic streptococcus proves to be bacitracin-resistant and SXT-susceptible, it is classified as being a **non-group-A or -B beta hemolytic** streptococcus. This means that the organism is probably a species within group C. Keep in mind that an occasional group A streptococcal strain is susceptible to both bacitracin and SXT disks. One must always remember that *exceptions to most tests do occur;* that is why this identification procedure leads us only to *presumptive* conclusions.

Record any zone of inhibition as positive for this test.

Hippurate Hydrolysis

Note in table 52.1 that hippurate hydrolysis and the CAMP test are grouped together as positive tests for *S. agalactiae.* If an organism is positive for both tests, or either one, one can assume with almost 100% certainty that the organism is *S. agalactiae.*

Proceed as follows to determine which of your isolates are able to hydrolyze sodium hippurate:

Materials:

serological test tubes
serological pipettes (1 ml size)
ferric chloride reagent
centrifuge

1. Centrifuge the culture for three to five minutes.
2. Pipette 0.2 ml of the supernatant and 0.8 ml of ferric chloride reagent into an empty serological test tube. Mix well.
3. Look for a **heavy precipitate** to form. If the precipitate forms and persists for 10 minutes or longer, the test is positive. If the culture proves to be weakly positive, incubate the culture for another 24 hours and repeat the test.
4. Record your results on the Laboratory Report.

Bile Esculin Hydrolysis

This is the best physiological test that we have for the identification of group D streptococci. Both enterococcal and nonenterococcal species of group D are able to hydrolyze esculin in the agar slant, causing the slant to blacken.

A positive BE test tells us that we have a group D streptococcus; differentiation of the two types of group D streptococci depends on the salt tolerance test.

Examine the BE agar slants, looking for **blackening of the slant,** as illustrated in figure 52.7. If less than half of the slant is blackened, or no blackening occurs within 24 to 48 hours, the test is negative.

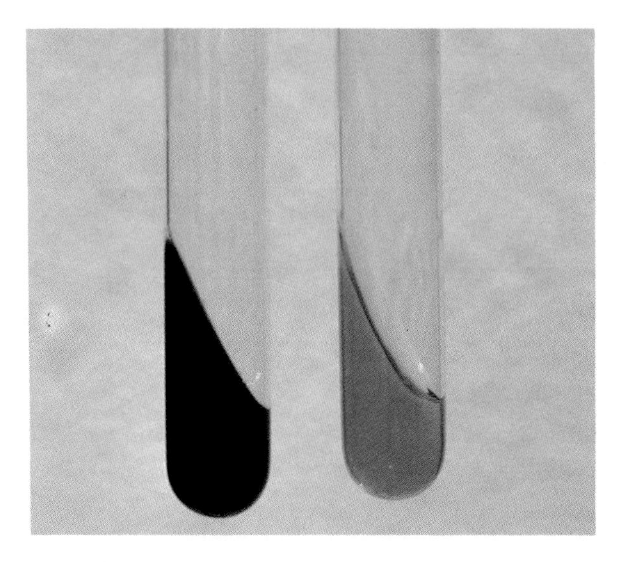

Figure 52.7 Positive bile esculin hydrolysis on left; negative on right.

Salt Tolerance (6.5% NaCl)

All enterococci of group D produce heavy growth in 6.5% NaCl broth. As indicated in table 52.1, none of the nonenterococci, group D, grow in this medium. This test, then, provides us with a good method for differentiating the two types of group D streptococci.

A positive result shows up as **turbidity** within 72 hours. A color change of purple to yellow *may* also be present. If the tube is negative at 24 hours, incubate it and check it again at 48 and 72 hours. *If the organism is salt tolerant and BE-positive,* it is considered to be an enterococcus.

Parenthetically, it should be added here that approximately 80% of group B streptococci will grow in this medium.

Optochin

Optochin susceptibility is used for differentiation of the alpha hemolytic viridans streptococci from the pneumococci. The pneumococci are sensitive to these disks; the viridans organisms are resistant.

Materials:

blood agar plates with optochin disks
plastic metric ruler

1. Measure the diameters of zones of inhibition that surround each disk, evaluating whether or not the zones are large enough to be considered positive. The standards are as follows:

   For 6 mm diameter disks, the zone must be at least 14 mm diameter to be considered positive.

   For 10 mm diameter disks, the zone must be at least 16 mm diameter to be considered positive.
2. Record your results on the Laboratory Report.

Bile Solubility

If a streptococcal organism is soluble in bile and positive on the optochin test, presumptive evidence indicates that the isolate is *S. pneumoniae.* Perform the bile solubility test on each of your alpha hemolytic isolates as follows:

Materials:

2 empty serological tubes (per test)
dropping bottle of phenol red indicator
dropping bottle of 1/20N NaOH
TSB culture of unknown
2% bile solution (sodium desoxycholate)
bottle of normal saline solution
2 serological pipettes (1 ml size)
water bath (37° C)

1. Mark one empty serological tube "bile" and the other "saline." Into their respective tubes, pipette 0.5 ml of 2% bile and 0.5 ml of saline.
2. Shake the TSB unknown culture to suspend the organisms and pipette 0.5 ml of the culture into each tube.
3. Add one or two drops of phenol red indicator to each tube and adjust the pH to 7.0 by adding drops of 1/20N NaOH.
4. Place both tubes in a 37° C. water bath and examine periodically for two hours. If the turbidity clears in the bile tube, it indicates that the cells have disintegrated and the organism is *S. pneumoniae.* Compare the tubes side-by-side.
5. Record your results on the Laboratory Report.

Final Confirmation

All the laboratory procedures performed so far lead us to *presumptive identification.* To confirm these conclusions it is necessary to perform serological tests on each of the unknowns. If commercial kits are available for such tests, they should be used to complete the identification procedures.

Laboratory Report

Complete the Laboratory Report for this exercise.

53 Gram-Negative Intestinal Pathogens

The enteric pathogens of prime medical concern are the **salmonella** and **shigella.** They cause enteric fevers, food poisoning, and bacillary dysentery. *Salmonella typhi,* which causes typhoid fever, is by far the most significant pathogen of the salmonella group. In addition to the typhoid organism, there are ten other distinct salmonella species and over 300 serotypes. The shigella, which are the prime causes of human dysentery, constitute four species and many serotypes. *Serotypes* within genera are organisms of similar biochemical characteristics that can be most easily differentiated by serological and phage typing.

Routine testing for the presence of these pathogens is a function of public health laboratories at various governmental levels. The isolation of these pathogenic enterics from feces is complicated by the fact that the colon contains a diverse population of bacteria. Species of such genera as *Escherichia, Proteus, Enterobacter, Pseudomonas,* and *Clostridium* exist in large numbers; hence it is necessary to use media that are selective and differential to favor the growth of the pathogens.

Figure 53.1 is a separation outline that is the basis for the series of tests that are used to demonstrate the presence of salmonella or shigella in a patient's blood, urine, or feces. Note that lactose fermentation separates the salmonella and shigella from most of the other Enterobacteriaceae. Final differentiation of the two enteric pathogens from *Proteus* relies on motility, hydrogen sulfide production, and urea hydrolyosis. The differentiation information of the positive lactose fermenters on the left side of the separation outline is provided here mainly for comparative references that can be used for the identification of other unknown enterics. See figures 41.2 and 41.3 on pages 153 and 154.

In this exercise you will be given a mixed culture containing a coliform, *Proteus,* and a salmonella or shigella. The pathogens will be of the less dangerous types, but their presence will, naturally, demand utmost caution in handling. Your problem will be to isolate the pathogen from the mixed culture and make a genus identification. There are five

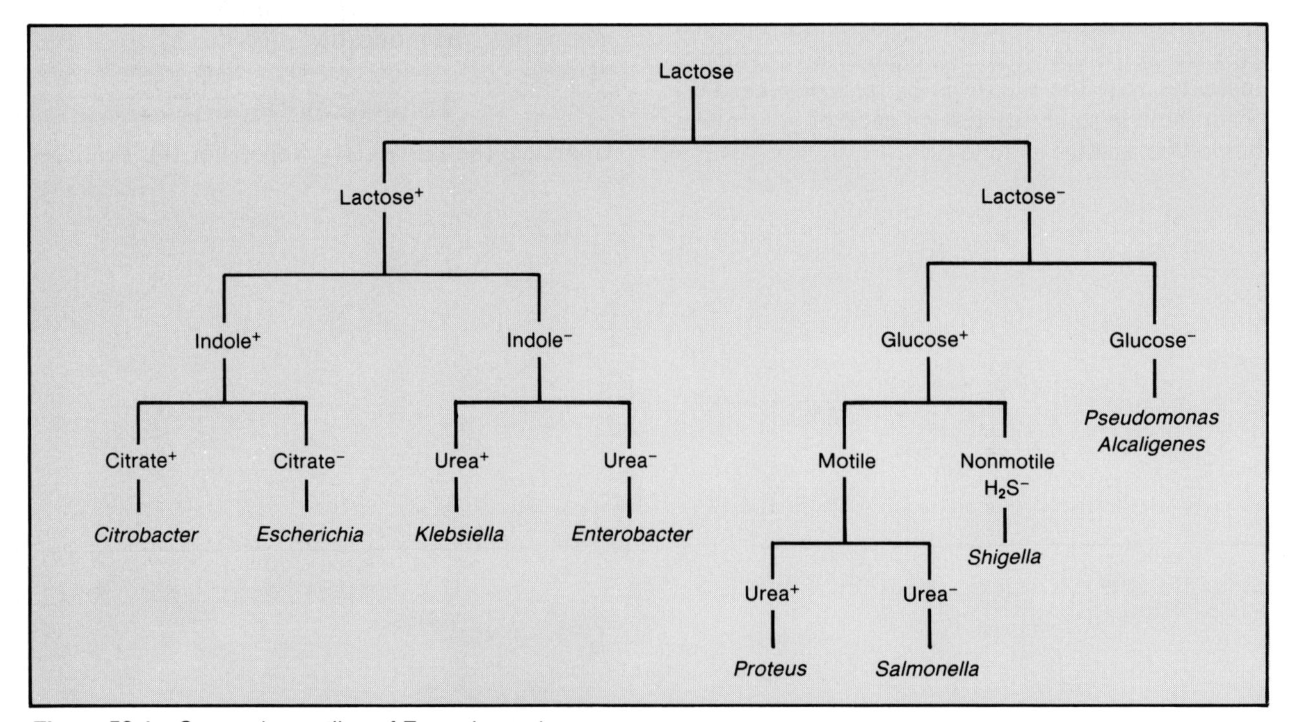

Figure 53.1 Separation outline of Enterobacteriaceae.

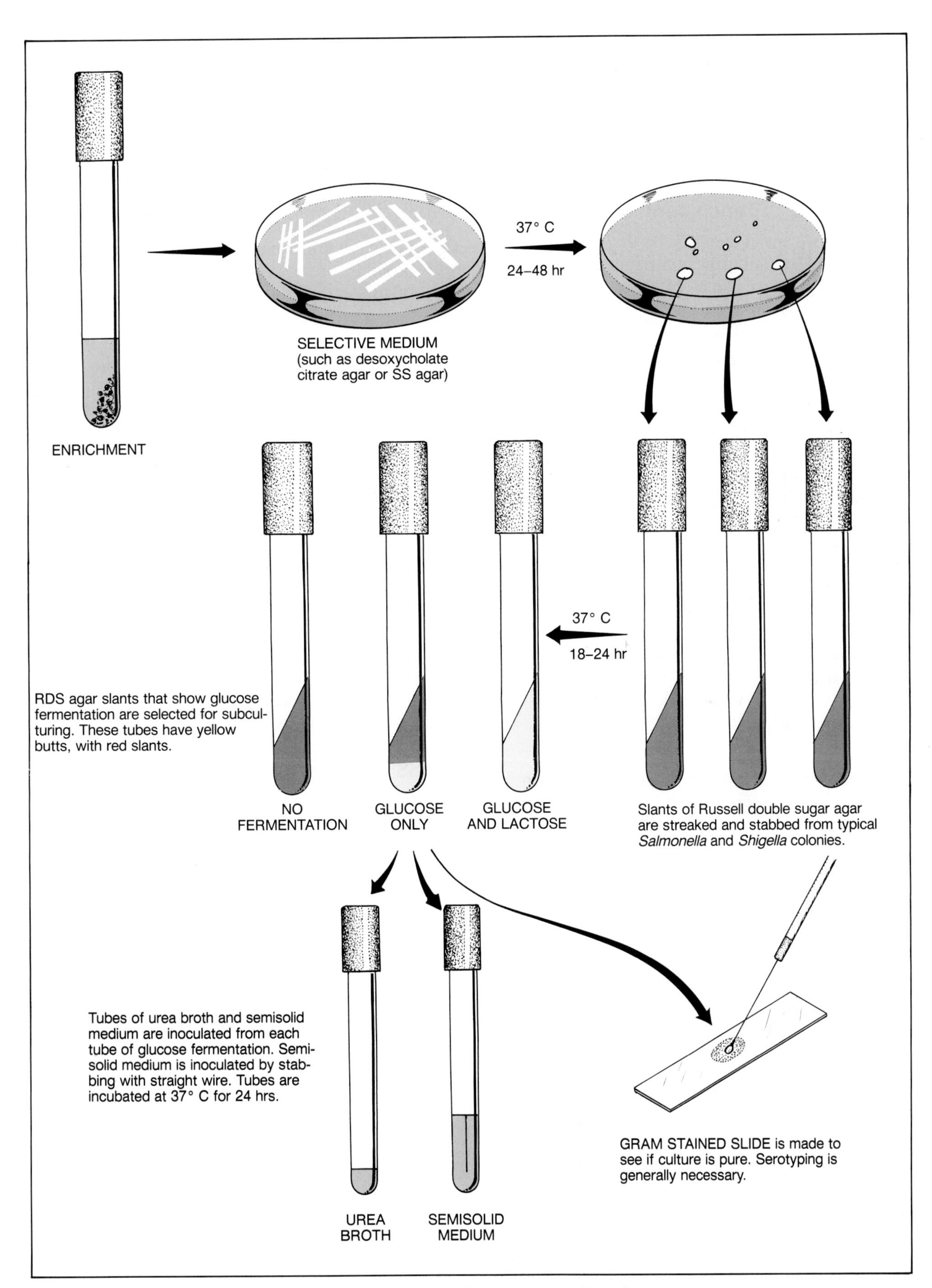

Figure 53.2 Isolation and presumptive identification of *Salmonella* and *Shigella*.

steps that are used to prove the presence of these pathogens in a stool sample: enrichment, isolation, fermentation tests, final tests, and serotyping. Figure 53.2 illustrates the general procedure.

Enrichment

There are two enrichment media that are most frequently used to inhibit the nonpathogens and favor the growth of pathogenic enterics. They are *selenite F* and *gram-negative (GN) broths.* While most salmonella grow unrestricted in these media, some of the shigella are inhibited; because of this, stool samples are usually plated directly on isolation media as well.

In actual practice, 1 to 5 grams of feces are placed in 10 ml of enrichment broth. In addition, plates of various kinds of selective media are inoculated directly. The broths are incubated for 12 to 18 hours.

Since we are not using stool samples in this exercise, the enrichment procedure is omitted. You will streak the isolation media directly from the unknown broth.

Isolation
(First Period)

There are several excellent selective differential media that have been developed for the isolation of these pathogens. Various inhibiting agents such as brilliant green, bismuth sulfite, sodium desoxycholate, and sodium citrate are included in them. For *Salmonella typhi,* bismuth sulfite agar appears to be the best medium. Colonies of *S. typhi* on this medium appear black due to the reduction of sulfite to sulfide.

Other widely used media are EMB, desoxycholate citrate (DCA), and salmonella-shigella (SS) agars. The bile salt, sodium desoxycholate, in DCA and SS agar inhibits gram-positive bacteria. The sodium citrate in these media inhibits coliforms and other gram-negative nonenterics. Both DCA and SS media contain lactose and neutral red, which causes the colonies of coliforms to appear red. The salmonella form large white opaque colonies on these media; the shigella produce ground-glass-appearing colonies with even margins.

After an enrichment culture has been incubated for 12 to 18 hours, one or more plates of these selective media are streaked. Proceed as follows to inoculate the selective media with your unknown:

Materials:

unknown culture (mixture of a coliform, *Proteus,* and a salmonella or shigella)
plate of each type of selective medium (DCA, EMB, and/or SS agar)

1. Label each plate with your name and unknown number.
2. With a loop, streak each plate with your unknown in a manner that will produce good isolation.
3. Incubate at 37° C for 24 to 48 hours.

Fermentation Tests
(Second Period)

The fermentation characteristic that separates the SS pathogens from the coliforms is their inability to ferment lactose. Once we have isolated colonies on differential media that look like salmonella or shigella, the next step is to determine whether or not the isolates ferment lactose. All media for this purpose contain at least two sugars, glucose and lactose. Some contain a third sugar, sucrose. They also contain phenol red to indicate when fermentation occurs. Russell double sugar agar (RDS) is one of the simpler media that works very well. It is the one we will use in this exercise.

Proceed as follows to inoculate three RDS slants from colonies that look like either salmonella or shigella. The reason for using three slants is that you may have difficulty distinguishing *Proteus* from the SS pathogens. By inoculating three tubes instead of only one, your chances of success are greater.

Materials:

3 RDS slants
streak plates of unknown

1. Label the three slants with your name and the number of your unknown.
2. Look for isolated colonies that look like salmonella or shigella. Remember that coliforms will appear pink-red on DCA or SS agar, and nucleated on EMB agar. If bismuth sulfite agar is used, the salmonella colonies usually appear black.
3. With a straight wire, inoculate the three RDS agar slants from separate SS-appearing colonies. Use the streak-stab technique. When streaking the surface of the slant before stabbing, move the wire over the entire surface for good coverage.
4. Incubate the slants at 37° C for 18 to 24 hours. Longer incubation time may cause *alkaline*

reversion. Even refrigeration beyond this time may cause reversion.

5. After incubation, examine the slants and *select those that have a yellow butt with a red slant.* These tubes contain organisms that ferment only glucose. If the slants are completely yellow, the organism is a coliform that ferments both lactose and glucose. If both the butt and slant are red, no fermentation has taken place. If tubes of glucose fermenters are incubated for 48 hours or longer, however, they often become completely red, losing the yellow butt due to alkaline reversion. That is why short incubation is desirable.

Final Tests
(Third Period)

Examination of the separation outline, figure 53.1, reveals that the next step in the differentiation of the SS pathogens is to determine whether or not the organisms are motile and produce hydrogen sulfide or urease. Two media will be used to complete this differentiation: semisolid medium and urea broth. Semisolid medium is used to determine the presence or absence of motility and hydrogen sulfide production. The urea broth determines whether or not urease is produced.

Materials (per RDS slant):

- 1 tube of semisolid medium (2 ml in serological tube)
- 1 tube of urea broth (0.5 ml in serological tube)

1. With a loop, inoculate one tube of urea broth from each RDS agar slant that is glucose-positive, lactose-negative.
2. With a straight wire, stab one tube of semisolid medium from each of the same RDS agar slants. Stab in the center to one-half of depth.
3. Incubate at 37° C for 24 to 48 hours.
4. After incubation, evaluate your cultures according to the following criteria:
   Semisolid medium: Motility will be evidenced as a spreading of growth through the medium from the line of inoculation, causing cloudiness. Hydrogen sulfide production is evidenced by a black precipitate.
   Urea broth: If the medium changes from yellow to red or cerise color, the organism produces urease.

   Record your results on the Laboratory Report.

Note: It should be kept in mind that biochemical testing of salmonella and shigella for species identification goes far beyond these tests. Carbohydrates such as xylose, maltose, rhamose, mannitol, dulcitol, and salicin are used in conjunction with indole, citrate, and other differential tests.

Miniaturized Multitests: If API 20E or Enterotube II tests are available to you, confirm your results with one of them. Consult Exercises 42 or 43.

Serotyping

In conjunction with biochemical testing, it is desirable to use serological tests to determine the antigens present in the organism. This is routine procedure in clinical and public health laboratories. The method is outlined in Exercise 55 and may be performed here as a means of confirmation of the biochemical results.

54 Urinary Tract Pathogens

Chronic or acute infections of the urinary tract may involve the kidneys, ureters, bladder, or urethra. Such infections may cause high blood pressure, kidney damage, uremia, or death. In some instances the infections are inapparent and may go unnoticed for some time. Most infections of this tract enter by way of the urethra; very few originate in the blood.

A multitude of organisms can cause urinary infections. The most common cause of such infections in women of childbearing age is *Escherichia coli.* In order of frequency after *E. coli,* the following are implicated: (1) *Proteus* spp., (2) *Pseudomonas aeruginosa* and other pseudomonas, (3) enterococci, (4) *Candida albicans,* (5) hemolytic streptococci, (6) *Staphylococcus aureus,* (7) *Mycobacterium tuberculosis,* and (8) salmonella and shigella.

The importance of performing microbial analyses of urine on patients with urinary infections cannot be overemphasized. While some physicians tend to treat patients with antimicrobics and watch only for symptomatic improvement without performing follow-up urinary tests, this practice is not reliable. Clinical testing of urine 48 to 78 hours after the start of chemotherapy should be performed to evaluate the effectiveness of the therapy. If the antimicrobics are effective, the urine will be free of bacteria at this time.

The sequence of steps in performing a complete study of microorganisms in urine includes (1) aseptic collection of specimen, (2) quantitative evaluation, (3) isolation of the pathogen, (4) identification, and (5) antimicrobic sensitivity testing.

Collection of Urine

The presence of bacteria in urine does not necessarily indicate a urinary tract infection since many bacteria exist within the urethra and near its external orifice. Catheterization is undesirable because it may dislodge bacteria within the urethra or even cause infection. Meaningful results, however, can be obtained with midstream voided specimens collected in sterile containers. For best results, the external genitalia should be cleansed with a liquid soap containing hexachlorophene. Even with midstream samples we still may expect to find the following organisms in normal urine: coagulase-negative staphylococci, diphtheroid bacilli, enterococci, *Proteus,* hemolytic streptococci, yeasts, and aerobic gram-positive spore-forming rods.

Specimens are most reliable when plated out immediately after collection. If bacterial tests cannot be performed immediately, refrigeration is mandatory. It must be kept in mind that urine is an ideal growth medium for many bacteria. Specimens not promptly refrigerated should be considered unsatisfactory for study.

Quantitative Evaluation

When it can be demonstrated that urine contains 100,000 organisms or more per ml in a clean-voided midstream sample, *significant bacteriuria* exists. Counts of between 1000 and 100,000 per ml of a single species represent possible or probable infections and should be repeated.

There are two ways that one can use to determine whether or not significant bacteriuria exists. One method is to do a plate count. The other method is to do a microscopic study of the fresh sample before centrifugation. It is customary to utilize both methods simultaneously.

Plate Count Method

Prepare two pour plates according to the routine shown in figure 54.1.

Materials:

First period:
- 1 sterile empty shake bottle
- 2 petri plates
- 2 trypticase soy agar pours
- 1 99-ml sterile water blank
- 2 1.1-ml dilution pipettes (with rubber bulbs or mechanical suction device)

Second period:
- Quebec colony counter
- mechanical hand counter

1. Liquefy two TSA pours and cool to 50° C.
2. Label one petri plate 1:100 and the other 1:1000.
3. Pour the urine into an empty sterile shake bottle, cap it tightly, and shake 25 times, as in figure 19.4, page 81.
4. Transfer 1 ml of the mixed urine to a 99-ml sterile water blank. Use a pipette with a rubber bulb or mechanical suction device.
5. Mix the water blank with 25 shakes and, with a fresh pipette, transfer 0.1 ml to the 1:1000 plate and 1.0 ml to the 1:100 plate.
6. Empty the tubes of TSA into the plates, swirl them, and let stand to cool.
7. Place the plates in a 37° C incubator for 24 hours.
8. After incubation, select the plate that contains between 30 and 300 colonies. Count all colonies on a Quebec colony counter, using a mechanical hand counter to tally.

Microscopic Examination

As soon as the plates have been poured, prepare a wet mount and a gram-stained slide of a well-mixed sample of the urine. If a phase microscope is available, use it for the wet mount. The presence of yeasts will show up well on such a preparation.

Examine the gram-stained slide under oil immersion. The presence of any bacteria on the slide is good presumptive evidence that more than 100,000 organisms per ml exist in the sample.

Isolation Procedure

Figure 54.2 reveals the procedure for isolation of a urinary pathogen. Note that the sediment from a centrifuged sample is used to inoculate a tube of thioglycollate medium and three plates of special media. Two slides also are made from the sediment. The three media used in the plates will provide adequate selectivity to separate out most of the different types of organisms that might be present.

Materials:

1 sterile centrifuge tube (with screw cap)
1 tube of thioglycollate medium-135C (BBL)
1 petri plate of blood agar (TSA base)
1 petri plate of desoxycholate lactose agar (DLA)
1 petri plate of phenylethyl alcohol medium with blood (PEA-B)

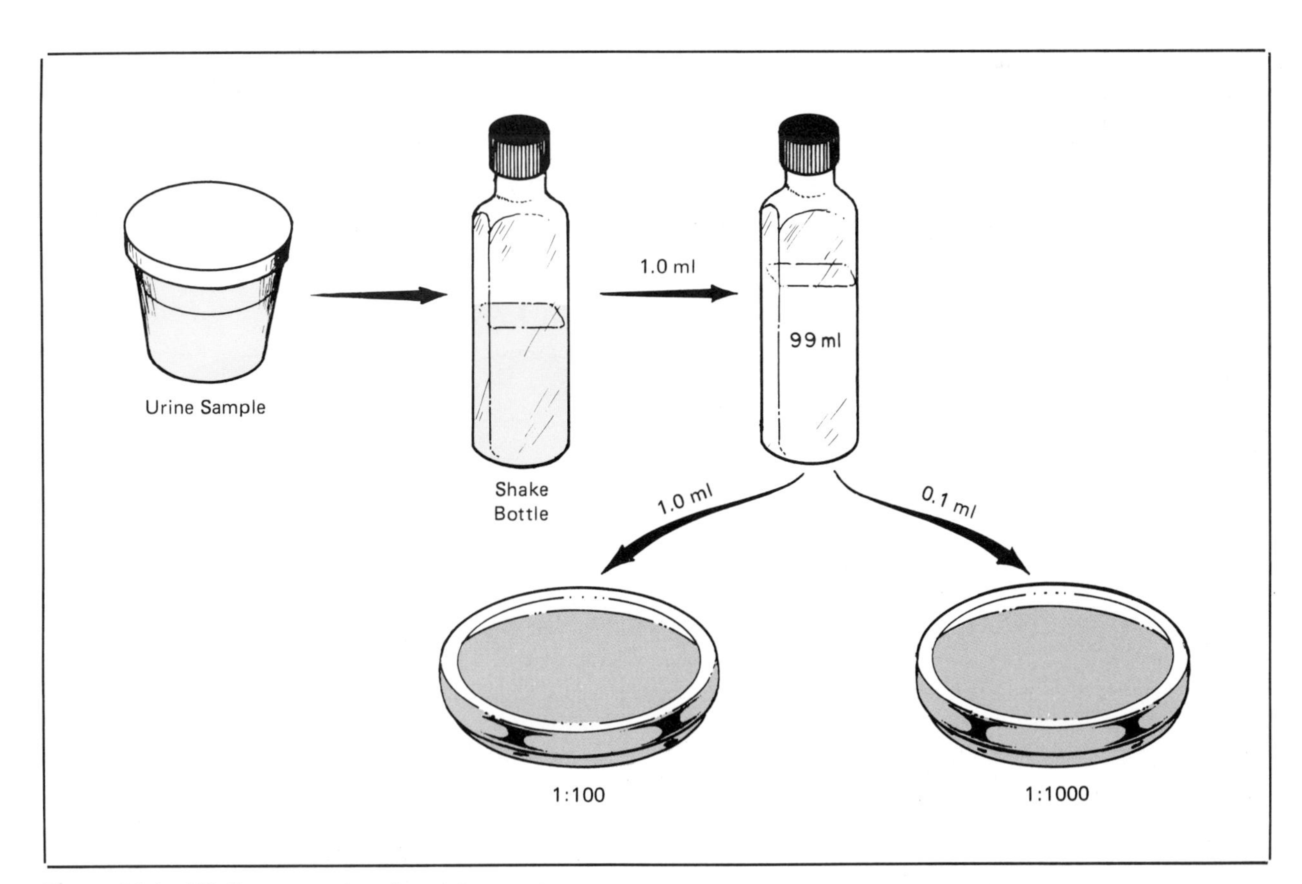

Figure 54.1 Dilution procedure for plate count.

1. Shake the sample to resuspend the organisms and decant 10 ml into a centrifuge tube. Keep tube capped.
2. Centrifuge for 10 minutes at 2000 rpm.
3. Decant all but 0.5 ml from the tube and resuspend the sediment with a sterile wire loop.
4. Inoculate a tube of thioglycollate medium with a loopful of the sediment.
5. Streak out a loopful of the sediment on each of the three agar plates (blood agar, DLA, and PEA-B). Use a good isolation technique.
6. Incubate the thioglycollate tube and three plates at 37° C for 18 to 24 hours.

Microscopic Examination

Although slides were made prior to centrifugation, new slides now must be made of centrifuged material to achieve greater concentration for evaluation. Make both wet mount and gram-stained slides. Examine the wet mount under high-dry, preferably with phase optics. Look for casts, pus cells, and other elements. Refer to figure 54.3 for help in identifying objects that are present. Normal urine will contain an occasional leukocyte, some epithelial cells, mucus, bacteria, and crystals of various kinds.

Examine the gram-stained slide under oil immersion. Determine the morphology and staining reaction of the predominant organism.

Presumptive Identification

Analyses of the three plates and thioglycollate medium after 24 hours of incubation should enable one to make a presumptive determination of

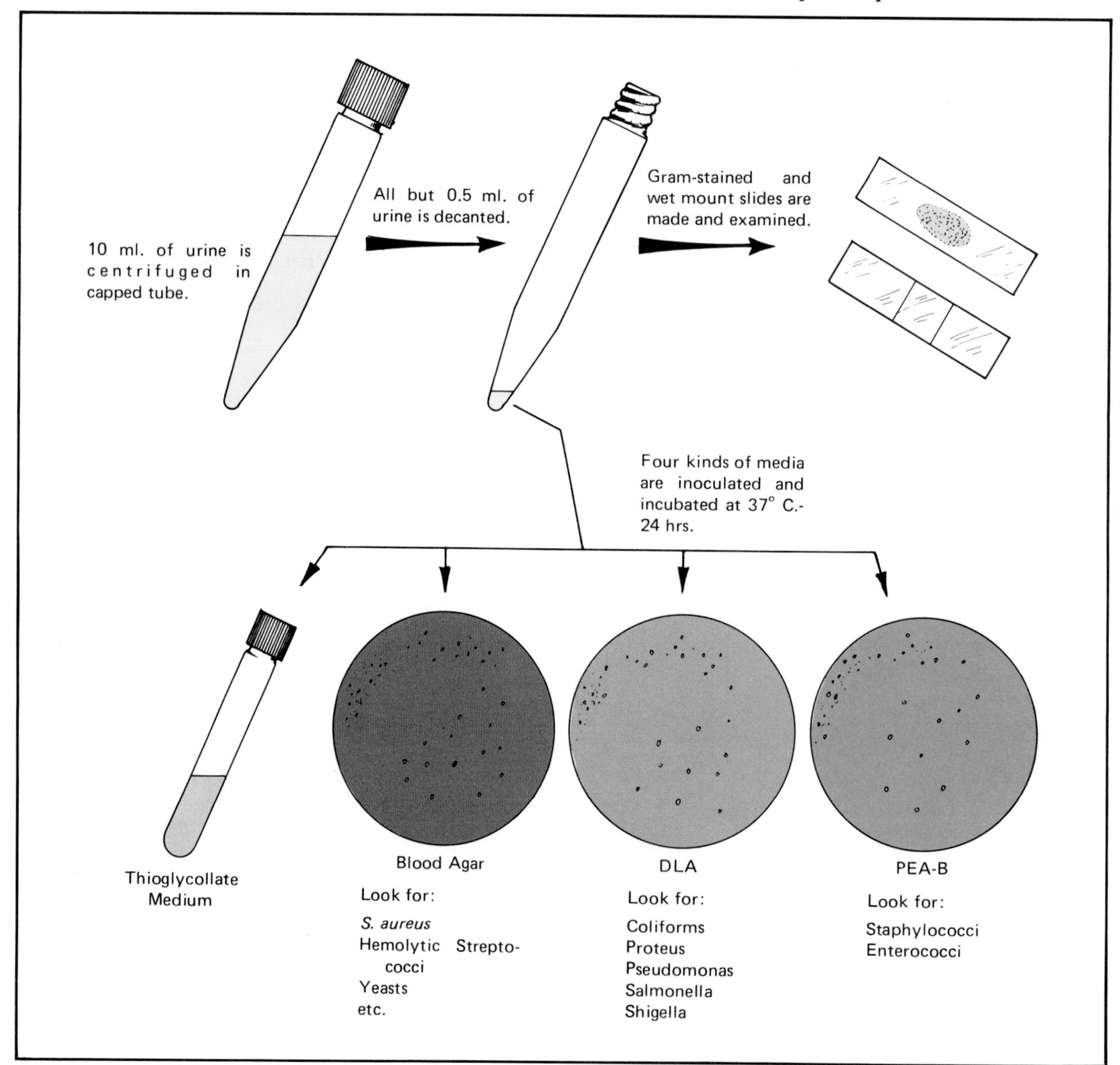

Figure 54.2 Procedure for presumptive identification of urinary microorganisms.

the cause of the urinary infection. If the DLA or PEA-B plates show only scanty growth, reincubate for another 24 hours. A discussion of the composition and functions of each of the media follows. Additional tests also are suggested.

Fluid Thioglycollate Medium

This medium was inoculated to promote growth from urine samples that have low bacterial counts or contain fastidious organisms. In the event that none of the plates produce colonies from the urine of the patient known to have a urinary infection, then this tube will be very useful.

Blood Agar

Practically all pathogens of the urinary tract will grow on this medium. This includes the coliforms, *Proteus, Pseudomonas, Candida,* staphylococci, streptococci, etc. Subculturing from this plate to trypticase soy broth can provide pure cultures of the pathogen for physiological testing or antimicrobic sensitivity testing. The presence or absence of hemolytic activity also can be determined at this time. If the pathogen appears to be a hemolytic, gram-positive coccus, one should follow the procedures outlined in Exercises 51 and 52 for identification.

Desoxycholate Lactose Agar

The presence of sodium desoxycholate and sodium citrate in this medium is inhibitory to gram-positive bacteria. Coliforms and other gram-negative bacteria grow well on it. If the predominant organism is gram-negative, some differentiation may be made at this point. If the colonies on this medium are flat and rose-red in color, the organism is *E. coli. Pseudomonas* and *Proteus,* which do not ferment the lactose in the medium, produce white colonies. *Proteus* can be confirmed with the

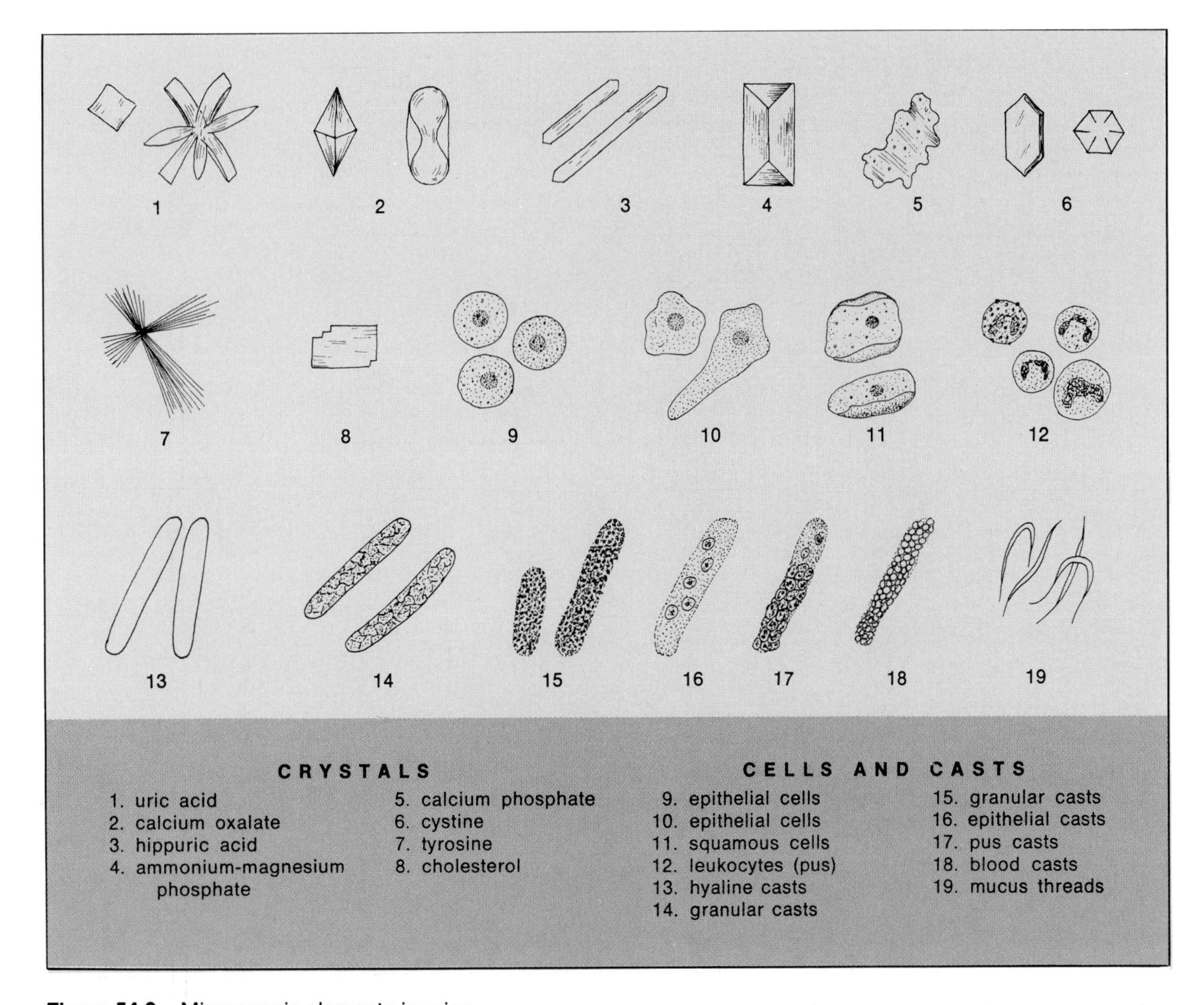

Figure 54.3 Microscopic elements in urine.

urease test, being positive for urease production. *Pseudomonas* species give a positive reaction with Taxo N disks. Fermentation and additional physiological testing may be necessary for species identification. Exercises 38, 39, and 53 should be consulted for further testing.

Phenylethyl Alcohol Medium

This medium, to which blood has been added, is highly inhibitory to gram-negative organisms. *Proteus,* in particular, is prevented from growing on it. If considerable growth occurs on the DLA plate, and very little or no growth occurs here, then one can assume that the disease is due to a gram-negative organism. This, of course, would be confirmed by the findings on the gram-stained slide. The findings on this plate should be correlated with those on blood agar. If enterococci (*S. faecalis*) are suspected, a plate of Mead agar should be streaked and incubated at 37° C. Enterococci produce pink colonies on this medium.

Laboratory Report

Record all findings on the Laboratory Report and answer all questions.

Slide Agglutination Test: Serological Typing 55

Organisms of different species not only differ morphologically and physiologically, but they also differ in protein makeup. The different proteins of bacterial cells that are able to stimulate antibody production when injected into an animal are **antigens.** The antigenic structure of each species of bacteria is unique to that species and, like the fingerprint of an individual, can be used to identify the organism. Many closely related microorganisms that are identical physiologically can be differentiated only by determining their antigenic nature. The method of determining the presence of specific antigens in a microorganism is called **serological typing** (serotyping). It consists of adding a suspension of the organisms to **antiserum,** which contains antibodies that are specific for the known antigens. If the antigens are present, the antibodies in the antiserum will combine with the antigens, causing **agglutination,** or clumping, of the bacterial cells. Serotyping is particularly useful in the identification of various organisms that cause salmonella and shigella infections. In the identification of the various serotypes of these two genera, the use of antisera is generally performed after basic biochemical tests have been utilized as in Exercise 53.

In this exercise you will be issued two unknown organisms, one of which is a salmonella. By following the procedure in figure 55.1, you will determine which one of the unknowns is a salmonella. Note that you will use two test controls. A **negative test control** will be set up in depression A on the slide to see what the absence of agglutination looks like. The negative control is a mixture of antigen and saline (antibody is lacking). A **positive test control** will be performed in depression C with standardized antigen and antiserum to give you a typical reaction of agglutination.

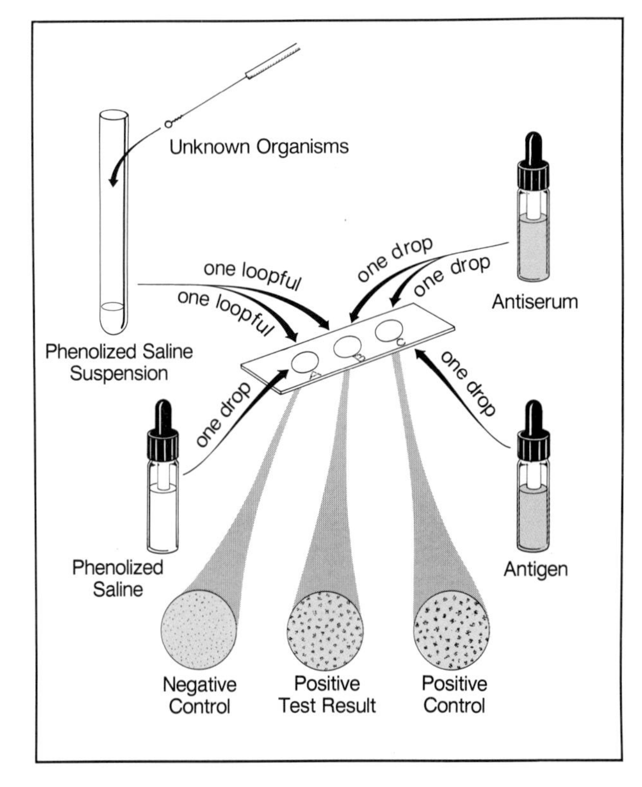

Figure 55.1 Slide agglutination technique.

Materials:

two numbered unknowns per student (slant cultures of a salmonella and a coliform)
salmonella O antigen, group B (Difco #2840–56)
salmonella O antiserum, poly A-I (Difco #2264–47)
depression slides or spot plates
dropping bottle of phenolized saline solution (0.85% sodium chloride, 0.5% phenol)
2 serological tubes per student
1-ml pipettes

1. Label three depressions on a spot plate or depression slide A, B, and C, as shown in figure 55.1.
2. Make a phenolized saline suspension of each unknown in separate serological tubes by suspending one or more loopfuls of organisms in one ml of phenolized saline. Mix the organisms sufficiently to ensure complete dispersion of clumps of bacteria. The mixture should be very turbid.
3. Transfer one loopful (0.05 ml) from the phenolized saline suspension of one tube to depressions A and B.
4. To depressions B and C add one drop of salmonella O polyvalent antiserum. To depres-

sion A, add one drop of phenolized saline, and to depression C, add one drop of salmonella O antigen, group B. Mix the organisms in each depression with a clean wire loop. Do not go from one depression to the other without washing the loop first. Compare the three mixtures.

Agglutination should occur in depression C (positive control) but not in depression A (negative control). If agglutination occurs in depression B, the organism is a salmonella.

5. Repeat this process on another slide for the other organism.
6. **Caution:** Deposit all slides and serological tubes in the container of disinfectant provided by the instructor.
7. Record your results on the Laboratory Report.

Tube Agglutination Test: The Widal Test 56

A tube test for determining the quantity of agglutinating antibodies, or **agglutinins,** in the serum of a patient with typhoid fever was described by Grunbaum and Widal in 1896. This technique is still in use today and has been adapted to many other diseases as well. The procedure involves adding a suspension of dead typhoid cells to a series of tubes containing the patient's serum, which has been diluted out to various concentrations. After the tubes have been incubated for 30 minutes at 37° C and have been centrifuged, they are examined to note the amount of agglutination that has occurred. The highest dilution at which agglutination is seen is designated as the **antibody titer** of the patient's serum. This is expressed at 1:40, 1:80, or whatever the dilution might be. If the titer is, say, 1:320, this is interpreted as 320 antibody units per milliliter of the patient's serum. Naturally, the higher the titer, the greater is the antibody response of the individual to the disease. The technique can be used clinically to determine whether or not a patient with typhoidlike symptoms actually has the disease. If successive daily tests on the patient's serum reveal no antibody titer, or a lower titer which does not increase from day to day, it can be assumed that some other disease is present. A daily increase in the titer, on the other hand, would be indicative of typhoid infection.

In this exercise you will be given a sample of blood serum that is known to contain antibodies for the typhoid organism. By using the Widal tube agglutination method, you will determine the antibody titer.

Materials:

- test-tube rack (Wassermann type) with ten clean serological tubes
- bottle of saline solution (0.85%), clear or filtered

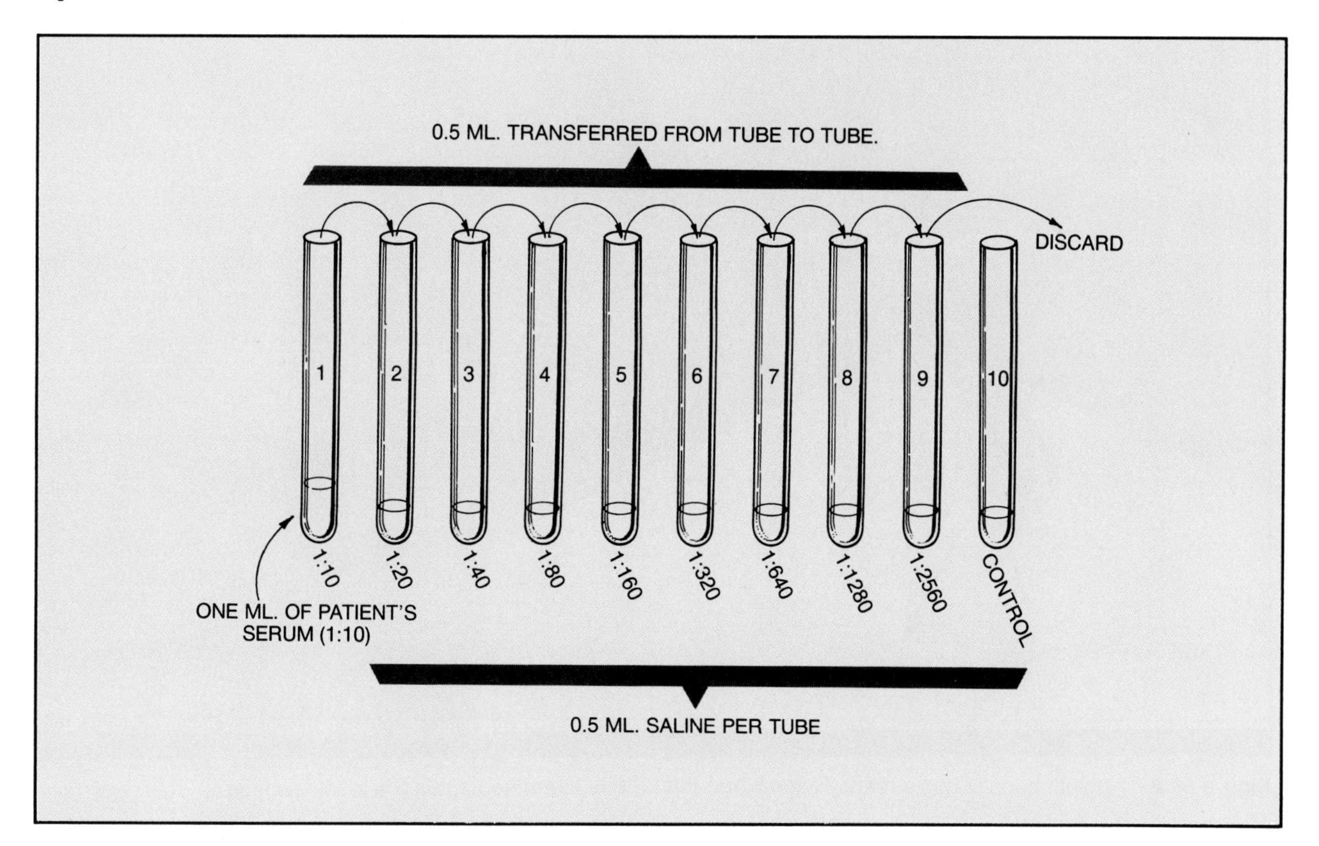

Figure 56.1 Procedure for dilution of serum.

1-ml pipettes
5-ml pipettes
water bath at 37° C
centrifuges
antigen (1:10 dilution) *Salmonella typhi* "O"
patient's serum (1:10 dilution), known positive "O"

1. Dilute the patient's serum as shown in figure 56.1. Follow this routine:
   a. Set up ten clean serological test-tubes in the front row of a test-tube rack and number them from one to ten (left to right) with a marking pencil.
   b. Into tube 1, pipette 1 ml of the patient's serum (1:10 dilution). For convenience, the instructor may wish to dispense this material to each student.
   c. With a 5-ml pipette, dispense 0.5 ml of saline to each of the remaining nine tubes.
   d. With a 1-ml pipette, transfer 0.5 ml of the serum from tube 1 to tube 2. Mix the serum and saline in tube 2 by carefully drawing the liquid up into the pipette and discharging it slowly back down into the tube three times.
   e. Repeat this process by transferring 0.5 ml from tube 2 to 3, tube 3 to 4, 4 to 5, etc. When you get to tube 9, discard the 0.5 ml drawn from it instead of adding it to tube 10; thus, tube 10 will contain only saline and can be used as a **negative test control** for comparing with the other tubes.
2. With a fresh 5-ml pipette, transfer 0.5 ml of antigen to each tube. Shake the rack to completely mix the antigen and diluted serum.
3. Place the rack in a water bath at 37° C for 30 minutes.
4. Centrifuge all tubes for three minutes at 2000 rpm. (If time permits, seven minutes centrifugation is preferable.)
5. Examine each tube for agglutination and record the titer as the highest dilution in which agglutination is seen. When examining each tube, jar it first by rapping the side of the tube with a snap of the finger to suspend the clumps of agglutinated cells. Hold it up against a desk lamp in the manner shown in figure 56.2. Do not look directly into the light. The reflection of the light off the particles is best seen against a dark background. *Compare each tube with tube 10, which is your negative test control.*
6. Record your results on the Laboratory Report.

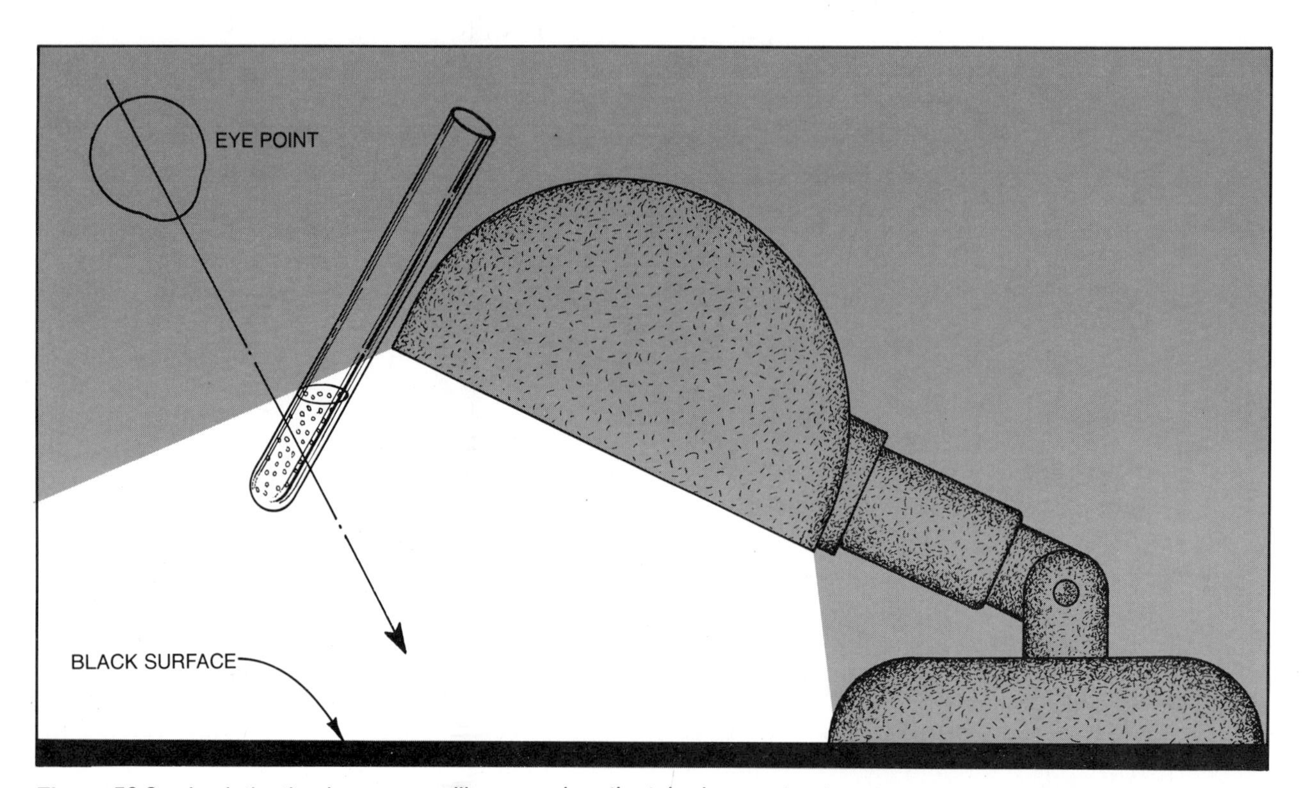

Figure 56.2 Agglutination is more readily seen when the tube is examined against a black surface.

Phage Typing 57

The host specificity of bacteriophage is such that it is possible to delineate different strains of individual species of bacteria on the basis of their susceptibility to various kinds of bacteriophage. In epidemiological studies, where it is important to discover the source of a specific infection, determining the phage type of the causative organism can be an important tool in solving the riddle. For example, if it can be shown that the phage type of *S. typhi* in a patient with typhoid fever is the same as the phage type of an isolate from a suspected carrier, chances are excellent that the two cases are epidemiologically related. Since all bacteria are probably parasitized by bacteriophages, it is theoretically possible, through research, to classify each species into strains or groups according to their phage type susceptibility. This has been done for *Staphylococcus aureus, Salmonella typhi,* and several other pathogens. The following table illustrates the lytic groups of *S. aureus* as proposed by M. T. Parker.

Lytic Group	Phages in Group
I	29 52 52A 79 80
II	3A 3B 3C 55 71
III	6 7 42E 47 53 54 75 77 83A
IV	42D
not allotted	81 187

In bacteriophage typing, a suspension of the organism to be typed is swabbed over an agar surface. The bottom of the plate is marked off in squares and labeled to indicate which phage types are going to be used. To the organisms on the surface, a small drop of each phage type is added to their respective squares. After incubation, the plate is examined to see which phages were able to lyse the organisms. This is the procedure to be used in this exercise. See figure 57.1.

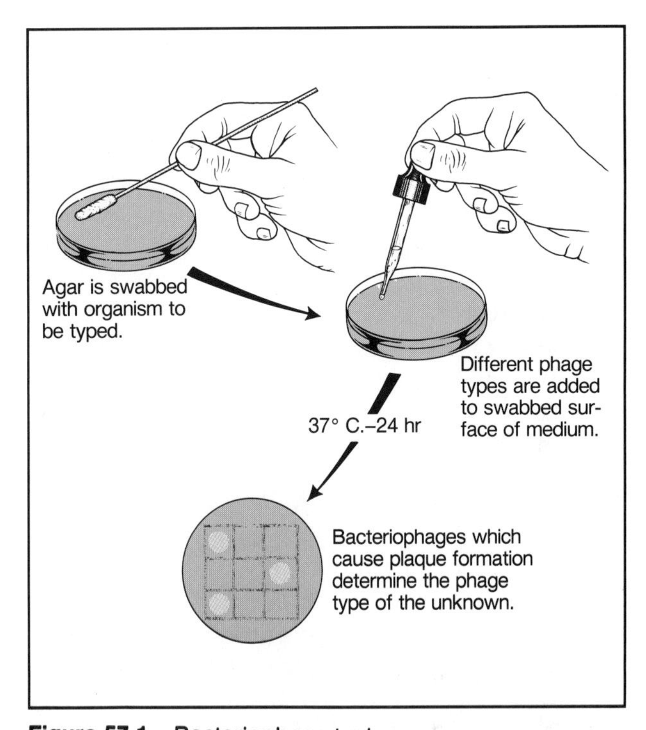

Figure 57.1 Bacteriophage typing.

Materials:

1 petri plate of tryptone yeast extract agar
bacteriophage cultures (available types)
nutrient broth cultures of *S. aureus* with swabs

1. Mark the bottom of a plate of tryptone yeast extract agar with as many squares as there are phage types to be used. Label each square with the phage type numbers.
2. Swab the entire surface of the agar with the organisms.
3. Deposit one drop of each phage in its respective square.
4. Incubate the plate at 37° C for 24 hours and record the lytic group and phage type of the culture.
5. Record your results on the Laboratory Report.

58 Blood Grouping

Another good example of the phenomenon of agglutination is seen in blood typing. Around 1900, Karl Landsteiner discovered that there are at least four different kinds of human blood with respect to the presence or absence of specific **agglutinogens** (agglutinating antigens) in the red blood cells. These antigens have been designated as A and B. The four groups are known as types A, B, AB, and O. The last type, which is characterized by the absence of A or B agglutinogens, is most common in the United States (45% of the population). Type A is next in frequency, found in 39% of the population. The incidences of types B and AB are 12% and 4%, respectively.

Blood typing is performed with antisera containing high titers of anti-A and anti-B agglutinins. The test may be performed by either slide or tube methods. In both instances, a drop of each kind of antiserum is added to separate samples of saline suspension of red blood cells. Figure 58.1 illustrates the slide technique. If agglutination occurs only in the suspension to which the anti-A serum was added, the blood is type A. If agglutination occurs only in the anti-B mixture, the blood is type B. Agglutination in both samples indicates that the blood is type AB. The absence of agglutination indicates that the blood is type O.

Between 1900 and 1940 a great deal of research was done to uncover the presence of other antigens in human red blood cells. Finally, in 1940, Landsteiner and Wiener reported that rabbit sera containing antibodies for the red blood cells of the Rhesus monkey would agglutinate the red blood cells of 5% of white humans. This antigen in humans, which was first designated as the **Rh factor** in due respect to the Rhesus monkey, was later found to exist as six antigens, which were given the letters C, c, D, d, E, and e by Fisher and Race. Of these six antigens, the D factor is responsible for the Rh-positive condition and is found in 85% of Caucasians, 94% of Blacks, and 99% of Orientals.

Typing blood for the Rh factor can also be performed by both tube and slide methods, but certain differences in the techniques are involved. First of all, the antibodies in the typing sera are of the incomplete, albumin variety which *will not agglutinate human red cells when they are diluted with saline.* Therefore, it is necessary to use whole blood or to dilute the cells with plasma. Another difference is that the *test must be performed at higher temperatures* (37° C for tube test, 45° C for slide test).

In this exercise, two separate slide methods are presented for typing blood. If only the Landsteiner ABO groups are to be determined, the first method may be preferable. If Rh typing is to be included, the second method, which utilizes a slide warmer, will be followed. The availability of materials will determine which method is to be used.

ABO Blood Typing

Materials:

small vial (10 mm dia. × 50 mm long)
disposable lancets (B-D Microlance, Serasharp, etc.)
70% alcohol and cotton
china marking pencil and microscope slides
typing sera (anti-A and anti-B)
applicators or toothpicks
saline solution (0.85% NaCl)
1-ml pipettes

1. Mark a slide down the middle with a marking pencil, dividing the slide into two halves (see figure 58.1). Write "anti-A" on the left side and "anti-B" on the right side.
2. Pipette 1 ml of saline solution into a small vial or test tube.
3. Scrub the middle finger with a piece of cotton saturated with 70% alcohol and pierce it with a sterile disposable lance. Allow two or three drops of blood to mix with the saline by holding the finger over the end of the vial and washing it with the saline by inverting the tube several times.
4. Place a drop of this red cell suspension on each side of the slide.
5. Add a drop of anti-A serum to the left side of the slide and a drop of anti-B serum to the right side. ***Do not contaminate the tips of the serum pipettes with the material on the slide.***

6. After mixing each side of the slide with separate applicators or toothpicks, look for agglutination. The slide should be held about 6″ above an illuminated white background and rocked gently for two or three minutes. Record your results on the Laboratory Report as of three minutes.

Combined ABO and Rh Typing

As stated, Rh typing must be performed with heat on blood that has not been diluted with saline. A warming box such as the one in figure 58.2 is essential in this procedure. In performing this test, two factors are of considerable importance: first,

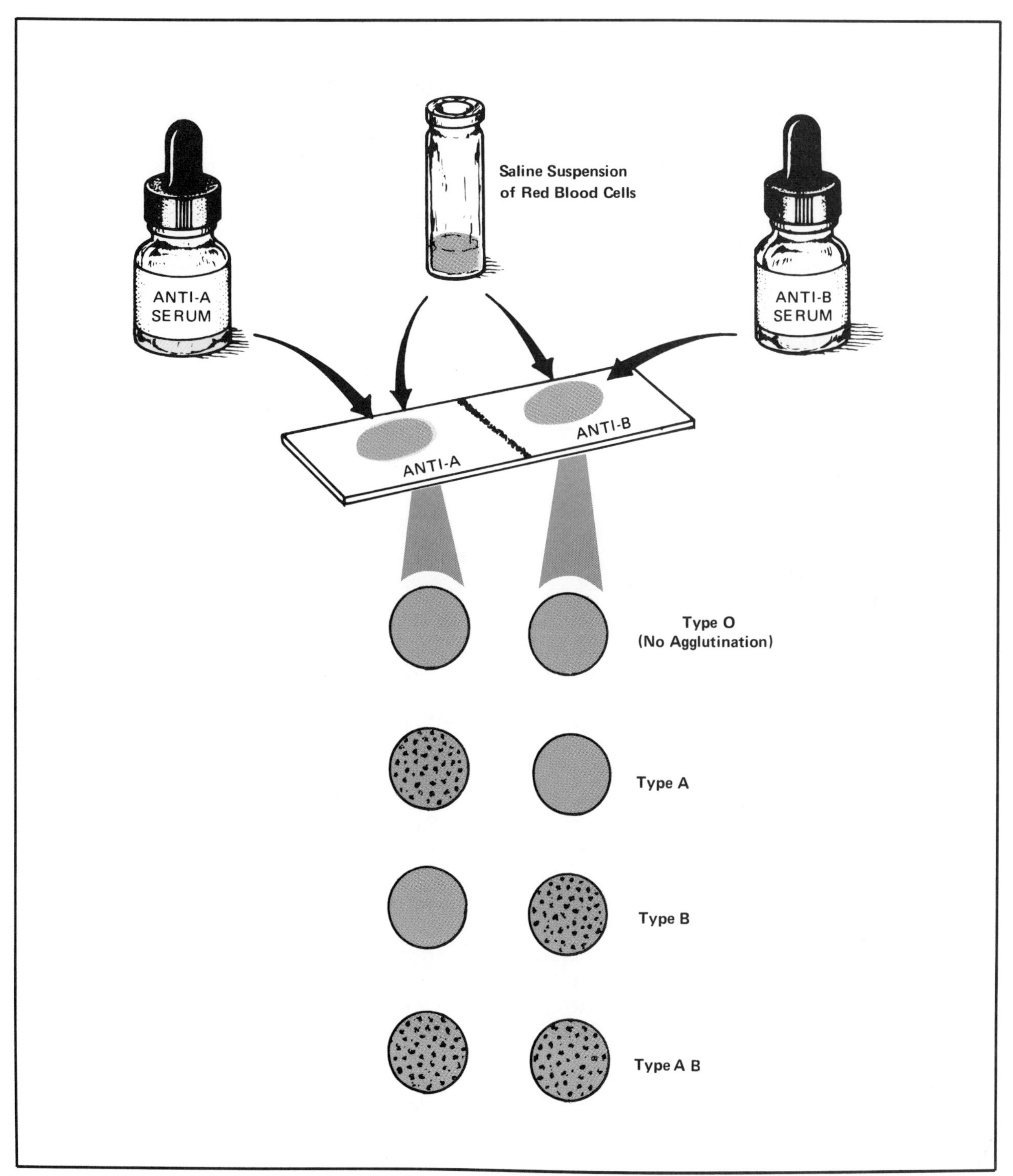

Figure 58.1 Typing of ABO blood groups.

only a small amount of blood must be used (a drop of about 3 mm diameter on the slide) and, secondly, proper agitation must be executed. The agglutination that occurs in this antibody-antigen reaction results in finer clumps; therefore, closer examination is essential. If the agitation is not properly performed, agglutination may not be as apparent as it should be. In this combined method we will use whole blood for the ABO typing, also. Although this method works satisfactorily as a classroom demonstration for the ABO groups, it is *not as reliable* as the previous method in which saline and room temperature are used *and is not recommended as a clinical method.*

Materials:

slide warming box with a special marked slide
anti-A, anti-B, and anti-D typing sera
applicators or toothpicks
70% alcohol and cotton
disposable sterile lancets (B-D Microlances, Serasharp, etc.)

1. Scrub the middle finger with a piece of cotton saturated with 70% alcohol and pierce it with a sterile disposable lance. Place a small drop in each of three squares on the marked slide on the warming box. To get the proper proportion of serum to blood, do not use a drop larger than 3 mm diameter on the slide.
2. Add a drop of anti-D serum to the blood in the anti-D square, mix with a toothpick, and note the time. ***Only two minutes should be allowed for agglutination.***
3. Add a drop of anti-B serum to the anti-B square and a drop of anti-A serum to the anti-A square. Mix the sera and blood in both squares with separate fresh toothpicks.
4. Agitate the mixtures on the slide by slowly rocking the box back and forth on its pivot. At the end of two minutes, examine the anti-D square carefully for agglutination. If no agglutination is apparent, consider the blood to be Rh negative. By this time the ABO type can also be determined. Record your results on the Laboratory Report.

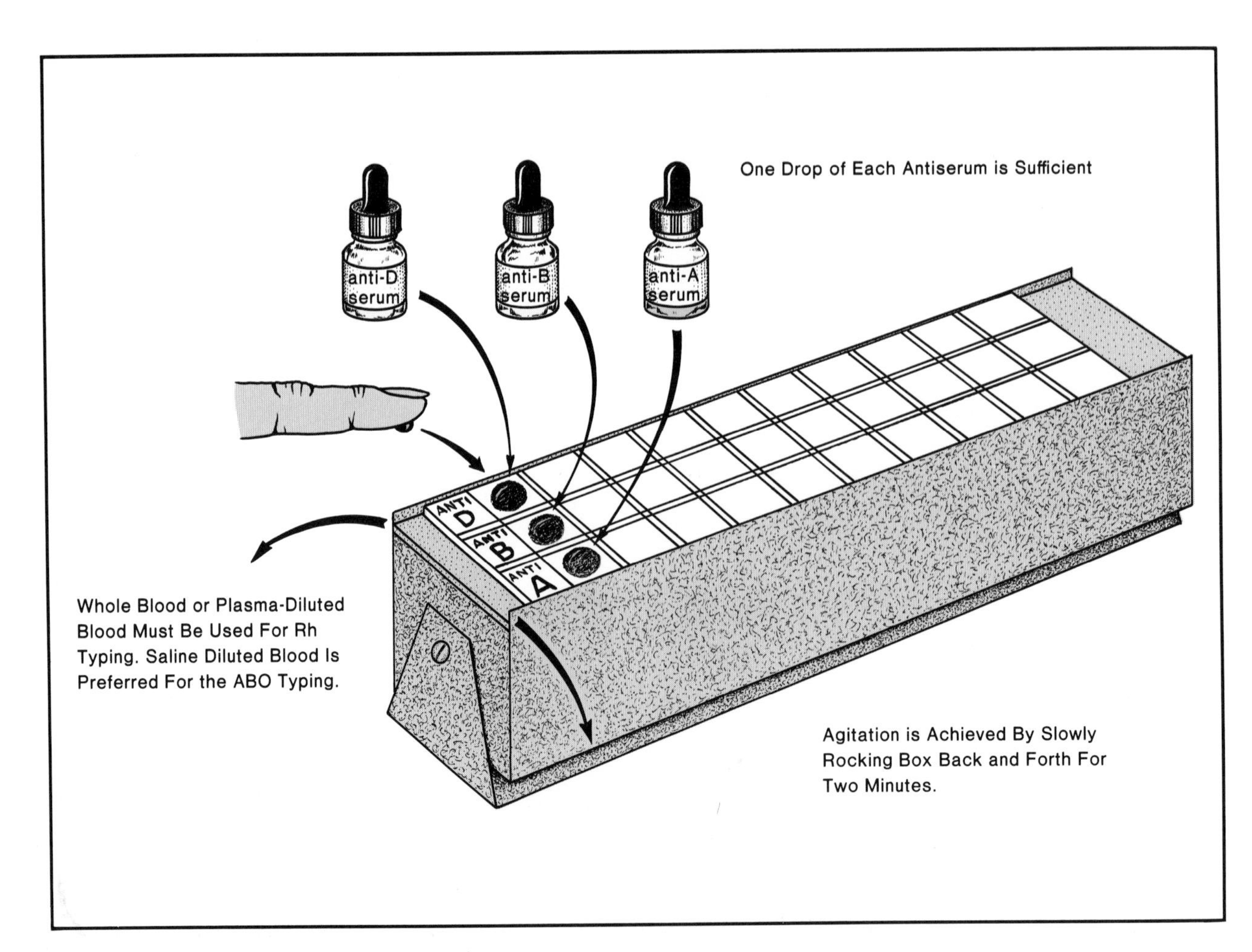

Figure 58.2 Blood typing with warming box.

White Blood Cell Study: Differential Blood Count 59

In 1883, at the Pasteur Institute in Paris, Metchnikoff published a paper proposing the **phagocytic theory of immunity.** On the basis of studies performed on transparent starfish larvae, he postulated that amoeboid cells in the tissue fluid and the blood of all animals are the major guardians of health against bacterial infection. He designated the large phagocytic cells of the blood as *macrophages* and the smaller ones as *microphages*. Today, Metchnikoff's macrophages are known as monocytes and his microphages as neutrophils or polymorphonuclear leukocytes.

Figure 59.1 illustrates the five types of leukocytes that are normally seen in the blood. Blood platelets and erythrocytes also are shown to present a complete picture of all formed elements in the blood. When observed as living cells under the microscope, they appear as refractile, colorless structures. As shown here, however, they reflect the dyes that are imparted by Wright's stain.

In this exercise, a smear of blood will be stained and examined under oil immersion to determine the relative percentages of each type of leukocyte. A total of 100 white blood cells will be recorded as to type. This method of white blood cell enumeration is called a **differential count.** Figure 59.1 will be used to help in identification of each kind of cell.

As you proceed with this count, it will become obvious that the **neutrophils** are most abundant (50–70%). The next most prominent cells are the **lymphocytes** (20–30%). **Monocytes** comprise about 2–6%; **eosinophils,** 1–5%; and **basophils,** less than 1%.

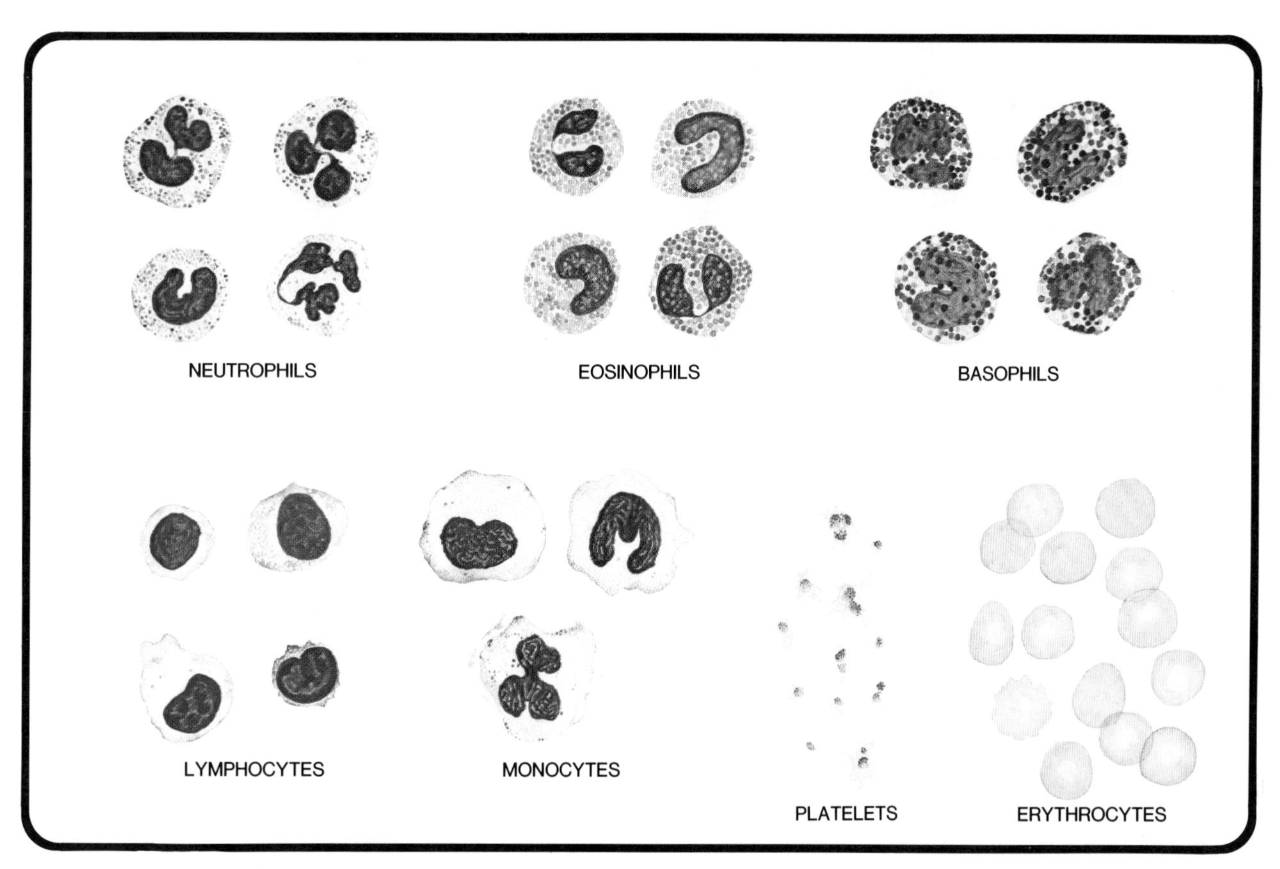

Figure 59.1 Formed elements of blood.

K. P. Talaro

The value of a differential count is immeasurable in the diagnosis of infectious diseases. High neutrophil counts, or *neutrophilia,* often signal localized infections such as appendicitis or abscesses in some other part of the body. *Neutropenia,* a condition in which there is a marked decrease in the numbers of neutrophils, occurs in typhoid fever, undulant fever, and influenza. *Eosinophilia* may indicate allergic conditions or invasions by parasitic roundworms such as *Trichinella spiralis,* the "pork worm." Counts of eosinophils may rise as much as 50% in cases of trichinosis. High lymphocyte counts, or *lymphocytosis,* are present in whooping cough and some viral infections. The normal white cell count is between 5000 and 10,000 cells per cubic millimeter. Elevated total white cell counts are referred to as *leukocytosis.* A count of 30,000 or 40,000 represents marked leukocytosis.

Preparation of the slide involves six steps: (1) spreading the blood, (2) drying, (3) staining, (4) addition of distilled water, (5) rinsing, and (6) blotting dry. The most difficult step is the first one: spreading the blood. If a good, even smear is not produced, the slide should be washed off and the process repeated. When staining the slide, make sure that the smear is completely covered with an ample amount of stain.

Preparation of Slide

In the preparation of a suitable stained blood slide, it is essential that the smear be thick at one end and drawn out to a feather-thin edge. This type of preparation will provide a gradient of cellular density that will make it possible to choose an area which is ideal for counting. The angle at which the spreading slide is held in making the smear will determine the thickness of the smear. It may be necessary for you to make more than one slide to get an ideal one.

Materials:

- clean microscope slides (polished edges)
- sterile disposable lancets
- sterile absorbent cotton
- Wright's stain
- distilled water in dropping bottle
- 70% alcohol
- wax pencil
- bibulous paper

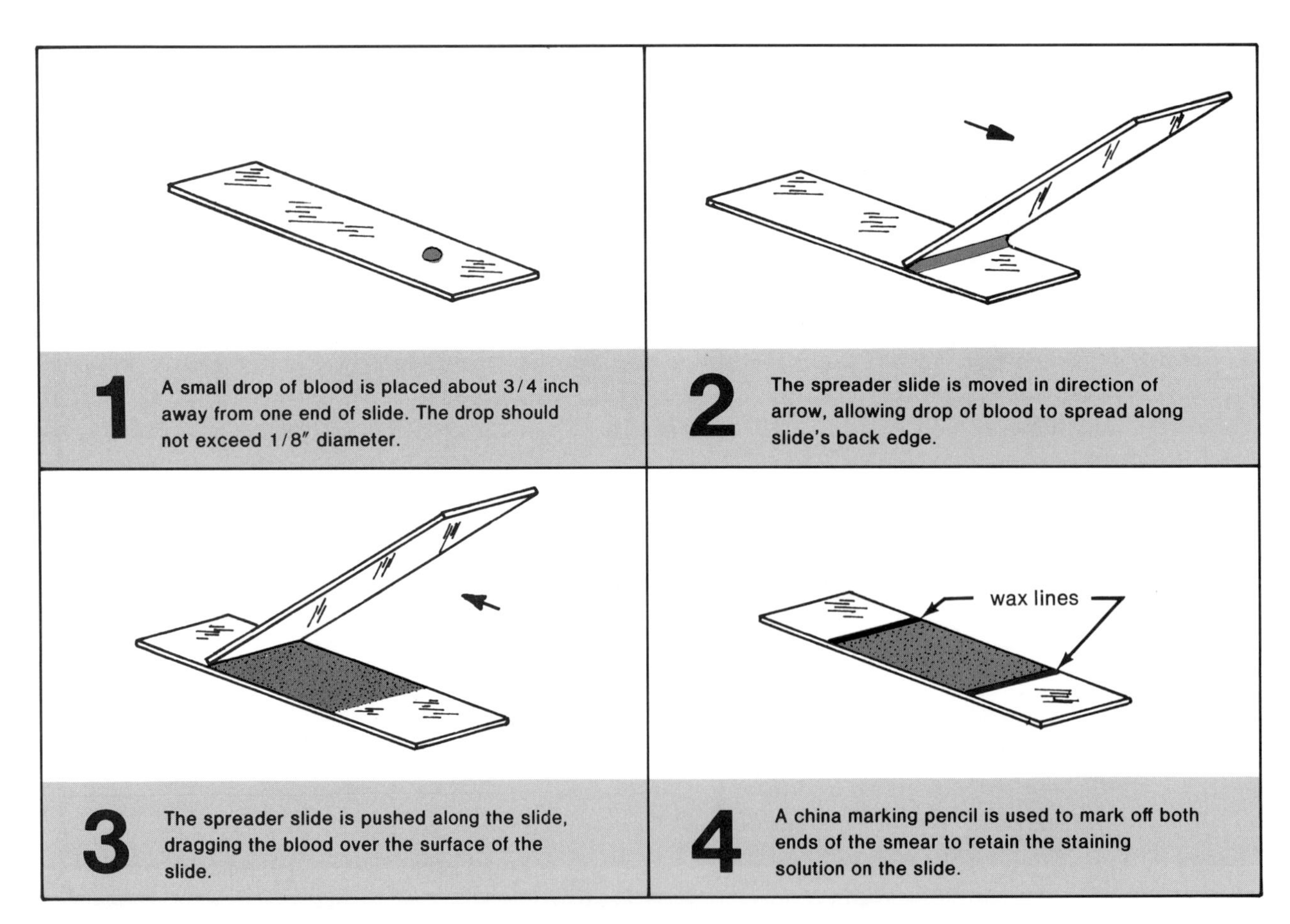

Figure 59.2 Smear preparation technique for differential WBC count.

1. Clean three or four slides with soap and water. Handle them with care to avoid getting their flat surfaces dirty with the fingers. Although only two slides may be used, it is often necessary to repeat the spreading process, thus the extra slides.
2. Scrub the middle finger with 70% alcohol and stick with a lancet. Put a drop of blood on the slide 1″ from one end and spread with another slide in the manner illustrated in figure 59.2. Note that the blood is dragged over the slide, not pushed. Do not pull the slide over the smear a second time. If you don't get an even smear the first time, repeat the process on a fresh, clean slide. To get a smear that will be of the proper thickness, hold the spreading slide at an angle greater than 45°.
3. Draw a line on each side of the smear with a wax pencil to confine the stain which is to be added.
4. Cover the film with **Wright's stain,** *counting the drops* as you add them. Stain for **4 minutes** and then add the same number of drops of **distilled water** to the stain and let stand for another **10 minutes.** Blow gently on the mixture every few minutes to keep the solutions mixed.
5. Gently wash off the slide under **running water** for **30 seconds** and shake off the excess. Blot dry with bibulous paper.

Performing the Cell Count

As soon as the slide is completely dry, scan it with the low-power objective to find that area of the slide that has the best distribution of cells. Avoid the excessively dense areas. Place a drop of oil near the edge of the smear in the selected area and examine with the oil immersion objective.

Remove your Laboratory Report sheet from the back of the manual and record each type of leukocyte encountered as you move the slide. Follow the path indicated in figure 59.3. For identification of each type of cell, refer to figure 59.1. Your Laboratory Report sheet indicates how the cells are to be tabulated.

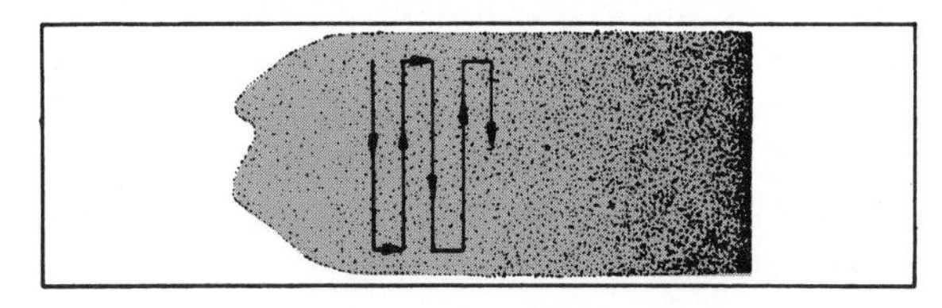

Figure 59.3 Examination path for differential count.

60 Fluorescent Antibody Technique

The science of immunology advanced rapidly in the latter part of the nineteenth century with the discovery and application of the principles of agglutination, precipitation, lysis, and opsonization to serological testing. By the turn of the century, most of the serological tests that are in use today were well advanced and being used in the better clinics and hospitals. Not until 1942, however, did Coons, Creek, Jones, and Berliner describe a new method which could be used for detecting antibody antigen reactions—the **fluorescent antibody technique.** In that year, they published a paper describing a method of using a pneumococcal antibody that had been coated with a fluorescent dye.

With the use of this type of antibody it becomes possible to determine the specific strain of pneumococcus by examining the organisms with ultraviolet light after they have been treated with fluorescent antibody serum. By 1950, Coons and Kaplan had improved these techniques to the point where they could be used for the identification of rickettsia and viruses as well as bacteria. Present-day procedures for fluorescent antibody (FA) techniques are based on the work of these men.

One of the main advantages of the FA technique over other serological tests is the rapidity with which results are obtained. In some cases, it is possible for a laboratory technician to report findings the very same day that a sample for testing is received. Such rapid interpretation of results is possible because *pure cultures of the pathogen are not essential.* Even when the pathogen is apparently obscured by large numbers of nonrelated organisms, the technique works very well.

The FA principle is utilized in several ways in diagnostic bacteriology. The **direct method** is the one we will use to identify a specific group of streptococcus. In this procedure, a smear of the organism is prepared in the usual manner. To the smear is added a drop of rabbit serum that contains fluorescent globulin specific for the pathogen. The slide is then placed in a moist chamber for 15 to 30 minutes to allow the antibody and antigen to combine. After the excess antibody has been washed off with a special buffered solution and blotted dry, it is examined with darkfield ultraviolet illumination. If the organisms fluoresce under the ultraviolet light, it is identified as belonging to the streptococcus group for which the antiserum is designated.

In this exercise you will be given a culture of either *Streptococcus pyogenes,* which belongs to group A, or *Streptococcus faecalis,* which belongs to group D. Fluorescent antibody sera for both of these groups will be available. A fluorescence microscope will be set up for the examination of your slide.

Materials:

numbered unknown (broth culture of
S. pyogenes or *S. faecalis*)
FA streptococcus group A (Difco #2318)
FA streptococcus group D (Difco #2321)
petri dish and filter paper
wood applicators (cut to 1¼″ lengths)
1% phosphate buffer solution, pH 7.2
glycerol, pH 7.2
microscope slides and cover glasses
fluorescence microscope

1. With a china marking pencil, divide a slide down the center with a line (see illustration 1, figure 60.1). On each half of the slide, prepare smears of the unknown. Fix the smears to the slide in the customary manner over a Bunsen burner flame. Label the left side A and the right side D.
2. Place the slide in a petri dish that has some wet filter paper in the bottom. Short lengths of wood applicator or a glass U-shaped rod should be placed between the slide and wet paper.
3. To the A side of the slide add a drop of serum containing fluorescent antibody for group A streptococci (FA streptococcus group A). To the D side of the slide add serum containing fluorescent antibody for the group D streptococci (FA streptococcus group D).
4. Cover the petri dish and let stand at room temperature for 15 to 30 minutes.

5. Rinse the smears with 1% buffer, pH 7.2, to remove the excess labeled antibody and then immerse in a Coplin jar containing the same buffered solution for 10 minutes.
6. Remove the slide from the jar and gently blot dry with bibulous paper.
7. Add a drop of glycerol, pH 7.2, to each smear and cover with clean cover glasses.
8. Examine under a fluorescence microscope and record your conclusions on the Laboratory Report.

Figure 60.1 Preparation of fluorescent antibody slide.

LABORATORY REPORT **1**

Student: ____________________

Desk No.: ________ Section: ____________

Brightfield Microscopy

A. Completion Questions

Record the answers to the following questions in the column at the right.

1. List three fluids that may be used for cleaning lenses.
2. How can one greatly increase the bulb life on a microscope lamp if voltage is variable?
3. What characteristic of a microscope enables one to switch from one objective to another without altering the focus?
4. What effect (*increase* or *decrease*) does closing the diaphragm have on the following?
   a. Image brightness
   b. Image contrast
   c. Resolution
5. In general, at what position should the condenser be kept?
6. Express the maximum resolution of the compound microscope in terms of micrometers (μm).
7. If you are getting 225× magnification with a 45× high-dry objective, what would be the power of the eyepiece?
8. What is the magnification of objects observed through a 100× oil immersion objective with a 7.5× eyepiece?
9. Immersion oil must have the same refractive index as ______ to be of any value.
10. Substage filters should be of a ____________ color to get the maximum resolution of the optical system.

B. True–False

Record these statements as *true* or *false* in the answer column.

1. Eyepieces are of such simple construction that almost anyone can safely disassemble them for cleaning.
2. Lenses can be safely cleaned with almost any kind of tissue or cloth.
3. When swinging the oil immersion objective into position after using high-dry, one should always increase the distance between the lens and slide to prevent damaging the oil immersion lens.
4. Instead of starting first with the oil immersion lens, it is best to use one of the lower magnifications first, and then swing the oil immersion into position.
5. The 45× and 100× objectives have shorter working distances than the 10× objective.

Answers
Completion
1a. ____________
b. ____________
c. ____________
2. ____________
3. ____________
4a. ____________
b. ____________
c. ____________
5. ____________
6. ____________
7. ____________
8. ____________
9. ____________
10. ____________
True–False
1. ____________
2. ____________
3. ____________
4. ____________
5. ____________

C. Multiple Choice

Select the best answer for the following statements.

1. The resolution of a microscope is increased by
   1. using blue light.
   2. stopping down the diaphragm.
   3. lowering the condenser.
   4. raising the condenser to its highest point.
   5. Both 1 and 4 are correct.
2. The magnification of an object seen through the 10× objective with a 10× ocular is
   1. ten times.
   2. twenty times.
   3. 1000 times.
   4. None of the above are correct.
3. The most commonly used ocular is
   1. 5×.
   2. 10×.
   3. 15×.
   4. 20×.
4. Microscope lenses may be cleaned with
   1. lens tissue.
   2. a soft linen handkerchief.
   3. an air syringe.
   4. Both 1 and 3 are correct.
   5. 1, 2, and 3 are correct.
5. When changing from low power to high power, it is generally necessary to
   1. lower the condenser.
   2. open the diaphragm.
   3. close the diaphragm.
   4. Both 1 and 2 are correct.
   5. Both 1 and 3 are correct.

Answers
Multiple Choice
1. ________
2. ________
3. ________
4. ________
5. ________

LABORATORY REPORT **2**

Student: ______________________________

Desk No.: __________ Section: ______________

Phase-contrast Microscopy

A. Questions

1. Which rays (*direct* or *diffracted*) are altered by the phase ring on the phase plate?

   ____

2. How much phase shift occurs in the light rays that emerge from a transparent object? __________

   ____

3. How does the diaphragm of a phase-contrast microscope differ from a diaphragm of a brightfield microscope? ______________________________

   ____

4. Where is the phase plate located in a phase-contrast microscope? ______________________________

   ____

5. How does one increase brightness in a phase-contrast microscope? ______________________________

   ____

   ____

6. Differentiate between the following:

   bright-phase microscope: ______________________________

   ____

   ____

   dark-phase microscope: ______________________________

   ____

   ____

7. List two items that can be used for observing the concentricity of the annulus and phase ring.

   (1) ______________________________

   (2) ______________________________

LABORATORY REPORT **4**

Student: ______________________

Desk No.: ________ Section: ____________

Protozoa, Algae, and Cyanobacteria

A. Tabulation of Observations

In this study of freshwater microorganisms, record your observations in the following tables. The number of organisms to be identified will depend on the availability of time and materials. Your instructor will indicate the number of each type that should be recorded.

Record the genera of each identifiable type. Also, indicate the phylum or division to which the organism belongs. Microorganisms that you are unable to identify should be sketched in the space provided. It is not necessary to draw those that are identified.

PROTOZOA

GENUS	PHYLUM	BOTTLE NO.	SKETCHES OF UNIDENTIFIED

ALGAE

GENUS	DIVISION	BOTTLE NO.	SKETCHES OF UNIDENTIFIED

CYANOBACTERIA

GENUS	BOTTLE NO.	SKETCHES OF UNIDENTIFIED

B. General Questions

Record the answers to the following questions in the answer column. It may be necessary to consult your text or library references for one or two of the answers.

1. Give the kingdom in which each of the following groups of organisms is found:
   a. protozoans
   b. algae
   c. cyanobacteria
   d. bacteria
   e. fungi
   f. microscopic invertebrates
2. Four kingdoms are represented by the organisms in the above question. Name the fifth kingdom.
3. What is the most significant characteristic seen in eukaryotes that is lacking in prokaryotes?
4. What characteristic in the microscopic invertebrates distinguishes them from protozoans?
5. Which protozoan phylum was not found in pond samples because phylum members are all parasitic?
6. Indicate whether the following are *present* or *absent* in the algae:
   a. cilia b. flagella c. chloroplasts
7. Indicate whether the following are *present* or *absent* in the protozoans:
   a. cilia
   b. chloroplasts
   c. mitochondria
   d. mitosis
8. Which photosynthetic pigment is common to all algae and cyanobacteria?
9. Name two photosynthetic pigments that are found in the cyanobacteria but not in the algae.
10. What photosynthetic pigment is found in bacteria but is lacking in all other photosynthetic organisms?
11. What type of movement is exhibited by the diatoms?

Answers

1a. ____________
b. ____________
c. ____________
d. ____________
e. ____________
f. ____________
2. ____________
3. ____________
4. ____________
5. ____________
6a. ____________
b. ____________
c. ____________
7a. ____________
b. ____________
c. ____________
d. ____________
8. ____________
9a. ____________
b. ____________
10. ____________
11. ____________

C. *Protozoan Characterization*

Select the protozoan groups in the right-hand column that have the following characteristics:

1. move with flagella
2. move with cilia
3. move with pseudopodia
4. have nuclear membranes
5. lack nuclear membranes
6. all species are parasitic
7. produce resistant cysts

1. Sarcodina
2. Mastigophora
3. Ciliophora
4. Sporozoa
5. all of above
6. none of above

D. *Characterization of Algae and Cyanobacteria*

Select the groups in the right-hand column that have the following characteristics:

Pigments

1. chlorophyll a
2. chlorophyll b
3. chlorophyll c
4. fucoxanthin
5. c-phycocyanin
6. c-phycoerythrin

Food Storage

7. fats
8. oils
9. starches
10. laminarin
11. leucosin
12. paramylum
13. mannitol

Other Structures

14. pellicle, no cell wall
15. cell walls, box in lid
16. chloroplasts
17. phycobilisomes
18. thylakoids

1. Euglenophycophyta
2. Chlorophycophyta
3. Chrysophycophyta
4. Phaeophycophyta
5. Pyrrophycophyta
6. Cyanobacteria
7. all of above
8. none of above

Answers

Protozoa

1. ____
2. ____
3. ____
4. ____
5. ____
6. ____
7. ____

Algae

1. ____
2. ____
3. ____
4. ____
5. ____
6. ____
7. ____
8. ____
9. ____
10. ____
11. ____
12. ____
13. ____
14. ____
15. ____
16. ____
17. ____
18. ____

LABORATORY REPORT **6**

Student: ______________________

Desk No.: ________ Section: ____________

Fungi: Yeasts and Molds

A. *Yeast Study*

Draw a few representative cells of *Saccharomyces cerevisiae* in the appropriate circles below. Blastospores (buds) and ascospores, if seen, should be shown and labeled.

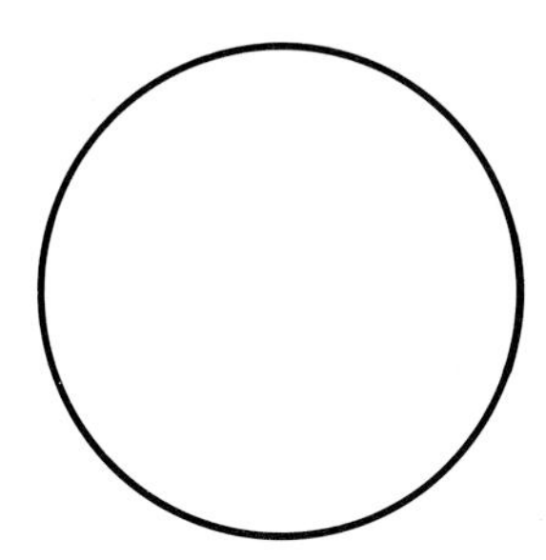

Prepared Slide

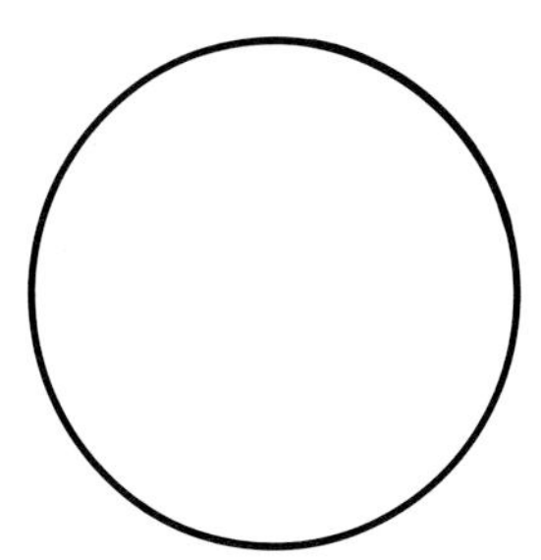

Living Cells

B. *Mold Study*

In the following table, list the genera of molds identified in this exercise. Under colony description, give the approximate diameter of the colony, its topside color and backside (bottom) color. For microscopic appearance, make a sketch of the organism as it appears on slide preparation.

GENUS	COLONY DESCRIPTION	MICROSCOPIC APPEARANCE (DRAWING)

C. Questions

Record the answers for the following questions in the answer column.

1. The science that is concerned with the study of fungi is called ______________ .
2. The kingdom to which the fungi belong is ______________ .
3. Microscopic filaments of molds are called ______________ .
4. A filamentlike structure formed by a yeast from a chain of blastospores is called a ______________ .
5. A mass of mold filaments, as observed by the naked eye, is called a ______________ .
6. Most molds have ______________ hyphae (*septate* or *non-septate*).
7. List three kinds of sexual spores that are the basis for classifying the molds.
8. What is the name of the rootlike structure that is seen in *Rhizopus?*
9. What type of hypha is seen in *Mucor* and *Rhizopus?*
10. What kind of asexual spores are seen in *Mucor* and *Rhizopus?*
11. What kind of asexual spores are seen in *Penicillium?*
12. What kind of asexual spores are seen in *Alternaria?*
13. Which subdivision of the Astigomycota contains individuals that lack sexual spores?
14. What division of Myceteae consists of slime molds?
15. Fungi that exist both as yeasts and molds are said to be ______________ .

Answers

1. ______________
2. ______________
3. ______________
4. ______________
5. ______________
6. ______________

7a. ______________

b. ______________

c. ______________

8. ______________
9. ______________
10. ______________
11. ______________
12. ______________
13. ______________
14. ______________
15. ______________

LABORATORY REPORT **7**

Student: ______________________

Desk No.: ________ Section: ____________

Bacteria

A. Tabulation

After examining your TSA and blood agar plates, record your results in the following table and on a similar table that your instructor has drawn on the chalkboard. With respect to the plates, we are concerned with a quantitative evaluation of the degree of contamination, and differentiation as to whether or not the organisms are bacteria or molds. Quantify your recording as follows:

0	no growth	+++	51 to 100 colonies
+	1 to 10 colonies	++++	over 100 colonies
++	11 to 50 colonies		

After shaking the tube of broth to disperse the organisms, look for cloudiness (turbidity). If the broth is clear, no bacterial growth occurred. Record no growth as 0. If tube is turbid, record + in last column.

STUDENT INITIALS	PLATE EXPOSURE METHOD		COLONY COUNTS		BROTH	
	TSA	Blood Agar	Bacteria	Fungi	Source	Result

B. Questions

1. Using the number of colonies as an indicator, which habitat sampled by the class appears to be the most contaminated one? ____________________

   Why do you suppose this habitat contains such a high microbial count? ____________________

   ____

2. In a few words, describe some differences in the macroscopic appearance of bacteria and fungal colonies: ____________________

   ____

   ____

   ____

   ____

3. How can you tell when a tube of broth contains bacterial growth? ____________________

   ____

   ____

4. a. Were any of the plates completely lacking in colonies? ____________________

   b. Do you think that the habitat sampled was really sterile? ____________________

   c. If your answer to *b* is *no,* then how can you account for the lack of growth on the plate? ____

   ____

   d. If your answer to *b* is *yes,* defend it: ____________________

   ____

   ____

   ____

LABORATORY REPORT **8–11**

Student: ______________________________

Desk No.: ________ Section: ______________

Negative Staining, Smear Preparation, Simple and Capsular Staining

A. *Negative Staining* (Exercise 8)

1. **Drawing:** Make a drawing in the circle at the right of some of the organisms as seen under oil immersion.
2. In addition to nigrosine, what other agent is often used for making negative-stained slides?

   ____

3. Other than bacteria, what other kinds of microorganisms might one encounter in the mouth?

   ____

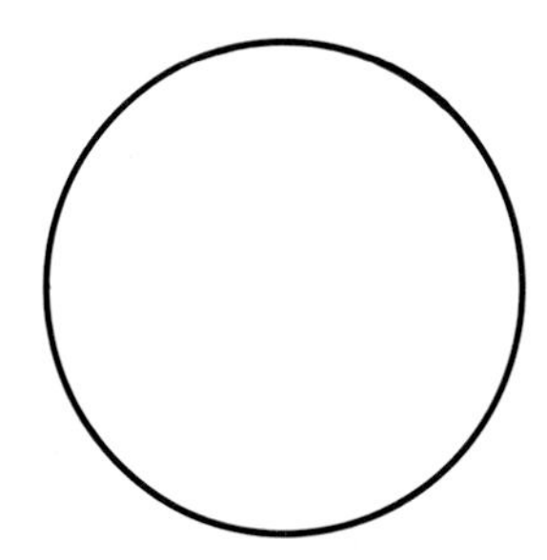

Oral organisms
(nigrosine)

B. *Smear Preparation* (Exercise 9)

1. Give two reasons for heating the slide after the smear is air-dried:

   a. ____

   b. ____

2. Why is an inoculating needle preferred to a wire loop when making smears from solid media?

   ____

   ____

   ____

C. *Simple Staining* (Exercise 10)

1. **Drawing:** Draw a few cells of *C. diphtheriae* from the portion of the slide that exhibits metachromatic granules and palisade arrangement.
2. Why are basic dyes more successful on bacteria than acidic dyes?

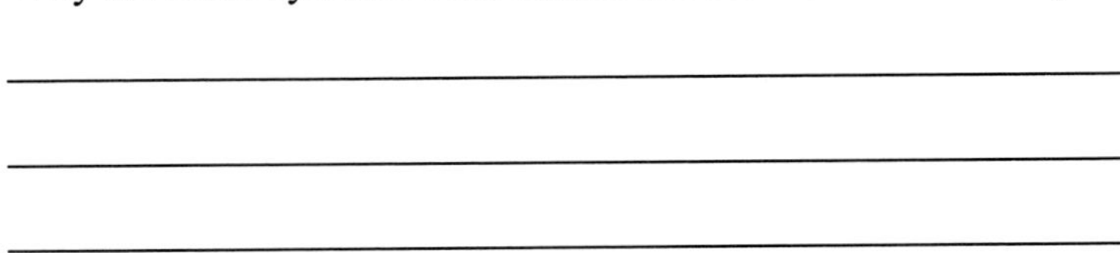

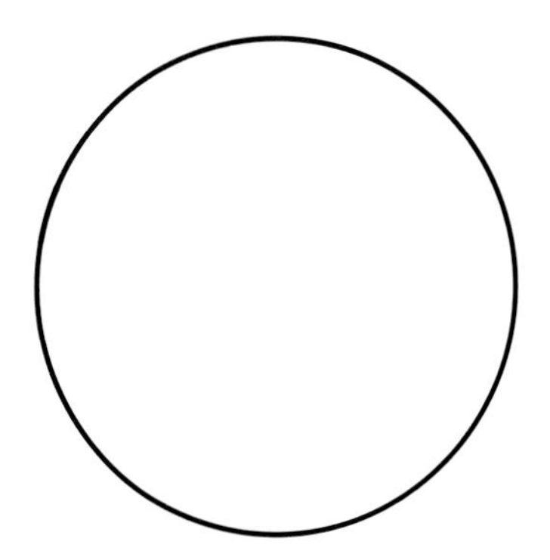

Corynebacterium diphtheriae

3. List three basic dyes that are used to stain bacteria: a. ______________________,

   b. ______________________, c. ______________________

D. *Capsular Staining* (Exercise 11)

1. List some of the chemical substances that have been identified in bacterial capsules.

   ____

   ____

   ____

Klebsiella pneumoniae
(capsular stain)

2. What relationship is there between capsules and bacterial virulence?

   ____

   ____

   ____

   ____

   ____

   ____

3. What is the **Quellung test?**

   ____

   ____

4. Of what value is the **Quellung test?**

   ____

   ____

   ____

   ____

   ____

LABORATORY REPORT 12–14

Student: ____________________

Desk No.: ________ Section: ____________

Differential Staining: Gram, Spore, and Acid-Fast

A. Drawings

With colored pencils, draw the various organisms as seen under oil immersion. Extra circles are provided for additional assignments, if needed.

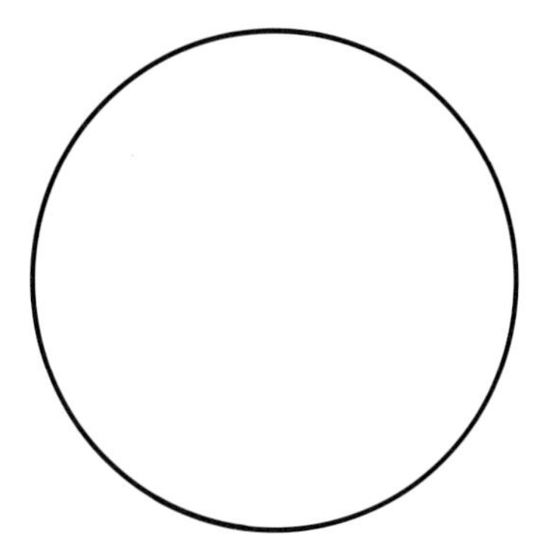

P. aeruginosa* & *S. aureus
(Gram stain)

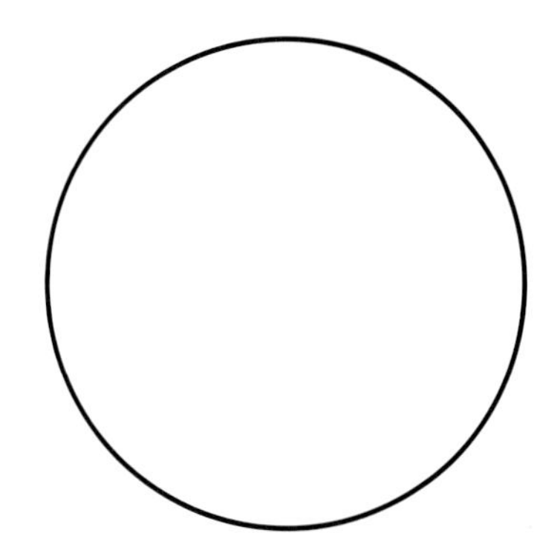

B. megaterium* & *M. B. catarrhalis
(Gram stain)

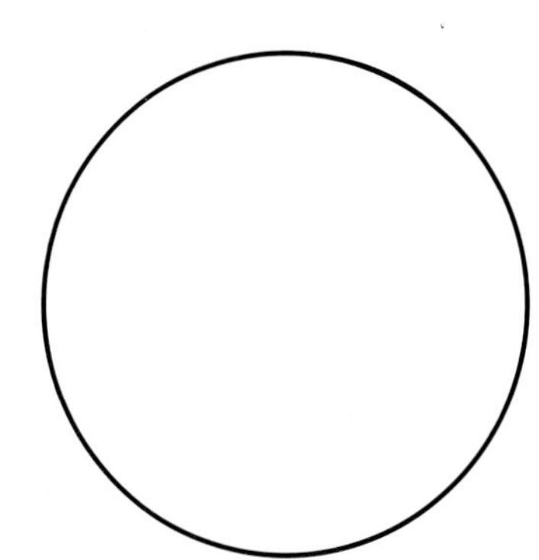

M. smegmatis
(Gram stain)

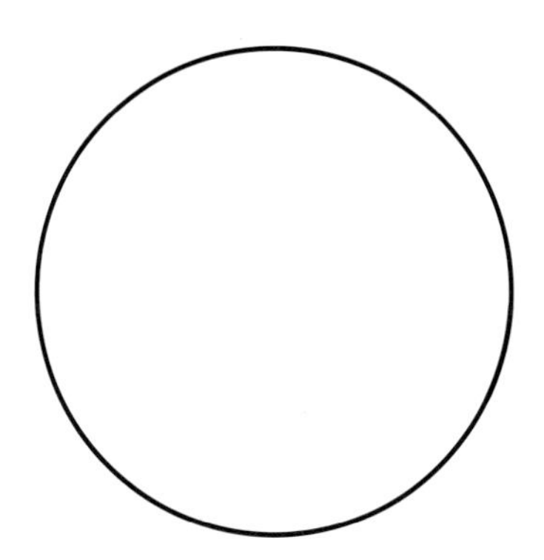

B. megaterium
(Schaeffer-Fulton method)

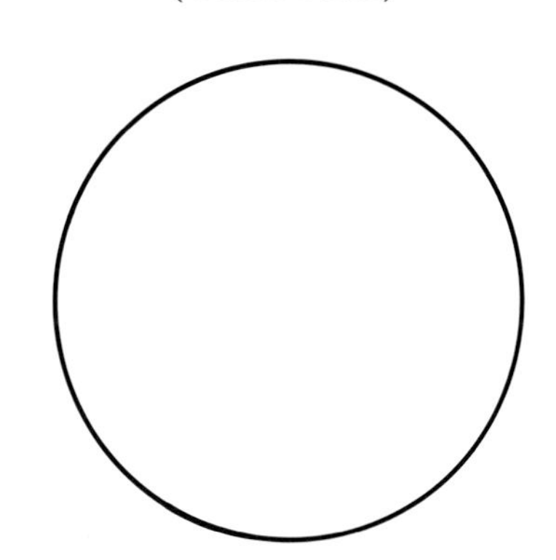

B. megaterium
(Dorner method)

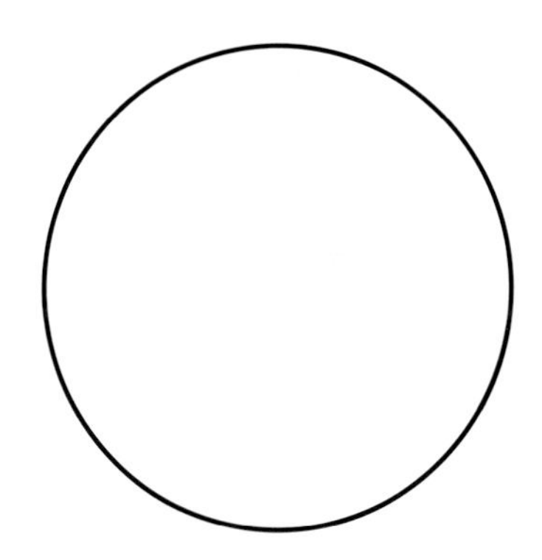

M. smegmatis* & *S. aureus
(Ziehl-Neelsen method)

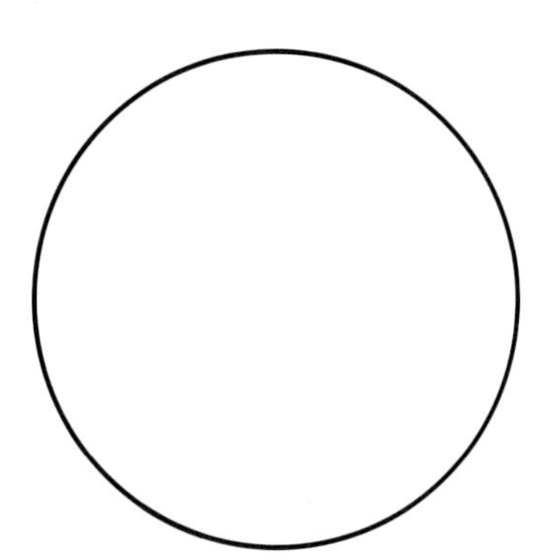

M. phlei* & *S. aureus
(Truant method)

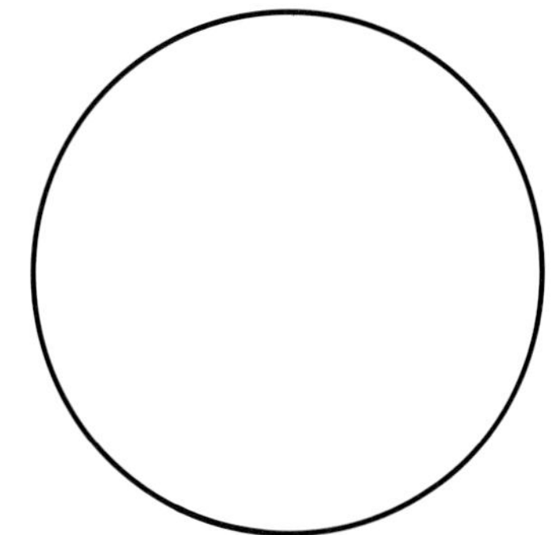

____

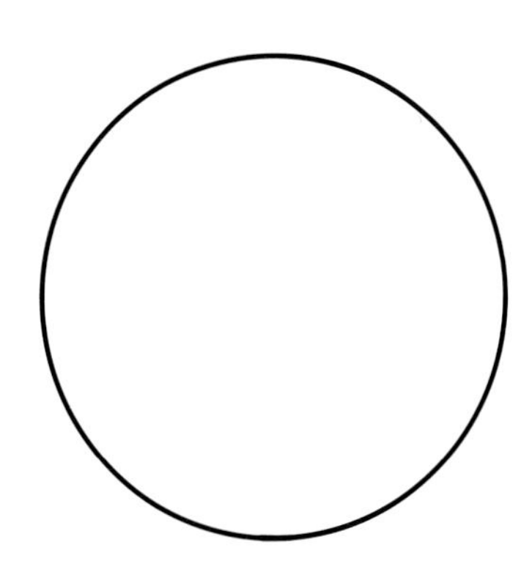

____

OPTIONAL STAINING

B. Completion Questions

1. What color would you expect *S. aureus* to be if the iodine step were omitted in the Gram staining procedure? ______

   Explanation: ______

2. What part of the bacterial cell (*cell wall* or *protoplast*) appears to play the most important role in determining whether or not an organism is gram-positive?

   ______

3. Why would methylene blue not work just as well as safranin for counterstaining in the Gram stain procedure? ______

4. Why are endospores so difficult to stain? ______

5. How do the following two genera of spore-formers differ physiologically?

   Bacillus: ______

   Clostridium: ______

6. How do you differentiate *S. aureus* and *M. B. catarrhalis* from each other on the basis of morphological characteristics? ______

7. Are the acid-fast mycobacteria gram-positive or gram-negative? ______
8. For what two diseases is acid-fast staining of paramount importance?

   a. ______ b. ______

9. Why is it desirable to combine *S. aureus* with acid-fast organisms such as *M. smegmatis* when applying an acid-fast staining technique?

   ______

LABORATORY REPORT **15**

Student: ____________________

Desk No.: ________ Section: ____________

Motility Determination

A. Test Results

1. Which of the two organisms exhibited true motility on the slides?

____

2. Did the semisolid medium inoculations confirm the results obtained from the slides?

____

3. Sketch in the appearance of the two tube inoculations:

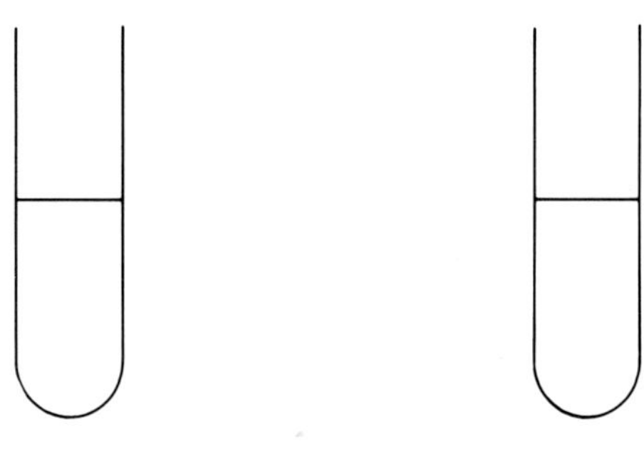

Micrococcus luteus *Proteus vulgaris*

B. Questions

1. How does brownian movement differ from true motility? ____________________

____

____

____

2. Make sketches that illustrate each of the following flagellar arrangements:

Monotrichic **Lophotrichic**

Amphitrichic **Peritrichic**

LABORATORY REPORT **16**

Student: ______________________

Desk No.: ________ Section: ____________

Culture Media Preparation

1. How do the following types of organisms differ in their carbon needs?

   Photoautotrophs: ______________________

   Photoheterotrophs: ______________________

2. Where do the above two types of organisms get their energy? ______________________

3. Where do **chemoheterotrophs** get their energy? ______________________

4. What is a growth factor? ______________________

   ____

5. Differentiate between the following two types of media:

   Synthetic medium: ______________________

   ____

   Nonsynthetic medium: ______________________

   ____

6. Briefly, list the steps that you would go through to make up a batch of nutrient agar slants:

   ____

   ____

   ____

   ____

LABORATORY REPORT **17**

Student: ____________________

Desk No.: ________ Section: ____________

Pure Culture Techniques

A. *Evaluation of Streak Plate*

Show within the circle the distribution of the colonies on your streak plate. To identify the colonies, use yellow for *Micrococcus luteus* and red for *Serratia marcescens*. If time permits, your instructor may inspect your plate and enter a grade where indicated.

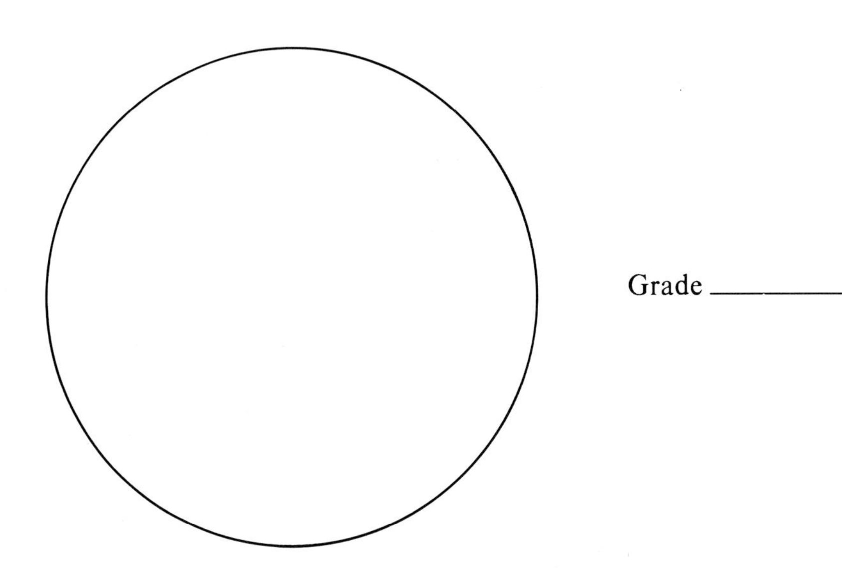

Grade ____________

B. *Evaluation of Pour Plates*

Show the distribution of colonies on plates II and III, using only the quadrant section for plate II. If plate III has too many colonies, follow the same procedure. Use colors.

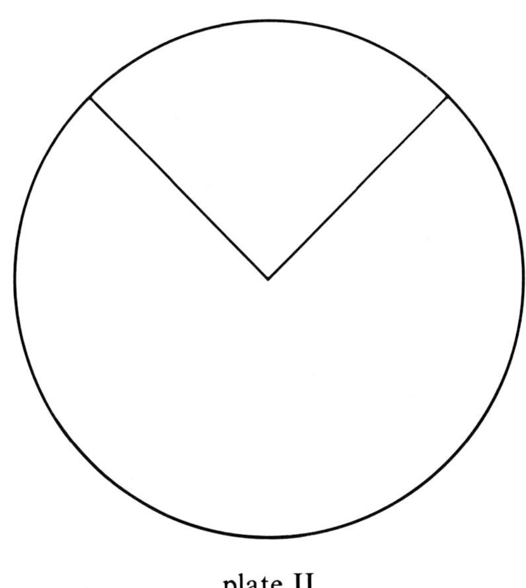

plate II

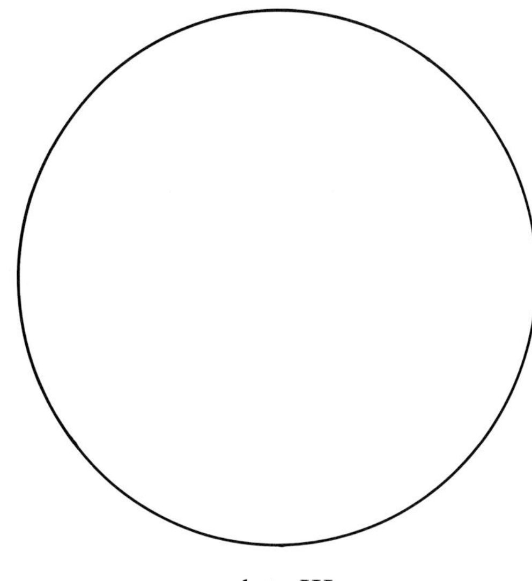

plate III

C. Subculture Evaluation

With colored pencils, sketch the appearance of the growth on the slant diagrams below. Also, draw a few cells of each organism as revealed by Gram staining in the adjacent circle.

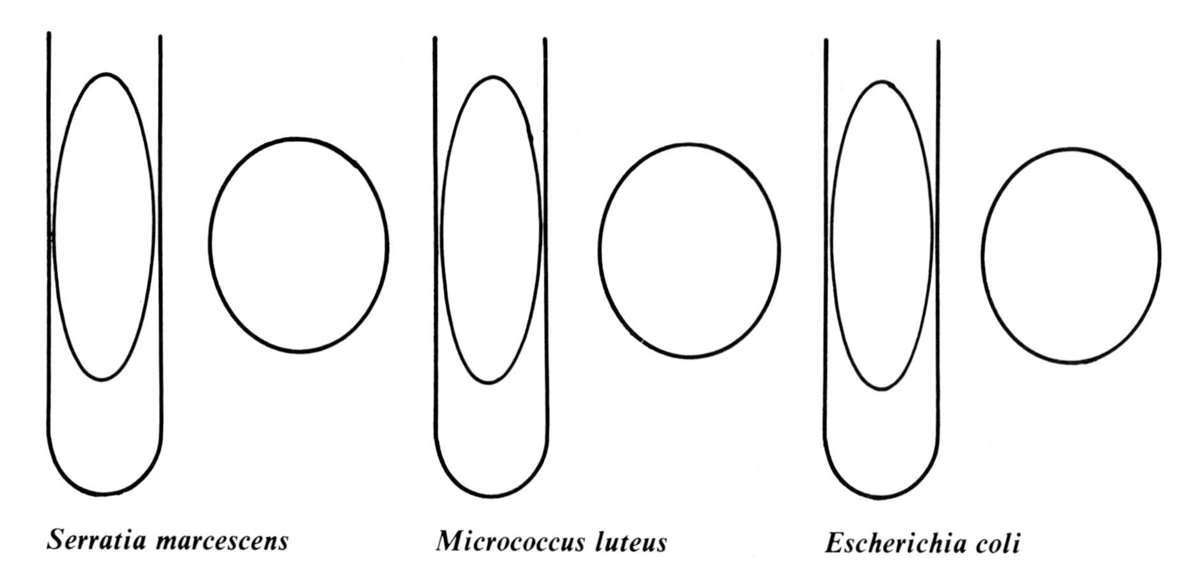

Serratia marcescens *Micrococcus luteus* *Escherichia coli*

D. Questions

1. Which method of separating organisms seems to achieve the best separation?

   ____

2. Which method requires the greatest skill? ______________________________

3. Do you think you have pure cultures of each organism on the slants? ______________

   Can you be absolutely sure by studying its microscopic appearance? ______________

   Explain: ______________________________

4. Give two reasons why the nutrient agar must be cooled to 50° C before inoculating and pouring.

   ____

   ____

   ____

5. Why should a petri plate be discarded if media is splashed up the side to the top?

   ____

   ____

6. Give two reasons why it is important to invert plates during incubation:

   ____

   ____

7. Why is it important not to dig into the agar with the loop? ______________________________

   ____

8. Why must loop be flamed before entering a culture? ______________________________

   Why must it be flamed after making an inoculation? ______________________________

   ____

LABORATORY REPORT **18**

Student: ____________________

Desk No.: ________ Section: ____________

Cultivation of Anaerobes

A. Evaluation of Plates

In the table below record the amount of growth for each streak according to this scale:

+++ good growth ++ fair growth + faint growth 0 no growth

ORGANISM	CONTROL	CANDLE JAR	GASPAK JAR
B. subtilis			
C. rubrum			
C. sporogenes			

B. Questions

1. On the basis of the results tabulated above, which organism appears to be

   aerobic? ____________________

   a facultative aerobe? ____________________

   a strict aerobe? ____________________

2. For what purpose is the candle method still used? ____________________

3. What toxic substance produced in a candle jar is harmful to some bacteria? ____________________

4. What agent in Brewer's anaerobic agar favors the growth of anaerobic bacteria? ____________________

   ____

C. Spore Study

Draw a few cells of each organism, illustrating the presence of spores.

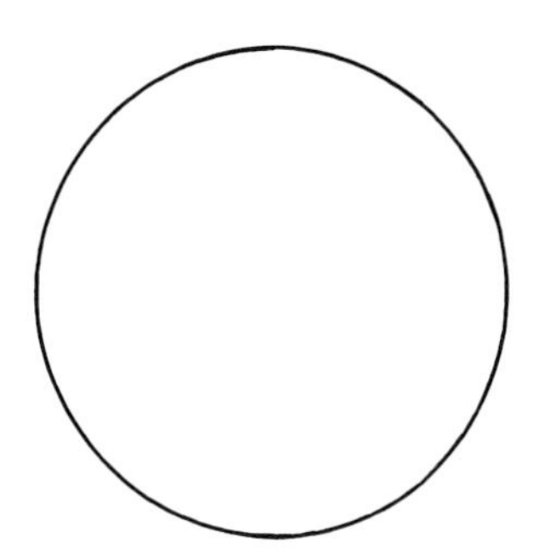

B. subtilis

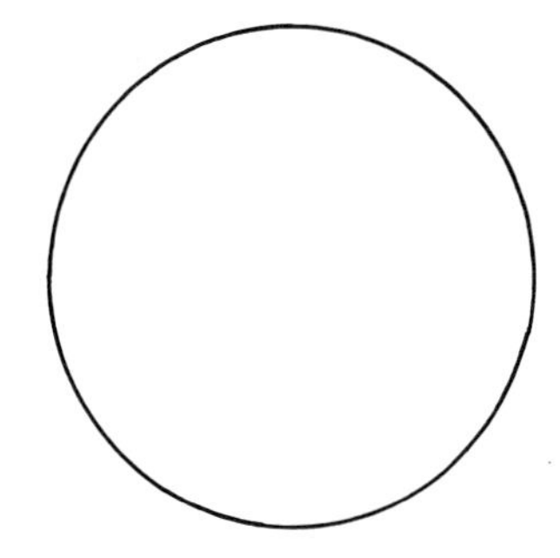

C. rubrum

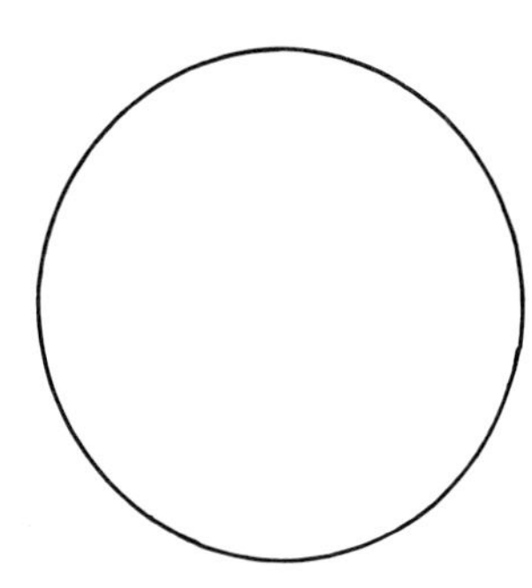

C. sporogenes

LABORATORY REPORT **19**

Student: ______________________

Desk No.: ________ Section: ____________

Bacterial Population Counts

A. Quantitative Plating Method

1. Which plate dilution gave you the best count? ______________________
2. How many colonies were there? ______________________
3. How many organisms per ml were present in the culture? ______________________
   Record this figure in the table below.*
4. Does each colony represent only one organism? ______________________
5. How would you inoculate a plate to get 1:100 dilution? ______________________

   ____
6. How would you inoculate a plate to get 1:10 dilution? ______________________

   ____
7. Give two reasons why it is necessary to shake the water blanks as recommended.

   a. ______________________

   b. ______________________

B. Turbidimetric Determinations

1. Record the percent transmittance and optical density values for your dilutions in the following table.

DILUTION	PERCENT TRANSMITTANCE	OPTICAL DENSITY	*NUMBER OF ORGANISMS PER ML
1:1			
1:2			
1:4			
1:8			
1:16			

2. Plot the optical densities versus the concentration of organisms. Complete the graph by drawing a line between plot points. Write in the number of organisms for each dilution.

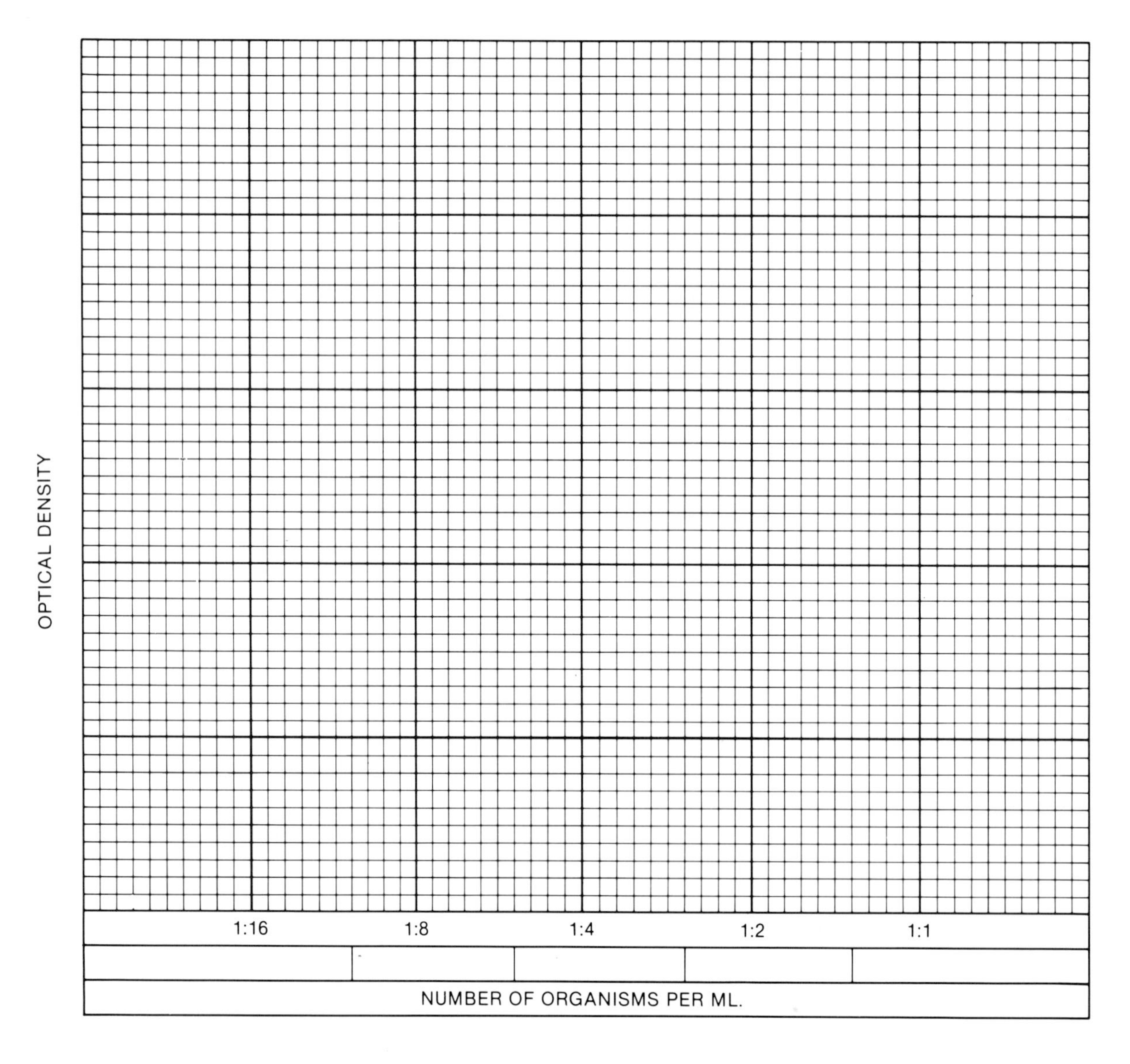

Evaluation: If your dilution techniques were flawless, you might expect a straight ascending line on this graph. However, it generally turns out that the line is bowed downward in the middle. This is due to the fact that these instruments do not function exactly the same at the extremes of percent transmittance.

3. Why is it necessary to perform a plate count in conjunction with the turbidimetry procedure?

____

____

4. If your medium was pale blue instead of amber-colored, as is the case of nutrient broth, would you set the wavelength control knob *higher* or *lower* than 686 nanometers?

____

LABORATORY REPORT **21**

Student: ______________________

Desk No.: ________ Section: ____________

Slide Culture: Autotrophs

A. *Microscopic Examination*

While examining the two slides, move them around to different areas to note the various types of organisms that are present. Draw representative types.

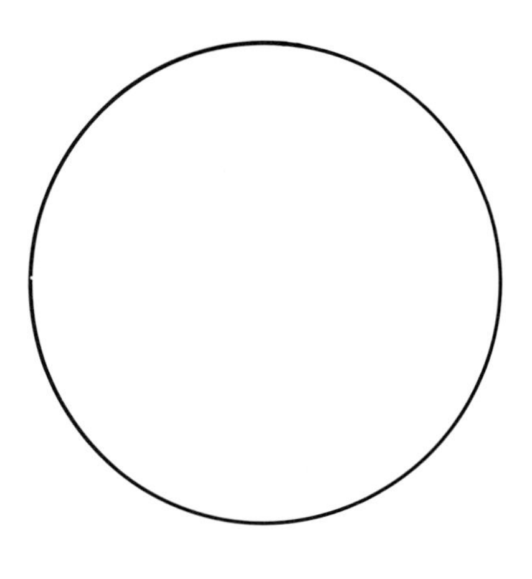

Gram Stain

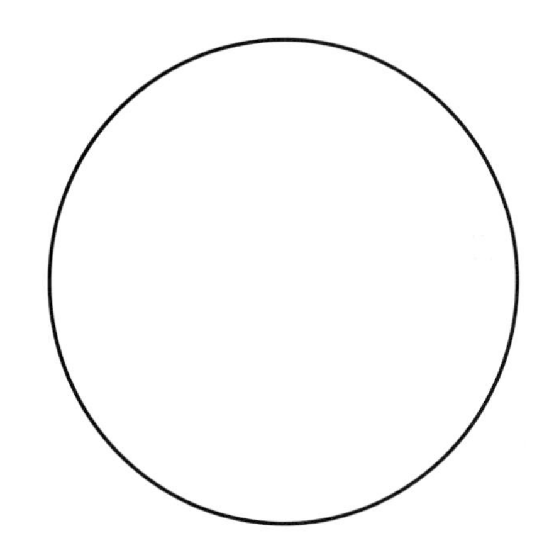

Living

B. *Questions*

1. With respect to Gram stain, which type (*gram-positive* or *gram-negative*) seems to predominate?

   ____

2. List as many different kinds of autotrophic protists as you can that can be cultured on this type of slide.

   ____

   ____

   ____

3. Some organisms that grow on this type of slide are chemosynthetic heterotrophs. What would be the source of their food?

   ____

   ____

LABORATORY REPORT **22**

Student: ____________________

Desk No.: __________ Section: __________

Bacteriophage: Isolation and Culture

A. *Plaque Size Increase*

With a china marking pencil, encircle and label three plaques on one of the plates and record their sizes in millimeters at one-hour intervals.

TIME	PLAQUE SIZE (millimeters)		
	Plaque No. 1	Plaque No. 2	Plaque No. 3
When first seen			
1 hour later			
2 hours later			
3 hours later			

B. *Questions*

1. Were any plaques seen on the negative control plate? ____________________
2. Do plates 1, 2, and 3 show a progressive increase in number of plaques with increased amount of sewage filtrate? ____________________
3. Did the phage completely "wipe out" all bacterial growth on any of the plates? __________

   If so, which plates? ____________________
4. Differentiate between:

   lysis: ____________________

   lysogeny: ____________________
5. Differentiate between:

   virulent phage: ____________________

   ____

   temperate phage: ____________________

   ____

LABORATORY REPORT **23**

Student: ______________________

Desk No.: __________ Section: ____________

Temperature: Effects on Growth

A. Pigment Formation and Temperature

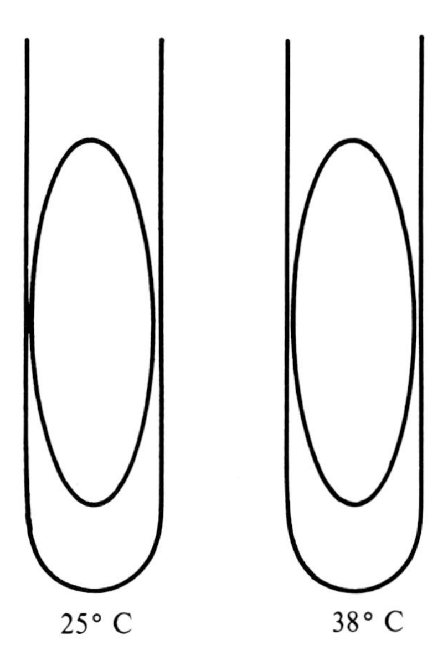

1. Draw the appearance of the growth of *Serratia marcescens* on the nutrient agar slants using colored pencils.
2. Which temperature seems to be closest to the optimum temperature for pigment formation?

   ____

3. What are the cellular substances that control pigment formation and are regulated by temperature?

   ____

B. Growth Rate and Temperature

If a photocolorimeter is available, dispense the cultures into labeled cuvettes and determine the percent transmittance of each culture. Calculate the O.D. values from the percent transmittance, using the formula given in Exercise 19.

If no photocolorimeter is available, record only the visual reading as +, + +, + + +, and none.

Temp. °C	*SERRATIA MARCESCENS*			*ESCHERICHIA COLI*		
	Visual Reading	Photocolorimeter		Visual Reading	Photocolorimeter	
		% T	O.D.		% T	O.D.
5						
25						
38						
42						

Growth curves of *Serratia marcescens* and *Escherichia coli* as related to temperature.

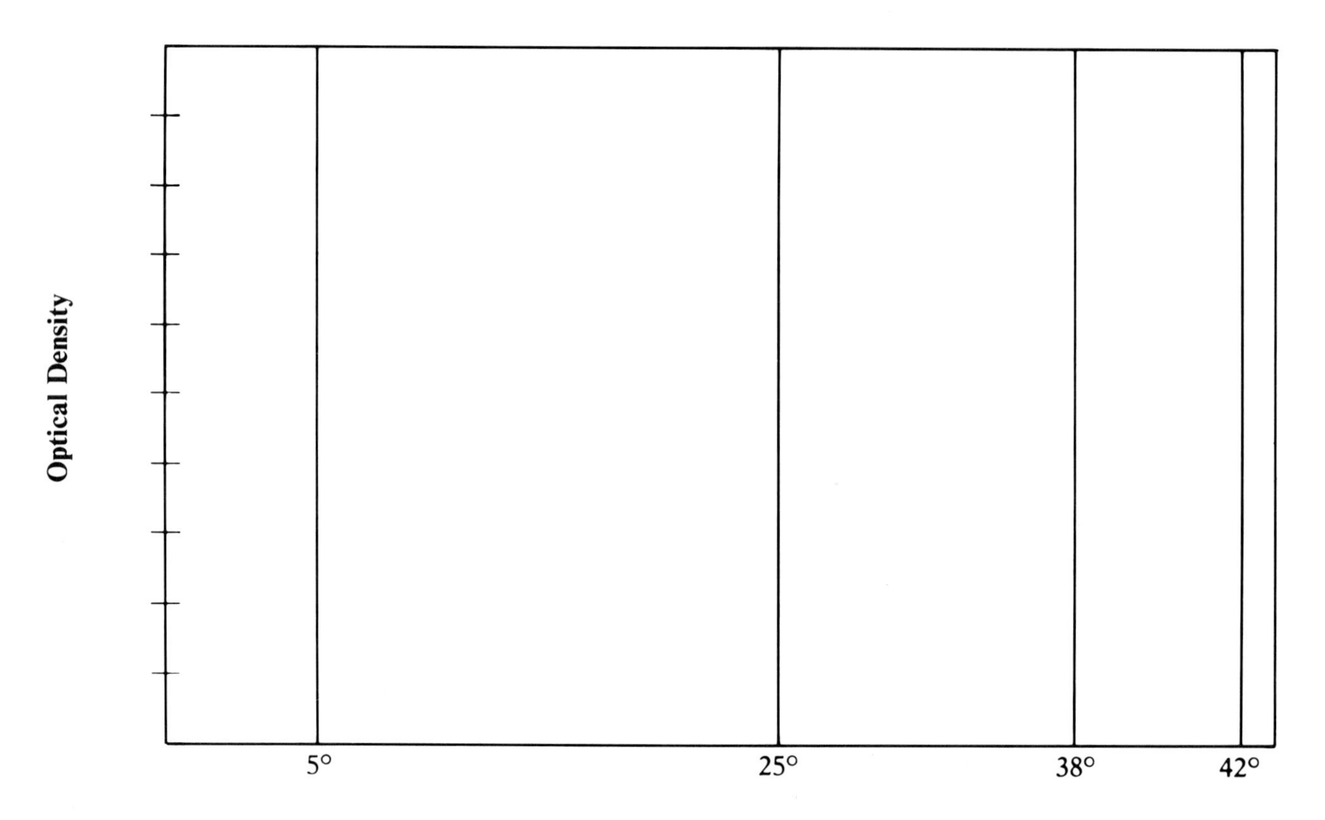

1. On the basis of the above graph, estimate the optimum growth temperature of the two organisms.

   Serratia marcescens: ______

   Escherichia coli: ______

2. To get more precise results for the above graph, what would you do?

   ______

   ______

3. Differentiate between the following:

   Thermophile: ______

   Mesophile: ______

   Psychrophile: ______

4. What is the optimum growth temperature range for most psychrophiles? ______

LABORATORY REPORT **24**

Student: ______________________

Desk No.: __________ Section: __________

Temperature: Lethal Effects

A. Tabulation of Results

Examine your five petri plates, looking for evidence of growth. Record on the chalkboard, using a chart similar to the one below, the presence or absence of growth as (+) or (−). When all members of the class have recorded their results, complete this chart.

ORGANISM	60° C					70° C					80° C					90° C					100° C				
	C	10	20	30	40	C	10	20	30	40	C	10	20	30	40	C	10	20	30	40	C	10	20	30	40
S. aureus																									
E. coli																									
B. megaterium																									

1. If they can be determined from the above information, record the **thermal death points** for each of the organisms.

   S. aureus: ______________ *E. coli:* ______________ *B. megaterium:* ______________

2. From the above table, determine the **thermal death times** for each organism at the tabulated temperatures.

ORGANISM	THERMAL DEATH TIME				
	60° C	70° C	80° C	90° C	100° C
S. aureus					
E. coli					
B. megaterium					

B. Questions

1. Give three reasons why endospores are much more resistant to heat than are vegetative cells.

   a. ______________________________

   b. ______________________________

   c. ______________________________

2. Differentiate between the following:

   Thermoduric: ____

   ____

   Thermophilic: ____

   ____

3. List four diseases caused by spore-forming bacteria.

   a. ____________________ b. ____________________

   c. ____________________ d. ____________________

4. Since boiling water is unreliable in destroying endospores, how should one use heat in medical applications to ensure spore destruction? (three ways)

   a. ____

   b. ____

   c. ____

LABORATORY REPORT **25**

Student: ____________________

Desk No.: __________ Section: ____________

Osmotic Pressure and Bacterial Growth

A. Results

Record the amount of growth of each organism at the different salt concentrations, using +, ++, +++, and none to indicate degree of growth.

ORGANISM	SODIUM CHLORIDE CONCENTRATION							
	0.5%		5%		10%		15%	
	48 hr	96 hr	48 hr	96 hr	48 hr	96 hr	48 hr	96 hr
Escherichia coli								
Staphylococcus aureus								
Halobacterium salinarium								

B. Questions

1. Evaluate the salt tolerance of the above organisms.

   Tolerates very little salt: ____________________

   Tolerates a broad range of salt concentration: ____________________

   Grows at only high salt concentration: ____________________

2. How would you classify *Halobacterium salinarium* as to salt needs? Check one.

   Obligate halophile __________

   Facultative halophile __________

   Give your reason for this decision: ____________________

3. Supply the following information concerning **mannitol salt agar** (see the Difco manual).

   Composition: ____________________

   ____

   For what organism is this medium selective? ____________________

   What ingredient makes it selective? ____________________

LABORATORY REPORT **26**

Student: ______________________

Desk No.: ________ Section: ______________

pH and Microbial Growth

A. *Tabulation of Results*

If a photocolorimeter is available, dispense the cultures into labeled cuvettes and determine the percent transmittance of each culture. Calculate the O.D. values from the percent transmittance, using the formula given in Exercise 19. To complete the tables, get the results of the other three organisms from other members of the class, and delete the substitution organisms in the tables that were not used.

If no photocolorimeter is available, record only the visual reading at +, ++, +++, and none.

pH	***S. CEREVISIAE* OR *CANDIDA GLABRATA***			***ESCHERICHIA COLI***		
	Visual Reading	Photocolorimeter		Visual Reading	Photocolorimeter	
		%T	O.D.		%T	O.D.
5						
7						
9						

pH	***SPOROSARCINA UREAE* OR *ALCALIGENES SPP.***			***STAPHYLOCOCCUS AUREUS***		
	Visual Reading	Photocolorimeter		Visual Reading	Photocolorimeter	
		%T	O.D.		%T	O.D.
5						
7						
9						

B. Growth Curves

Once you have computed all the O.D. values on the two tables, plot them on the following graph. Use different colored lines for each species.

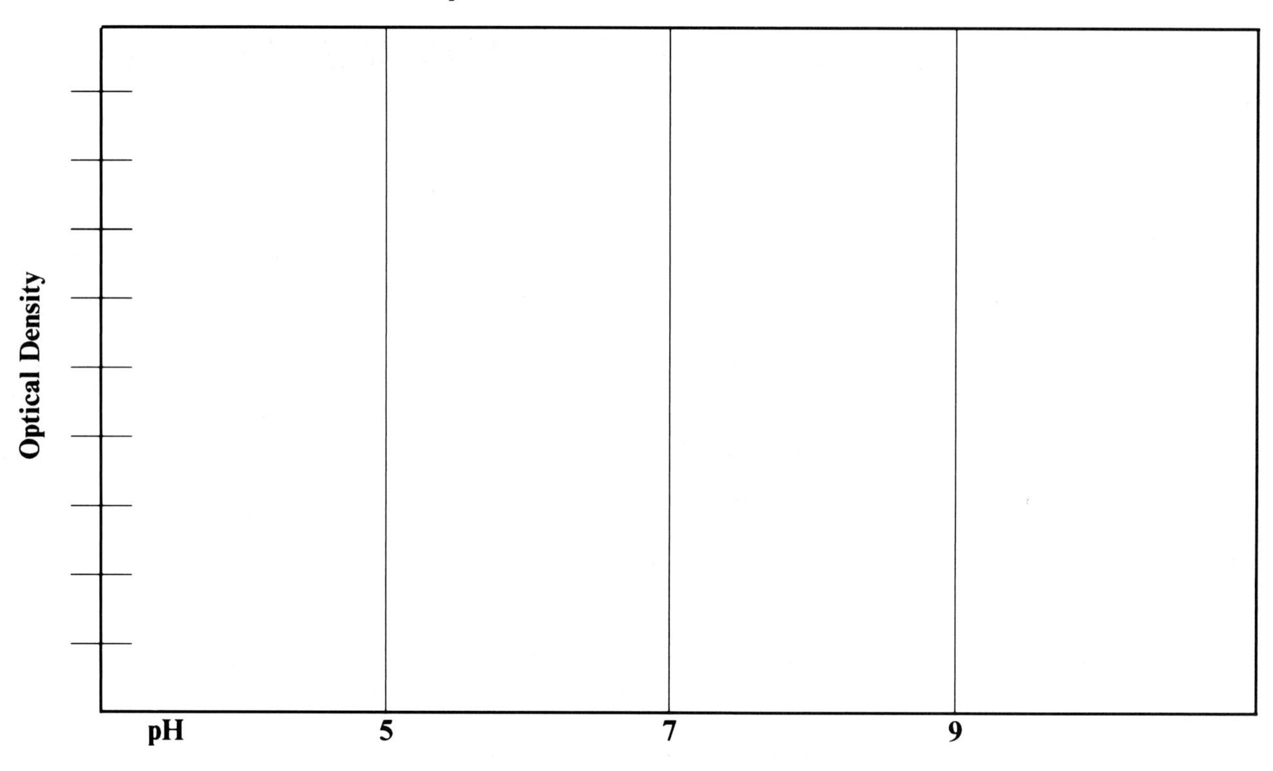

C. Questions

1. Which organism seems to grow best in acid media? ______
2. Which organism seems to grow best in alkaline media? ______
3. Which organism seems to tolerate the broadest pH range? ______

LABORATORY REPORT **27**

Student: ____________________

Desk No.: ________ Section: ____________

Ultraviolet Light: Lethal Effects

A. Tabulation of Results

Your instructor will construct a table similar to the one below on the chalkboard for you to record your results. If substantial growth is present in the exposed area, record your results as +++. If three or fewer colonies survived, record +. Moderate survival should be indicated as ++. No growth should be recorded as −. Record all information in the table.

ORGANISMS	EXPOSURE TIMES						
Staphylococcus aureus	10 sec	20 sec	40 sec	80 sec	2½ min	5 min	10 min
Survival:							
Bacillus megaterium	1 min	2 min	4 min	8 min	15 min	30 min	60 min
Survival:							

B. Questions

1. What length of time is required for the destruction of non-spore-forming bacteria such as *Staphylococcus aureus?*

   ____

2. Can you express, quantitatively, how much more resistant *B. megaterium* spores are to ultraviolet light than *S. aureus* vegetative cells (i.e., how many *times* more resistant are they)?

   ____

   ____

   ____

3. Why is it desirable to remove the cover from the petri dish when making exposures?

   ____

4. In what specific way does ultraviolet light destroy microorganisms?

   ____

   ____

5. What adverse effect can result from overexposure of human tissues to ultraviolet light?

____

____

____

____

____

6. What wavelength of ultraviolet is most germicidal? ______________________________

7. List several practical applications of ultraviolet light to microbial control.

____

____

____

____

____

____

____

____

LABORATORY REPORT **28**

Student: ____________________

Desk No.: ________ Section: ____________

Disinfectants: Evaluation Technique

A. *Tabulation of Results*

The instructor will draw a table on the chalkboard similar to the one below. Examine your tubes of nutrient broth and pins by shaking them and looking for growth (turbidity). If you are doubtful as to whether or not growth is present, compare the tubes with a tube of sterile nutrient broth. Record on the chalkboard a plus (+) sign if growth is present and a minus (−) sign if no growth is visible. After all students have recorded their results, complete the following chart.

DISINFECTANT		**MINUTES**											
		Staphylococcus aureus						*Bacillus megaterium*					
	Substitution	**C***	**1**	**5**	**10**	**30**	**60**	**C***	**1**	**5**	**10**	**30**	**60**
1:750 Zephiran													
5% phenol													
8% formaldehyde													

C* = control tube

B. *Questions*

1. What conclusions can be drawn from this experiment?

2. Distinguish between the following:

   Disinfectant: ______________________________

   ____

   Antiseptic: ______________________________

   ____

3. What factors other than time influence the action of a chemical agent on bacteria?

   ____

   ____

   ____

   ____

   ____

   ____

   ____

   ____

   ____

   ____

4. Fill in the equation which explains how the **phenol coefficient** is determined:

   P.C. = ______________________________

5. What are some drawbacks that one encounters when attempting to apply the phenol coefficient to all disinfectants?

   ____

   ____

   ____

   ____

   ____

   ____

LABORATORY REPORT **29**

Student: ____________________

Desk No.: ________ Section: ____________

Antiseptics: Evaluation Techniques

A. *Tabulation of Results*

With a millimeter scale, measure the zones of inhibition between the edge of the filter paper disk and the organisms. Record this information. Exchange your plates with other students' plates to complete the measurements for all chemical agents.

DISINFECTANT	MILLIMETERS OF INHIBITION	
	Staphylococcus aureus	*Pseudomonas aeruginosa*
5% phenol		
5% formaldehyde		
5% iodine		

B. *Questions*

1. What conclusions can be derived from these results? ____________________

2. What factors influence the size of the zone of inhibition? ____________________

LABORATORY REPORT **30**

Student: ____________________

Desk No.: __________ Section: ____________

Alcohol Evaluation: Its Effectiveness as a Skin Degerming Agent

A. *Tabulation of Results*

Count the number of colonies that appear on each of the thumbprints and record them in the following table. If the number of colonies has increased in the second press, record a 0 in percent reduction. Calculate the percentages of reduction and record this data in the appropriate column. Use this formula:

$$\text{Percent Reduction} = \frac{(\text{Colony Count 1st press}) - (\text{Colony Count 2nd press})}{(\text{Colony Count 1st press})} \times 100$$

LEFT THUMB (Control)			RIGHT THUMB (Dipped)			RIGHT THUMB (Swabbed)		
Colony Count 1st Press	Colony Count 2nd Press	Percent Reduction	Colony Count 1st Press	Colony Count 2nd Press	Percent Reduction	Colony Count 1st Press	Colony Count 2nd Press	Percent Reduction
Av. % Reduction, Left (C)			Av. % Reduction, Right (D)			Av. % Reduction, Right (S)		

B. Questions

1. In general, what effect does alcohol have on the level of skin contaminants? ______

______

2. Is there any difference between the effects of dipping versus swabbing? ______

   Which method appears to be more effective? ______

3. There is definitely survival of some microorganisms even after alcohol treatment. Without staining or microscopic scrutiny, predict what types of microbes are growing on the medium where you made the right thumb impression after treatment. ______

______

______

______

______

LABORATORY REPORT **31**

Student: ____________

Desk No.: ______ Section: ______

Antimicrobic Sensitivity Testing
(Kirby-Bauer Method)

A. Tabulation

List the antimicrobics that were used for each organism. Consult tables 31.1 and 31.2 to identify the various disks. After measuring and recording the zone diameters, consult table VII in appendix A for interpretation. Record the degrees of sensitivity (R, I, or S) in the sensitivity column. Exchange data with other class members to complete the entire chart.

	ANTIMICROBIC	ZONE DIA.	RATING (R, I, S)	ANTIMICROBIC	ZONE DIA.	RATING (R, I, S)
S. aureus						
P. aeruginosa						
Proteus vulgaris						
E. coli						

B. Questions

1. Which antimicrobics would be suitable for the control of the following organisms?

   S. aureus: ______________________________

   E. coli: ______________________________

   P. vulgaris: ______________________________

   P. aeruginosa: ______________________________

2. Differentiate between the following:

   Narrow spectrum antibiotic: ______________________________

   ____

   Broad spectrum antibiotic: ______________________________

   ____

3. Which antimicrobics used in this experiment would qualify as being excellent broad spectrum antimicrobics?

   ____

   ____

4. Differentiate between the following:

   Antibiotic: ______________________________

   ____

   Antimicrobic: ______________________________

   ____

5. How can drug resistance in microorganisms be circumvented? ______________________________

   ____

   ____

   ____

LABORATORY REPORT **32**

Student: ____________________

Desk No.: __________ Section: __________

Bacterial Mutagenicity and Carcinogenesis: The Ames Test

A. Tabulation of Results

Record the results of your tests in the following table and on a similar table on the chalkboard. Also record the results of substances tested by other students. A positive result will exhibit a zone of colonies similar to the zone shown on the plate in illustration 5, figure 32.1.

TEST SUBSTANCE	RESULT (+ or −)	TEST SUBSTANCE	RESULT (+ or −)	TEST SUBSTANCE	RESULT (+ or −)

B. Questions

1. Did you observe a zone of inhibition between the growing colonies and the impregnated disk on your positive plates? ____________________

   What is the cause of such a zone? ____________________

   ____

2. Differentiate between the following:

   Prototroph: ____________________

   Auxotroph: ____________________

3. Define **back mutation:** ____________________

   ____

   ____

4. List two characteristics of the Ames test that made this test so much superior to previous mutagenesis tests: ____________________

   ____

   ____

LABORATORY REPORT **33**

Student: ____________________

Desk No.: ________ Section: ____________

Effectiveness of Hand Scrubbing

A. Tabulation of Results

The instructor will draw a table on the chalkboard similar to the one below. Examine the six plates which your group inoculated from the basin of water. Select the two plates of a specific dilution which have approximately 30 to 300 colonies and count all of the colonies of each plate with a Quebec colony counter. Record the counts for each plate and their averages on the chalkboard. Once all the groups have recorded their counts, record the dilution factors for each group in the proper column. To calculate the organisms per milliliter multiply the average count by the dilution factor.

GROUP	0.1 ML COUNT		0.2 ML COUNT		0.4 ML COUNT		DILUTION FACTOR*	ORGANISMS PER MILLILITER
	Per Plate	Average	Per Plate	Average	Per Plate	Average		
A								
B								
C								
D								
E								

*Dilution factors: 0.1 ml = 10; 0.2 ml = 5; 0.4 ml = 2.5

B. Graph

After you have completed this tabulation, plot the number of organisms per milliliter that were present in each basin.

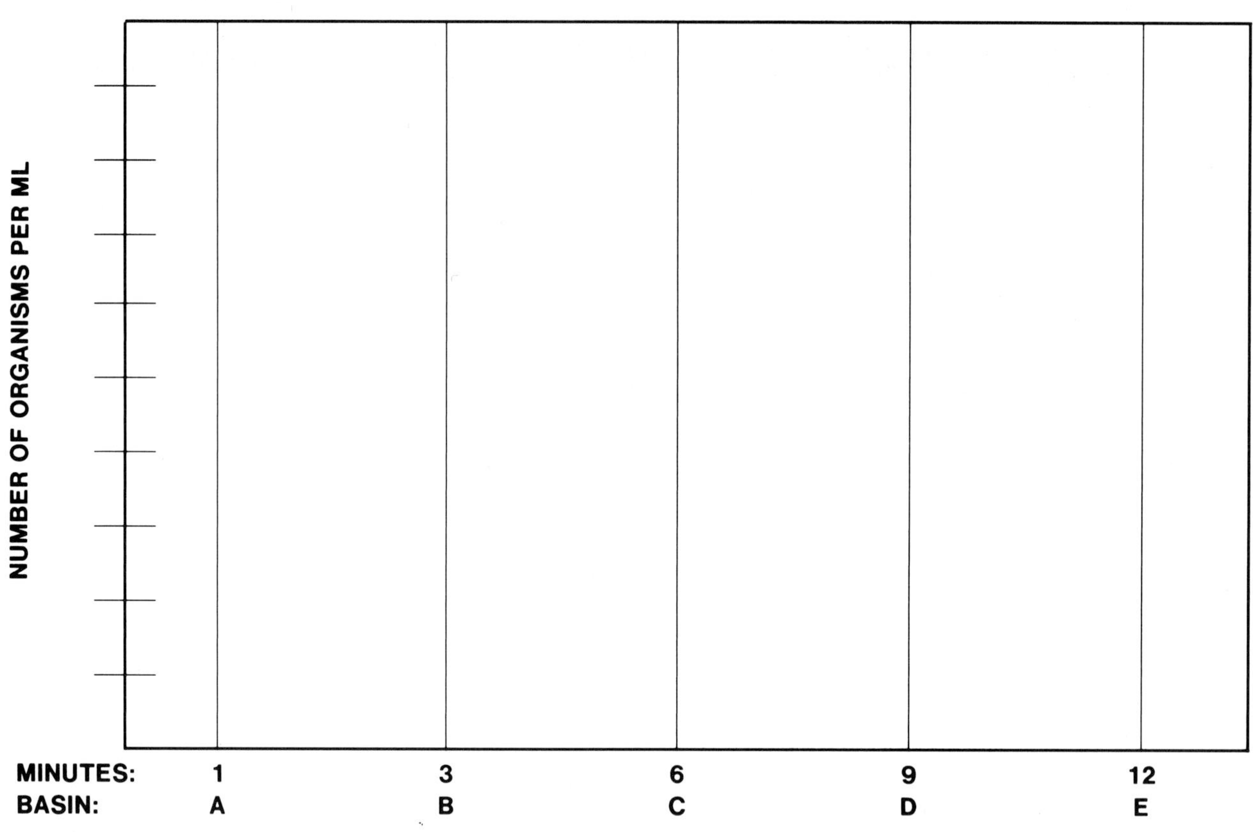

C. Questions

1. What conclusions can be derived from this exercise?

___

___

___

2. What might be an explanation of a higher count in basin D than in B, ruling out contamination or faulty techniques?

___

___

3. Why is it so important that surgeons scrub their hands prior to surgery even though they wear rubber gloves?

___

___

LABORATORY REPORT **34**

Student: ____________________

Desk No.: ________ Section: ____________

Preparation and Care of Stock Cultures

1. Why shouldn't cultures be stored at room temperature in your locker?

____

____

____

____

2. For what types of inoculations do you use your

reserve stock culture? ______________________________

____

working stock culture? ______________________________

____

3. What is **lyophilization?** ______________________________

____

____

____

____

What advantage does this procedure have over the method we are using for maintaining stock cultures?

____

____

____

____

____

LABORATORY REPORT **36**

Student: ____________________

Desk No.: ________ Section: ____________

Determination of Oxygen Requirements

A. *Observations*

After carefully comparing the appearance of the six cultures belonging to you and your laboratory partner, sketch in their characteristic growth.

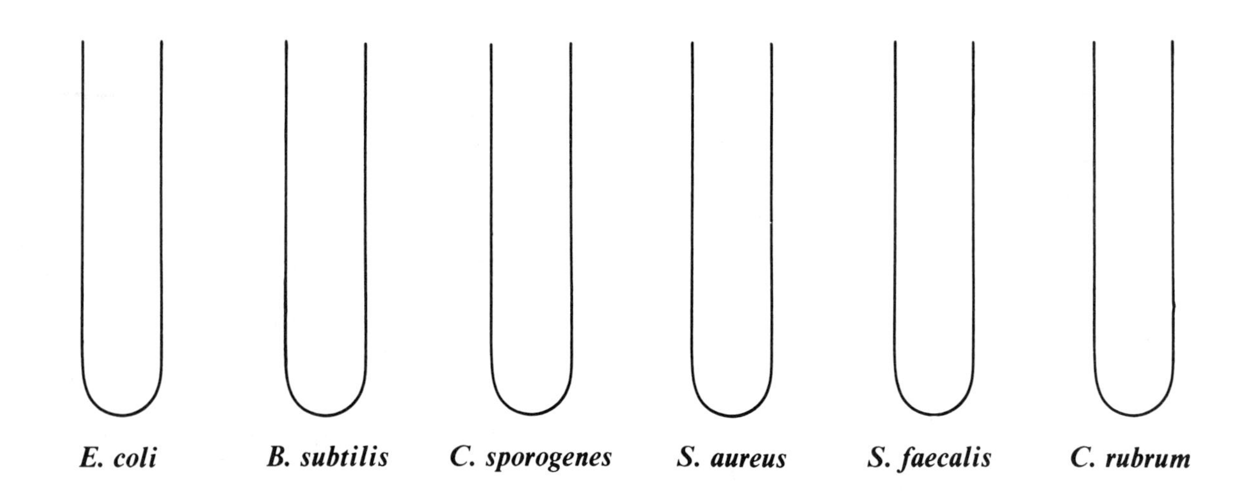

B. *Classification*

Classify each organism as to its oxygen requirements.

Escherichia coli: ____________________

Bacillus subtilis: ____________________

Clostridium sporogenes: ____________________

Staphylococcus aureus: ____________________

Streptococcus faecalis: ____________________

Clostridium rubrum: ____________________

Unknown No.: ( ) ____________________

Unknown No.: ( ) ____________________

LABORATORY REPORT **38–40**

Student: ____________________

Desk No.: ________ Section: ____________

Physiological Characteristics of Bacteria

A. *Media*

List the media that are used for the following tests:

1. Butanediol production
2. Hydrogen sulfide production
3. Indole production
4. Starch hydrolysis
5. Urease production
6. Citrate utilization
7. Fat hydrolysis
8. Casein hydrolysis
9. Catalase production
10. Mixed acid fermentation
11. Glucose fermentation
12. Nitrate reduction

B. *Reagents*

Select the reagents that are used for the following tests:

1. Indole test	Barritt's reagent—1
2. Voges-Proskauer test	Gram's iodine—2
3. Catalase test	Hydrogen peroxide—3
4. Starch hydrolysis	Kovacs' reagent—4
	None of these—5

C. *Ingredients*

Select the ingredients of the reagents for the following tests. Consult appendix B. More than one ingredient may be present in a particular reagent.

1. Oxidase test	α-naphthol—1
2. Voges-Proskauer test	Dimethyl-α-naphthylamine—2
3. Indole test	Dimethyl-ρ-phenylenediamine hydrochloride—3
4. Nitrite test	ρ-dimethylamine benzaldehyde—4
	Potassium hydroxide—5
	Sulfanilic acid—6

D. *Enzymes*

What enzymes are involved in the following reactions?

1. Urea hydrolysis
2. Hydrogen gas production from formic acid
3. Casein hydrolysis
4. Indole production
5. Nitrate reduction
6. Starch hydrolysis
7. Fat hydrolysis
8. Gelatin hydrolysis (Ex. 39)
9. Hydrogen sulfide production

Answers

Media

1. ____________
2. ____________
3. ____________
4. ____________
5. ____________
6. ____________
7. ____________
8. ____________
9. ____________
10. ____________
11. ____________
12. ____________

Reagents	Ingredients
1. ______	1. ______
2. ______	2. ______
3. ______	3. ______
4. ______	4. ______

Enzymes

1. ____________
2. ____________
3. ____________
4. ____________
5. ____________
6. ____________
7. ____________
8. ____________
9. ____________

E. Test Results

Indicate the appearance of the following positive test results.

1. Glucose fermentation, no gas
2. Citrate utilization
3. Urease production
4. Indole production
5. Acetoin production
6. Hydrogen sulfide production
7. Coagulation of milk
8. Peptonization in milk
9. Litmus reduction in milk
10. Nitrate reduction
11. Catalase production
12. Casein hydrolysis
13. Fat hydrolysis

Answers

1. ____________
2. ____________
3. ____________
4. ____________
5. ____________
6. ____________
7. ____________
8. ____________
9. ____________
10. ____________
11. ____________
12. ____________
13. ____________

F. General Questions

1. Differentiate between the following:

Respiration: ____

____

Fermentation: ____

____

Oxidation: ____

____

Reduction: ____

____

Catalase: ____

____

Peroxidase: ____

____

2. Differentiate between the following:

   Oxidative organism: ______________________________

   ____

   Fermentative organism: ______________________________

   ____

3. List two or three difficulties one encounters in trying to differentiate bacteria on the basis of physiological characteristics.

   ____

   ____

   ____

4. Now that you have determined the morphological, cultural, and physiological characteristics of your unknown, what other kinds of tests might you perform on the organism to assist in identification?

   ____

   ____

LABORATORY REPORT **42**

Student: ______________________

Desk No.: __________ Section: ______________

Enterobacteriaceae Identification: The API 20E System

A. *Tabulation of Results*

By referring to charts I and II, appendix D, determine the results of each test and record these results as positive (+) or negative (−) in the table below. Note that the results of the oxidase test must be recorded in the last column on the right side of the table.

ONPG	ADH	LDC	ODC	CIT	H_2S	URE	TDA	IND	VP	GEL	GLU	MAN	INO	SOR	RHA	SAC	MEL	AMY	ARA	OXI
1	2	4	1	2	4	1	2	4	1	2	4	1	2	4	1	2	4	1	2	4
	☐			☐			☐			☐			☐			☐			☐	

NO_2	N_2 GAS	MOT	MAC	OF-O	OF-F
1	2	4	1	2	4
	☐			☐	

Additional Digits

B. *Construction of Seven-Digit Profile*

Note in the above table that each test has a value of 1, 2, or 4. To compute the seven-digit profile for your unknown, total up the positive values for each group.

Example:
5 144 572 = ***E. coli.***

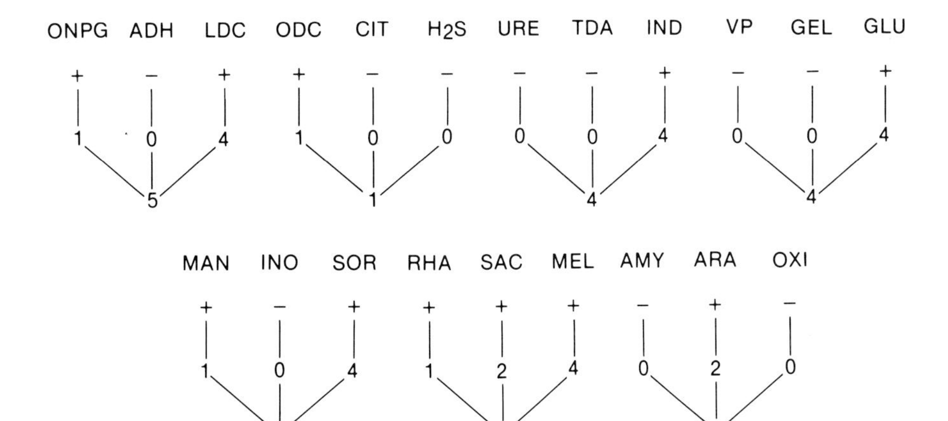

C. *Using the* API 20 E Analytic Index *or the API Characterization Chart*

If the *API 20E Analytical Index* is available on the demonstration table, utilize it to identify your unknown, using the seven-digit profile number that has been computed.

If no *Analytical Index* is available, use characterization chart III in appendix D.

Name of Unknown: ______________________

D. *Additional Tabulation Blanks*

Use the two forms below for recording information pertaining to additional unknowns.

api® 20E

Reference Number ____________ Patient ______________________ Date ____________

Source/Site ____________ Physician ______________________ Dept./Service ____________

	ONPG 1	ADH 2	LDC 4	ODC 1	CIT 2	H_2S 4	URE 1	TDA 2	IND 4	VP 1	GEL 2	GLU 4	MAN 1	INO 2	SOR 4	RHA 1	SAC 2	MEL 4	AMY 1	ARA 2	OXI 4
5 h																					
24 h																					
48 h																					
Profile Number																					

	NO_2 1	N_2 GAS 2	MOT 4	MAC 1	OF-O 2	OF-F 4
5 h						
24 h						
48 h						
Additional Digits						

Additional Information

Identification

00-42-012 E-3 (7/80)

api® 20E

Reference Number ____________ Patient ______________________ Date ____________

Source/Site ____________ Physician ______________________ Dept./Service ____________

	ONPG 1	ADH 2	LDC 4	ODC 1	CIT 2	H_2S 4	URE 1	TDA 2	IND 4	VP 1	GEL 2	GLU 4	MAN 1	INO 2	SOR 4	RHA 1	SAC 2	MEL 4	AMY 1	ARA 2	OXI 4
5 h																					
24 h																					
48 h																					
Profile Number																					

	NO_2 1	N_2 GAS 2	MOT 4	MAC 1	OF-O 2	OF-F 4
5 h						
24 h						
48 h						
Additional Digits						

Additional Information

Identification

00-42-012 E-3 (7/80)

LABORATORY REPORT **43**

Student: ____________________

Desk No.: ________ Section: ____________

Enterobacteriaceae Identification: The Enterotube II System

A. Tabulation of Results

Record the results of each test in the following table with a plus (+) or minus (−).

GLUCOSE	GAS PRODUCTION	LYSINE	ORNITHINE	H_2S	INDOLE	ADONITOL	LACTOSE	ARABINOSE	SORBITOL	VOGES-PROSKAUER	DULCITOL	PHENYLALANINE DEAMINASE	UREA	CITRATE

B. Identification by Chart Method

If no *Interpretation Guide* is available, apply the above results to chart IV, appendix D, to find the name of your unknown. Note that the spacing of the above table matches the size of the spaces on chart IV. If this page is removed from the manual, folded, and placed on chart IV, the results on the above table can be moved down the chart to make a quick comparison of your results with the expected results for each organism.

C. Using the Enterotube II Interpretation Guide

If the *Interpretation Guide* is available, determine the five-digit code number by encircling the numbers (4, 2, or 1) under each test that is positive, and then totaling these numbers within each group to form a digit for that group. Note that there are three tally charts on the back side of this Laboratory Report for your use.

Example:

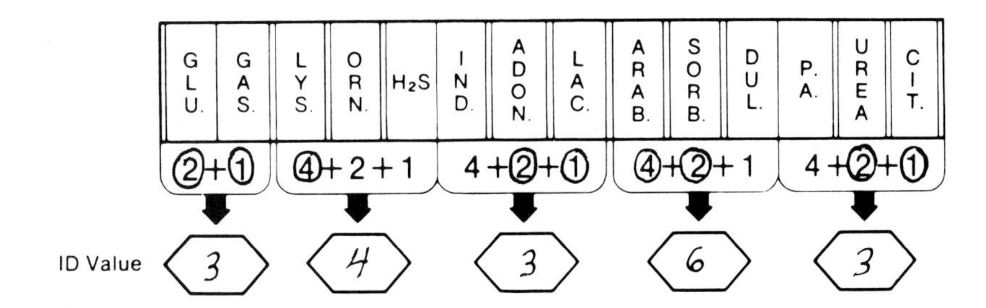

The "ID Value" 34363 can be found by thumbing the pages of the *Interpretation Guide*. You will find it on page 35. The listing is as follows:

ID Value	**Organism**	**Atypical Test Results**
34363	*Klebsiella pneumoniae*	None

Conclusion: Organism was correctly identified as *Klebsiella pneumoniae*. In this case, the identification was made independent of the V-P test.

ENTEROTUBE® II*

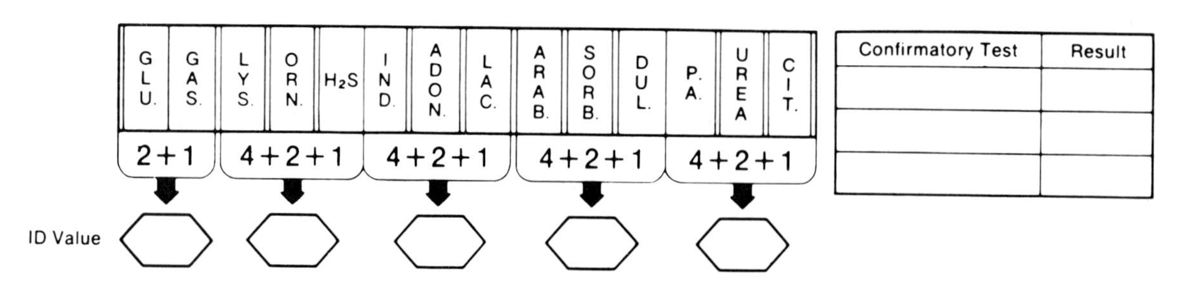

Culture Number, Case Number or Patient Name

Date

Organism Identified

*VP utilized as confirmatory test only.

ENTEROTUBE® II*

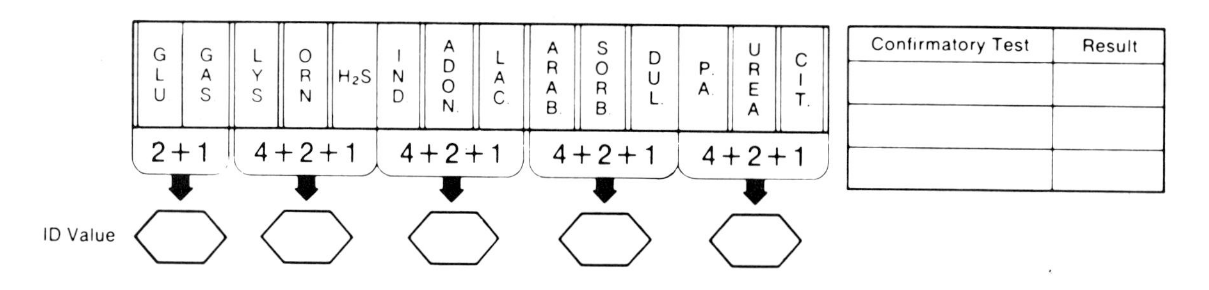

Culture Number, Case Number or Patient Name

Date

Organism Identified

*VP utilized as confirmatory test only.

ENTEROTUBE® II*

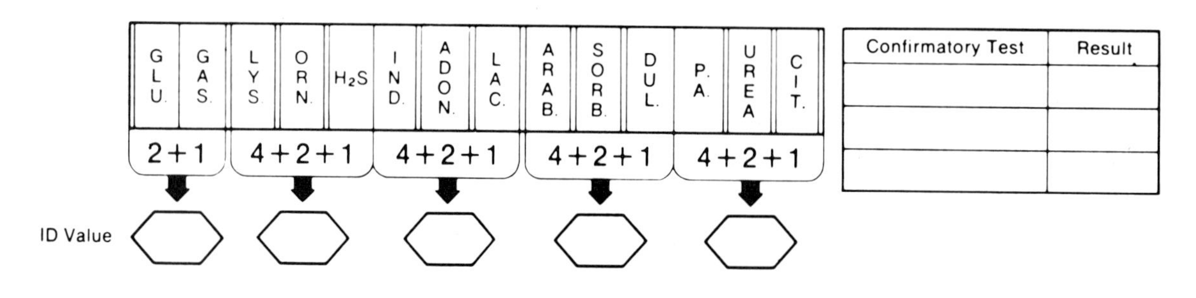

Culture Number, Case Number or Patient Name

Date

Organism Identified

*VP utilized as confirmatory test only.

LABORATORY REPORT **44**

Student: ______________________

Desk No.: ________ Section: ______________

O/F Gram-Negative Rods Identification:
The Oxi/Ferm Tube System

A. *Tabulation of Results*

Record the results of each test in the following table with a plus (+) or minus (−).

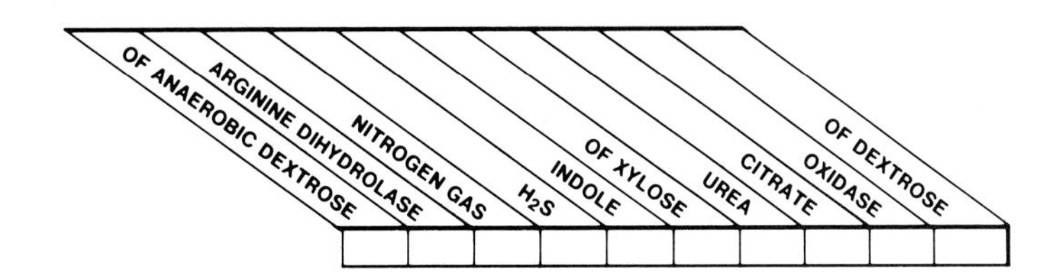

B. *Identification by Chart Method*

If no coding manual is available, apply the above results to chart V, appendix D, to find the name of your unknown. Note that the spacing of the above table matches the size of the spaces on chart V. If this page is removed from the manual, folded, and placed on chart V, the results on the above table can be moved down the chart to make a quick comparison of your results with the expected results for each organism.

C. *Computer Coding Method*

If the coding manual (*Oxi/Ferm Tube CCIS*) is available, determine the four-digit code number by encircling the numbers (4, 2, or 1) under each test that is positive, and then totaling these numbers within each group to form a digit for that group. Note that there are three tally charts on the reverse side of this Laboratory Report.

EXAMPLE

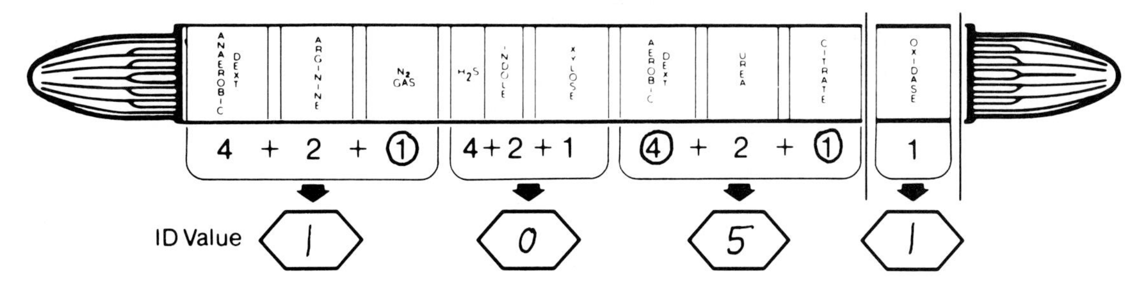

The "ID Value" 1051 is located in Part B of the coding manual. The listing is as follows:

ID Value	Organism
1051	*Pseudomonas stutzeri*

Conclusion: Organism was correctly identified as *Pseudomonas stutzeri.*

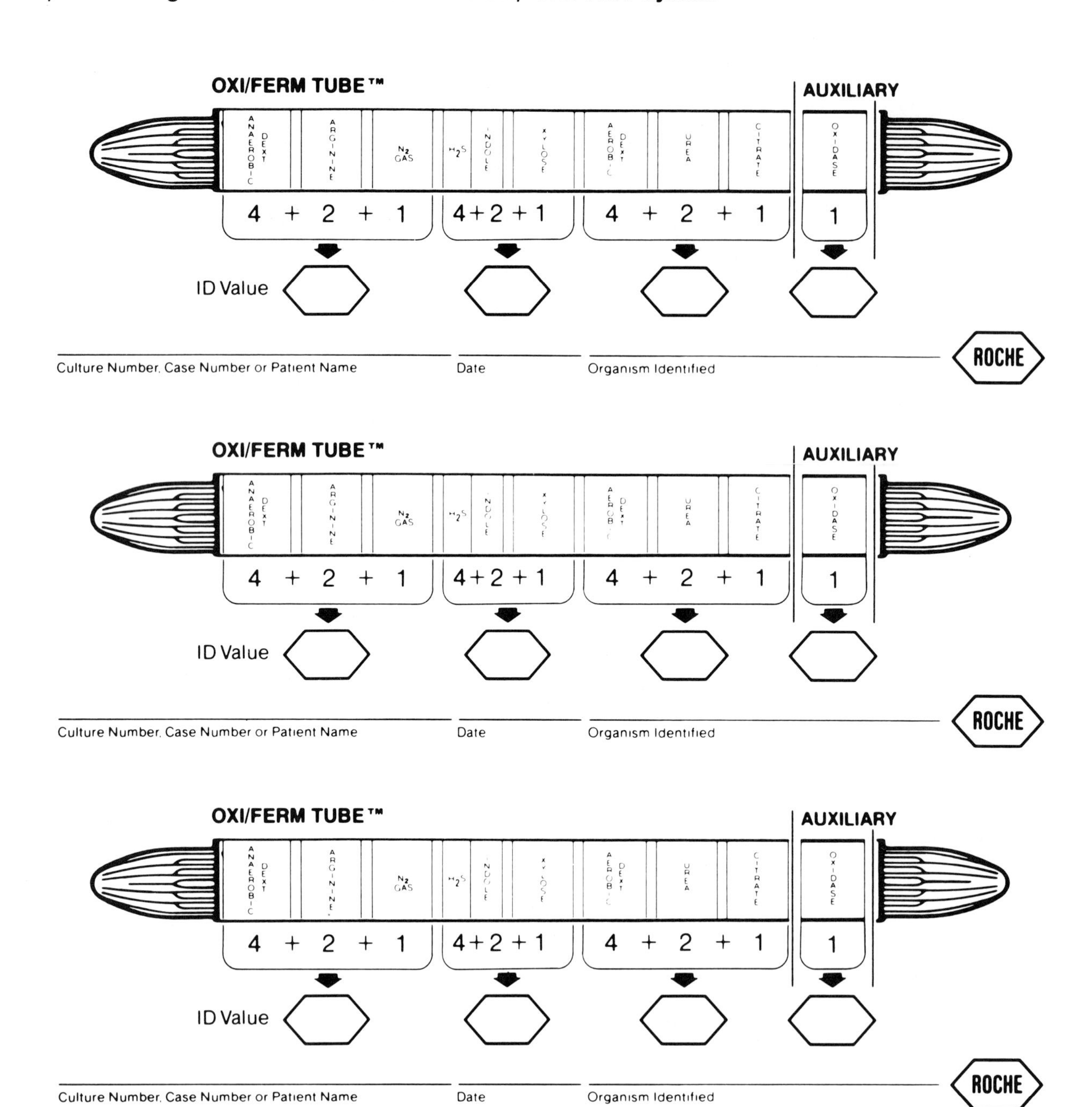
OXI/FERM TUBE™
AUXILIARY
ANAEROBIC DEXT
ARGININE
N2 GAS
H2S
INDOLE
XYLOSE
AEROBIC DEXT
UREA
CITRATE
OXIDASE
4 + 2 + 1
4+2+1
4 + 2 + 1
1
ID Value
Culture Number, Case Number or Patient Name
Date
Organism Identified
ROCHE

LABORATORY REPORT **45**

Student: ______________________

Desk No.: ________ Section: ______________

Staphylococcus Identification: The API Staph-Ident System

A. Tabulation of Results

By referring to charts VI and VII, appendix D, determine the results of each test, and record these results as positive (+) or negative (−) in the profile determination table below. Note that three of these tables have been printed on the back of this page for tabulation of additional organisms.

	PHS 1	URE 2	GLS 4	MNE 1	MAN 2	TRE 4	SAL 1	GLC 2	ARG 4	NGP 1
RESULTS										
PROFILE NUMBER										

GRAM STAIN [ ] COAGULASE [ ]

MORPHOLOGY [ ] CATALASE [ ]

Additional Information: [ ]

Identification: [ ]

B. Construction of Four-Digit Profile

Note in the above table that each test has a value of 1, 2, or 4. To compute the four-digit profile for your unknown, total up the positive values for each group.

Example: 7 700 = ***Staphylococcus aureus***

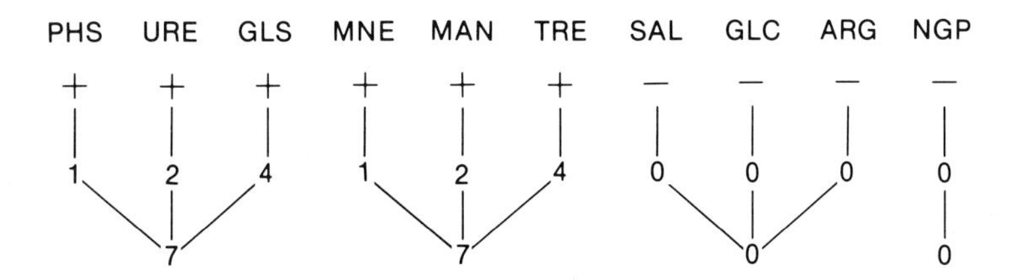

C. Final Determination

Refer to the Staph-Indent Profile Register (chart VIII, appendix D) to find the organism that matches your profile number. Write the name of your unknown in the space below and list any additional tests that are needed for final confirmation. If the materials are available for these tests, perform them.

Name of Unknown: ______________________________

Additional Tests: ______________________________

	PHS	URE	GLS	MNE	MAN	TRE	SAL	GLC	ARG	NGP
	1	2	4	1	2	4	1	2	4	1
RESULTS										

PROFILE NUMBER

GRAM STAIN
MORPHOLOGY
COAGULASE
CATALASE
Additional Information
Identification

	PHS	URE	GLS	MNE	MAN	TRE	SAL	GLC	ARG	NGP
	1	2	4	1	2	4	1	2	4	1
RESULTS										

PROFILE NUMBER

GRAM STAIN
MORPHOLOGY
COAGULASE
CATALASE
Additional Information
Identification

	PHS	URE	GLS	MNE	MAN	TRE	SAL	GLC	ARG	NGP
	1	2	4	1	2	4	1	2	4	1
RESULTS										

PROFILE NUMBER

GRAM STAIN
MORPHOLOGY
COAGULASE
CATALASE
Additional Information
Identification

LABORATORY REPORT **46**

Student: ______________________

Desk No.: ________ Section: ______________

Bacteriological Examination of Water (Multiple Tube Method)

A. Results of Presumptive Test (MPN Determination)

Record the number of positive tubes on the chalkboard and on the following table. When all students have recorded their results with the various water samples, complete this tabulation. Determine the MPN according to the instructions on page 178.

WATER SAMPLE (SOURCE)	NUMBER OF POSITIVE TUBES				MPN
	3 tubes DSLB 10 ml	3 tubes SSLB 1.0 ml	3 tubes SSLB 0.1 ml	3 tubes SSLB 0.01 ml	

B. Results of Confirmed Test

Record the results of the confirmed tests for each water sample that was positive on the presumptive test.

WATER SAMPLE (SOURCE)	POSITIVE	NEGATIVE

C. *Results of Completed Test*

Record the results of completed tests for each water sample that was positive on the confirmed test.

WATER SAMPLE (SOURCE)	LACTOSE FERMENTATION RESULTS	MORPHOLOGY	EVALUATION

D. *Questions*

1. Does a positive presumptive test mean that the water is absolutely unsafe to drink?

   ____

   Explain: ______________________________

   ____

   ____

2. What might be the explanation of false positive presumptive results? ______________________________

   ____

   ____

   ____

3. List three characteristics that an organism must have to qualify it as a sewage indicator.

   ____

   ____

   ____

4. What enteric **bacterial** diseases are transmitted in polluted water? ______________________________

   ____

   ____

   ____

5. What **protozoan** enteric disease is transmitted by polluted water? ______________________________

   ____

6. Why don't health departments routinely test for pathogens instead of using a sewage indicator?

7. Give the functions of the various media used in these tests:

   Lactose broth:

   Levine's EMB agar:

   Nutrient agar slant:

8. What other media are used for confirmatory tests?

LABORATORY REPORT **47**

Student: ____________

Desk No.: ________ Section: ____________

Bacteriological Examination of Water Membrane Filter Method

A. Tabulation

A table similar to the one below will be provided for you on the chalkboard. Record your coliform count on it. After all students have recorded their results, complete this table.

SAMPLE	SOURCE	COLIFORM COUNT	AMOUNT OF WATER FILTERED	MPN*
A				
B				
C				
D				
E				
F				
G				
H				

$$\text{MPN*} = \frac{\text{Coliform Count} \times 100}{\text{Amount of Water Filtered}}$$

B. Questions

1. Give two limitations of the membrane filter technique.

a. ____________

____

____

b. ____________

____

____

____

2. List some additional applications of the membrane filter in microbiology.

LABORATORY REPORT **48**

Student: ______________________

Desk No.: __________ Section: __________

Standard Plate Count of Milk

A. Tabulation of Results

After you have made your plate count, record your results on the following table. Get the results of the other milk sample from some member of the class.

TYPE OF MILK	PLATE COUNT	DILUTION	ORGANISMS PER ML
High-quality			
Poor-quality			

B. Questions

1. Do plate count figures actually represent numbers of organisms or numbers of clumps of bacteria?

2. What are some factors that will produce errors in the SPC technique?

3. What might be the explanation of a very high count in raw milk which has been properly refrigerated from the time of collection?

4. What is the most common source of bacteria in milk? ______________________

5. What is grade A raw milk? ______________________________

____

____

____

6. What is certified milk? ______________________________

____

____

____

____

7. Why is milk a more suitable vector of disease than water? ______________________________

____

____

____

8. What infectious diseases of cows can be transmitted to man via milk? ______________________________

____

____

____

LABORATORY REPORT **49**

Student: ____________________

Desk No.: ________ Section: ____________

Bacterial Counts of Foods

A. Tabulation of Results

Record your count and the bacterial counts of various other foods made by other students.

TYPE OF FOOD	PLATE COUNT	DILUTION	ORGANISMS PER ML

What generalizations can be drawn from this exercise?

____

____

____

____

____

____

B. Questions

1. What dangers and undesirable results may occur from ground meats of high bacterial counts?

____

____

2. What bacterial pathogens might be present in frozen foods?

3. What harm can result from repeated thawing and freezing of foods?

LABORATORY REPORT **51**

Student: ____________________

Desk No.: ________ Section: ____________

The Staphylococci: Isolation and Identification

A. Tabulation

At the beginning of the third laboratory period, the instructor will construct a chart similar to this one on the chalkboard. After examining your mannitol salt agar and staphylococcus medium 110 plates, record the presence (+) or absence (−) of staphylococcus growth in the appropriate columns. After performing coagulase tests on the various isolates record the results also as (+) or (−) in the appropriate columns.

STUDENT INITIALS	NOSE			FOMITES			
	Staph Colonies		Coagulase	Item	Staph Colonies		Coagulase
	MSA	SM110			MSA	SM110	

B. *Microscopy*

Provide drawings here of the various isolates as seen under oil immersion (Gram staining).

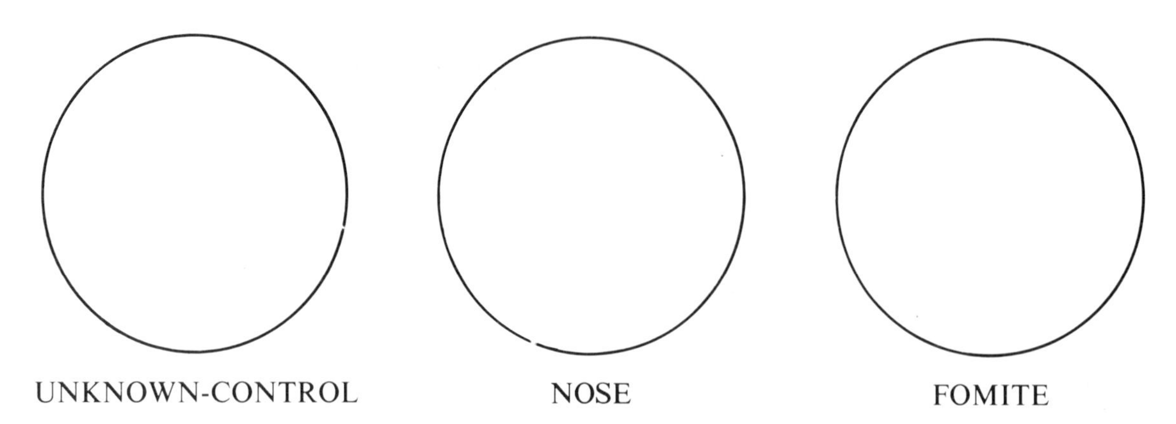

C. *Percentages*

From the data in the table on the previous page, determine the incidence (percentage) of individuals and fomites that harbor coagulase-positive and coagulase-negative staphylococci in this experiment.

SOURCE	TOTAL TESTED	TOTAL POSITIVE	PERCENTAGE POSITIVE	TOTAL NEGATIVE	PERCENTAGE NEGATIVE
Humans (Nose)					
Fomites					

D. *Record of Test Results*

Record here the results of each test performed in this experiment. Under GRAM STAIN indicate cellular arrangement as well as Gram reaction.

ISOLATE	GRAM STAIN	ALPHA TOXIN	MANNITOL (ACID)	COAGULASE
Unknown Control No.____				
Nose Isolate No. 1				
Nose Isolate No. 2				
Fomite Isolate				

E. *Final Determination*

Record here the name of your unknown-control. If API Staph-Ident miniaturized multitest strips are available, confirm your conclusions by testing each isolate. See Exercise 45.

Name of unknown-control: ______________________________

Staph-Ident results: ______________________________

LABORATORY REPORT **52**

Student: ______________________________

Desk No.: __________ Section: ______________

The Streptococci:
Isolation and Identification

A. *Tabulation of Pharynx Isolates*

The instructor will construct a chart similar to this one on the chalkboard. After examining the blood agar plates that were inoculated with pharynx organisms, record the types and size range of colonies that are present on your plates. Record this data first on this table, then on the chalkboard. After all students have recorded their results on the board, complete the tabulation of their results here, also. The names of the organisms will not be recorded until all tests are completed.

STUDENT INITIALS	TYPE OF HEMOLYSIS (ALPHA, ALPHA-PRIME, BETA)	SIZE RANGE OF COLONIES (MM)	NAMES OF ORGANISMS

B. Microscopy

Provide drawings here of the various pharyngeal isolates as seen under oil immersion (Gram staining).

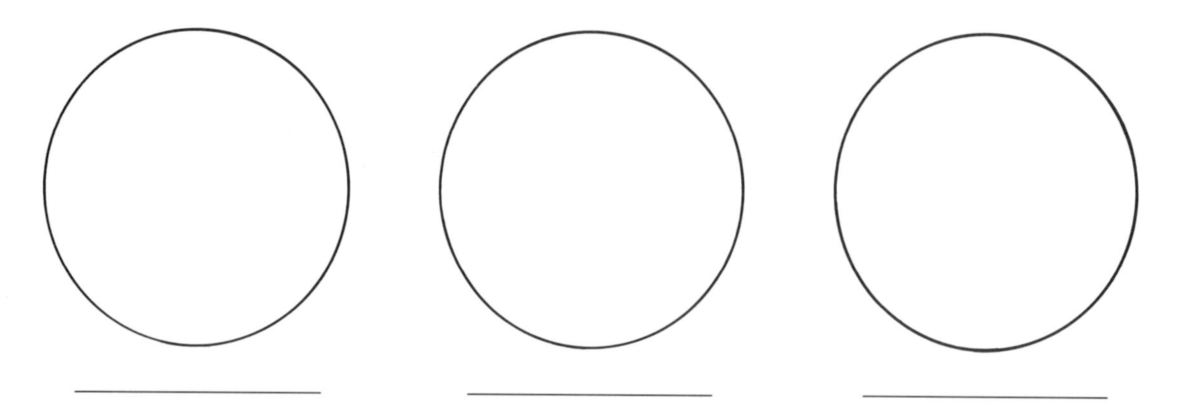

C. Percentages

From the data in the table on the previous page, calculate the percentages for each type of streptococci that were isolated from classmates.

S. pyogenes: ________________

S. agalactiae: ________________

S. pneumoniae: ________________

Viridans streptococci: ________

Group C streptococci: ____________

Group D enterococci: ____________

Group D nonenterococci: _________

D. Record of Test Results

Record here all information pertaining to the identification of pharyngeal isolates and unknowns.

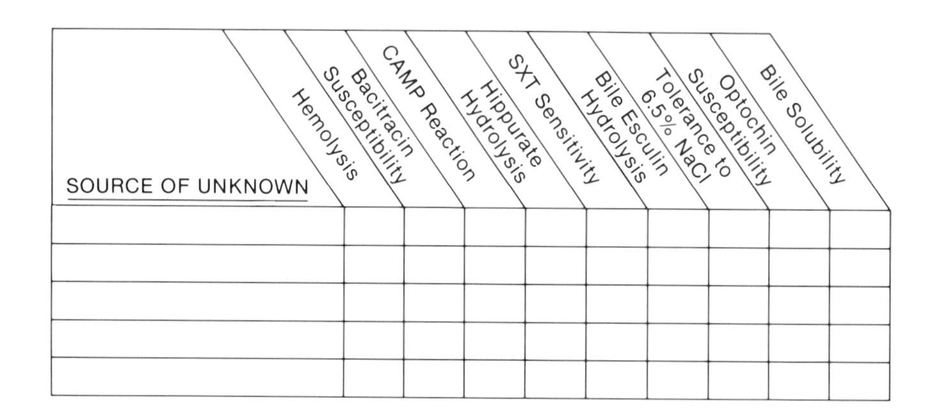

SOURCE OF UNKNOWN	Hemolysis	Bacitracin Susceptibility	CAMP Reaction	Hippurate Hydrolysis	SXT Sensitivity	Bile Esculin Hydrolysis	Tolerance to 6.5% NaCl	Optochin Susceptibility	Bile Solubility

E. Final Determination

Record here the identities of your various isolates and unknowns:

Pharyngeal isolates: ____________________________

____

Unknowns: ____________________________

____

____

F. Questions

Record the answers for the following questions in the answer column.

1. What two physiological tests are significant in the identification of *S. agalactiae?*
2. If an alpha hemolytic streptococcus is able to hydrolyze bile esculin, what test can be used to tell whether or not the organism is an enterococcus?
3. What test is used for differentiating group A from group C streptococci if both organisms are bacitracin susceptible?
4. What two tests are used to differentiate pneumococci from the viridans group?
5. What test is used for differentiating *S. pyogenes* from other beta hemolytic streptococci?
6. Hemolysis in streptococci can only be evaluated when the colonies develop ____________ (*aerobically* or *anaerobically*) in blood agar.
7. Which streptococcal species is frequently present in the vagina of third-trimester pregnant women?
8. Only one beta hemolytic streptococcus is primarily of human origin. Which one is it?
9. Who developed the system of classifying streptococci into groups A, B, C, etc.?
10. Who is credited with grouping streptococci according to the type of hemolysis?
11. Which streptococcal species is seen primarily as paired cells (diplococci)?
12. Name two species of streptococci that are implicated in dental caries.
13. Where in the body can *S. bovis* be found?
14. After performing all physiological tests, what type of tests must be performed to confirm identification?
15. Which hemolysin produced by *S. pyogenes* is responsible for the beta-type hemolysis that is characteristic of this organism?

Answers

1. a. ____________
   b. ____________
2. ____________
3. ____________
4. a. ____________
   b. ____________
5. ____________
6. ____________
7. ____________
8. ____________
9. ____________
10. ____________
11. ____________
12. a. ____________
    b. ____________
13. ____________
14. ____________
15. ____________

LABORATORY REPORT **53**

Student: ______________________

Desk No.: ________ Section: ____________

Gram-Negative Intestinal Pathogens

A. Unknown Identification

1. What was the genus of your unknown?

   ____________________ ________________
   Genus No.

2. What problems, if any, did you encounter?

3. Now that you know the genus of your unknown, what steps would you follow to determine the species?

B. General Questions

1. What ingredient in DCA and SS agar makes them selective?

2. What ingredients in DCA and SS agar cause coliforms to appear red while salmonella and shigella appear white?

3. What characteristics do the salmonella and shigella have in common?

4. How do the salmonella and shigella differ?

5. What restrictions might be placed on a person who is a typhoid carrier?

LABORATORY REPORT **54**

Student: ____________________

Desk No.: __________ Section: __________

Urinary Tract Pathogens

A. *Quantitative Evaluation*

After counting the colonies on the TSA plate, record the count as follows:

Number of colonies: ____________________

Dilution: ____________________

No. of organisms/ml of urine: ____________________

Gram-Stained Slide. If organisms are seen on a gram-stained slide of an uncentrifuged sample, sketch in color in the circle below.

Conclusion: Do the plate count and gram-stained slide of the uncentrifuged sample provide presumptive evidence of a urinary infection?

____

B. *Microscopic Study (Centrifuged Sample)*

Illustrate in the circle below the microscopic appearance of a centrifuged sample.

Conclusion: Describe here the morphological appearance of the predominant organism seen:

____

____

____

____

C. *Culture Analyses*

After studying the organisms on the three plates and thioglycollate medium, what organism do you believe is causing the infection?

Organism: ____________________

What further testing should be performed for confirmation?

____

____

____

LABORATORY REPORT **55**

Student: ______________________

Desk No.: __________ Section: __________

Slide Agglutination Test: Serological Typing

1. Record the unknown number which proved to be a salmonella. ______________
2. Why was phenolized saline used instead of plain physiological saline?

____

____

3. If your results were negative for both cultures, what might be the explanation?

____

____

____

____

____

LABORATORY REPORT **56**

Student: ____________________

Desk No.: ________ Section: ________

Tube Agglutination Test: The Widal Test

1. What was the titer of the serum which you tested?

 ____

2. Differentiate between the following:

 Serum: ____________________

 Antiserum: ____________________

 Antitoxin: ____________________

3. How would you prepare antiserum for an organism such as *E. coli?*

 ____

4. Indicate the type of antigen (*soluble protein, red blood cells,* or *bacteria*) that is used for each of the following serological tests.

 Agglutination: ____________________

 Precipitation: ____________________

 Hemolysis: ____________________

5. For which one of the above tests is complement necessary?

 ____

LABORATORY REPORT **57**

Student: ______________________

Desk No.: ________ Section: ____________

Phage Typing

1. To which phage types was this strain of *S. aureus* susceptible?

____

2. To what lytic group does this strain of staphylococcus belong?

____

3. In what way can bacteriophage alter the genetic structure of a bacterium?

____

____

____

____

____

LABORATORY REPORT **58**

Student: ______________________

Desk No.: __________ Section: __________

Blood Grouping

1. Record your blood type: ______________________

2. What agglutinins are present in each type of blood?

   Type A: ______________ Type AB: ______________

   Type B: ______________ Type O: ______________

3. Why does a person of type A blood go into a transfusion reaction when given type B blood?

   ____

   ____

   ____

4. Why can Rh-positive blood be given only once to a person who is Rh-negative?

   ____

   ____

   ____

   ____

   ____

LABORATORY REPORT **59**

Student: ____________________

Desk No.: ________ Section: ____________

White Blood Cell Study: Differential Blood Count

A. Tabulation of Results

As you encounter each type of leukocyte, record it on the following chart. After counting 50 or 100 cells, as time permits, calculate the percentages from the totals of each type.

NEUTROPHILS	LYMPHOCYTES	MONOCYTES	EOSINOPHILS	BASOPHILS
Totals				
Percent				

B. Questions

1. Were your percentages for each type within the normal ranges?

____

____

____

____

____

2. What errors might one be likely to make when doing this count for the first time?

____

____

____

____

____

____

3. Differentiate between the following:

   Cellular immunity: ______________________________

   Humoral immunity: ______________________________

4. Do cellular and humoral immunity work independently? ______________

   Explain: ______________________________

   ____

   ____

   ____

   ____

   ____

   ____

LABORATORY REPORT **60**

Student: ____________________

Desk No.: ________ Section: ____________

Fluorescent Antibody Technique

1. Record the number and species of your unknown. ____________________

2. Why is it necessary to let the slide stand in a moist chamber for 15 to 30 minutes? ____________

3. What would result if the slide were not washed off in the phosphate buffer solution?

4. What organisms other than streptococci are being routinely identified today with this technique?

Descriptive Chart

STUDENT: ______________________

LAB SECTION: ______________________

Habitat: __________ Culture No.: __________

Source: ______________________

Organism: ______________________

MORPHOLOGICAL CHARACTERISTICS

Cell Shape:

Arrangement:

Size:

Spores:

Gram stain:

Motility:

Capsules:

Special Stains:

CULTURAL CHARACTERISTICS

Colonies:

Nutrient Agar:

Blood Agar:

Agar Slant:

Nutrient Broth:

Gelatin Stab:

Oxygen Requirements:

Optimum Temp.:

PHYSIOLOGICAL CHARACTERISTICS

	TESTS	RESULTS
Fermentation	Glucose	
	Lactose	
	Sucrose	
	Mannitol	
Hydrolysis	Gelatin Liquefaction	
	Starch	
	Casein	
	Fat	
IMViC	Indole	
	Methyl Red	
	V-P (acetylmethylcarbinol)	
	Citrate Utilization	
	Nitrate Reduction	
	H_2S Production	
	Urease	
	Catalase	
	Oxidase	

	REACTION	TIME
Litmus Milk	Acid	______
	Alkaline	______
	Coagulation	______
	Reduction	______
	Peptonization	______
	No Change	______

Descriptive Chart

STUDENT: ____________________

LAB SECTION: ____________________

Habitat: __________ Culture No.: __________

Source: ____________________

Organism: ____________________

MORPHOLOGICAL CHARACTERISTICS

Cell Shape:

Arrangement:

Size:

Spores:

Gram stain:

Motility:

Capsules:

Special Stains:

CULTURAL CHARACTERISTICS

Colonies:

Nutrient Agar:

Blood Agar:

Agar Slant:

Nutrient Broth:

Gelatin Stab:

Oxygen Requirements:

Optimum Temp.:

PHYSIOLOGICAL CHARACTERISTICS

	TESTS	RESULTS
Fermentation	Glucose	
	Lactose	
	Sucrose	
	Mannitol	
Hydrolysis	Gelatin Liquefaction	
	Starch	
	Casein	
	Fat	
IMViC	Indole	
	Methyl Red	
	V-P (acetylmethylcarbinol)	
	Citrate Utilization	
	Nitrate Reduction	
	H_2S Production	
	Urease	
	Catalase	
	Oxidase	

	REACTION	TIME
Litmus Milk	Acid	______
	Alkaline	______
	Coagulation	______
	Reduction	______
	Peptonization	______
	No Change	______

Descriptive Chart

STUDENT: ______________________

LAB SECTION: ______________________

Habitat: __________ Culture No.: __________

Source: ______________________

Organism: ______________________

MORPHOLOGICAL CHARACTERISTICS

Cell Shape:

Arrangement:

Size:

Spores:

Gram stain:

Motility:

Capsules:

Special Stains:

CULTURAL CHARACTERISTICS

Colonies:

Nutrient Agar:

Blood Agar:

Agar Slant:

Nutrient Broth:

Gelatin Stab:

Oxygen Requirements:

Optimum Temp.:

PHYSIOLOGICAL CHARACTERISTICS

	TESTS	RESULTS
Fermentation	Glucose	
	Lactose	
	Sucrose	
	Mannitol	
Hydrolysis	Gelatin Liquefaction	
	Starch	
	Casein	
	Fat	
IMViC	Indole	
	Methyl Red	
	V-P (acetylmethylcarbinol)	
	Citrate Utilization	
	Nitrate Reduction	
	H_2S Production	
	Urease	
	Catalase	
	Oxidase	

	REACTION	TIME
Litmus Milk	Acid	________
	Alkaline	________
	Coagulation	________
	Reduction	________
	Peptonization	________
	No Change	________

Appendix Tables A

Table I International Atomic Weights

Element	Symbol	Atomic Number	Atomic Weight
Aluminum	Al	13	26.97
Antimony	Sb	51	121.76
Arsenic	As	33	74.91
Barium	Ba	56	137.36
Beryllium	Be	4	9.013
Bismuth	Bi	83	209.00
Boron	B	5	10.82
Bromine	Br	35	79.916
Cadmium	Cd	48	112.41
Calcium	Ca	20	40.08
Carbon	C	6	12.010
Chlorine	Cl	17	35.457
Chromium	Cr	24	52.01
Cobalt	Co	27	58.94
Copper	Cu	29	63.54
Fluorine	F	9	19.00
Gold	Au	79	197.2
Hydrogen	H	1	1.0080
Iodine	I	53	126.92
Iron	Fe	26	55.85
Lead	Pb	82	207.21
Magnesium	Mg	12	24.32
Manganese	Mn	25	54.93
Mercury	Hg	80	200.61
Nickel	Ni	28	58.69
Nitrogen	N	7	14.008
Oxygen	O	8	16.0000
Palladium	Pd	46	106.7
Phosphorus	P	15	30.98
Platinum	Pt	78	195.23
Potassium	K	19	39.096
Radium	Ra	88	226.05
Selenium	Se	34	78.96
Silicon	Si	14	28.06
Silver	Ag	47	107.880
Sodium	Na	11	22.997
Strontium	Sr	38	87.63
Sulfur	S	16	32.066
Tin	Sn	50	118.70
Titanium	Ti	22	47.90
Tungsten	W	74	183.92
Uranium	U	92	238.07
Vanadium	V	23	50.95
Zinc	Zn	30	65.38
Zirconium	Zr	40	91.22

Table II Four-Place Logarithms

N	0	1	2	3	4	5	6	7	8	9
10	0000	0043	0086	0128	0170	0212	0253	0294	0334	0374
11	0414	0453	0492	0531	0569	0607	0645	0682	0719	0755
12	0792	0828	0864	0899	0934	0969	1004	1038	1072	1106
13	1139	1173	1206	1239	1271	1303	1335	1367	1399	1430
14	1461	1492	1523	1553	1584	1614	1644	1673	1703	1732
15	1761	1790	1818	1847	1875	1903	1931	1959	1987	2014
16	2041	2068	2095	2122	2148	2175	2201	2227	2253	2279
17	2304	2330	2355	2380	2405	2430	2455	2480	2504	2529
18	2553	2577	2601	2625	2648	2672	2695	2718	2742	2765
19	2788	2810	2833	2856	2878	2900	2923	2945	2967	2989
20	3010	3032	3054	3075	3096	3118	3139	3160	3181	3201
21	3222	3243	3263	3284	3304	3324	3345	3365	3385	3404
22	3424	3444	3464	3483	3502	3522	3541	3560	3579	3598
23	3617	3636	3655	3674	3692	3711	3729	3747	3766	3784
24	3802	3820	3838	3856	3874	3892	3909	3927	3945	3962
25	3979	3997	4014	4031	4048	4065	4082	4099	4116	4133
26	4150	4166	4183	4200	4216	4232	4249	4265	4281	4298
27	4314	4330	4346	4362	4378	4393	4409	4425	4440	4456
28	4472	4487	4502	4518	4533	4548	4564	4579	4594	4609
29	4624	4639	4654	4669	4683	4698	4713	4728	4742	4757
30	4771	4786	4800	4814	4829	4843	4857	4871	4886	4900
31	4914	4928	4942	4955	4969	4983	4997	5011	5024	5038
32	5051	5065	5079	5092	5105	5119	5132	5145	5159	5172
33	5185	5198	5211	5224	5237	5250	5263	5276	5289	5302
34	5315	5328	5340	5353	5366	5378	5391	5403	5416	5428
35	5441	5453	5465	5478	5490	5502	5514	5527	5539	5551
36	5563	5575	5587	5599	5611	5623	5635	5647	5658	5670
37	5682	5694	5705	5717	5729	5740	5752	5763	5775	5786
38	5798	5809	5821	5832	5843	5855	5866	5877	5888	5899
39	5911	5922	5933	5944	5955	5966	5977	5988	5999	6010
40	6021	6031	6042	6053	6064	6075	6085	6096	6107	6117
41	6128	6138	6149	6160	6170	6180	6191	6201	6212	6222
42	6232	6243	6253	6263	6274	6284	6294	6304	6314	6325
43	6335	6345	6355	6365	6375	6385	6395	6405	6415	6425
44	6435	6444	6454	6464	6474	6484	6493	6503	6513	6522
45	6532	6542	6551	6561	6571	6580	6590	6599	6609	6618
46	6628	6637	6646	6656	6665	6675	6684	6693	6702	6712
47	6721	6730	6739	6749	6758	6767	6776	6785	6794	6803
48	6812	6821	6830	6839	6848	6857	6866	6875	6884	6893
49	6902	6911	6920	6928	6937	6946	6955	6964	6972	6981
50	6990	6998	7007	7016	7024	7033	7042	7050	7059	7067
51	7076	7084	7093	7101	7110	7118	7126	7135	7143	7152
52	7160	7168	7177	7185	7193	7202	7210	7218	7226	7235
53	7243	7251	7259	7267	7275	7284	7292	7300	7308	7316
54	7324	7332	7340	7348	7356	7364	7372	7380	7388	7396
N	0	1	2	3	4	5	6	7	8	9

Table II *(continued)*

N	0	1	2	3	4	5	6	7	8	9
55	7404	7412	7419	7427	7435	7443	7451	7459	7466	7474
56	7482	7490	7497	7505	7513	7520	7528	7536	7543	7551
57	7559	7566	7574	7582	7589	7597	7604	7612	7619	7627
58	7634	7642	7649	7657	7664	7672	7679	7686	7694	7701
59	7709	7716	7723	7731	7738	7745	7752	7760	7767	7774
60	7782	7789	7796	7803	7810	7818	7825	7832	7839	7846
61	7853	7860	7868	7875	7882	7889	7896	7903	7910	7917
62	7924	7931	7938	7945	7952	7959	7966	7973	7980	7987
63	7993	8000	8007	8014	8021	8028	8035	8041	8048	8055
64	8062	8069	8075	8082	8089	8096	8102	8109	8116	8122
65	8129	8136	8142	8149	8156	8162	8169	8176	8182	8189
66	8195	8202	8209	8215	8222	8228	8235	8241	8248	8254
67	8261	8267	8274	8280	8287	8293	8299	8306	8312	8319
68	8325	8331	8338	8344	8351	8357	8363	8370	8376	8382
69	8388	8395	8401	8407	8414	8420	8426	8432	8439	8445
70	8451	8457	8463	8470	8476	8482	8488	8494	8500	8506
71	8513	8519	8525	8531	8537	8543	8549	8555	8561	8567
72	8573	8579	8585	8591	8597	8603	8609	8615	8621	8627
73	8633	8639	8645	8651	8657	8663	8669	8675	8681	8686
74	8692	8698	8704	8710	8716	8722	8727	8733	8739	8745
75	8751	8756	8762	8768	8774	8779	8785	8791	8797	8802
76	8808	8814	8820	8825	8831	8837	8842	8848	8854	8859
77	8865	8871	8876	8882	8887	8893	8899	8904	8910	8915
78	8921	8927	8932	8938	8943	8949	8954	8960	8965	8971
79	8976	8982	8987	8993	8998	9004	9009	9015	9020	9025
80	9031	9036	9042	9047	9053	9058	9063	9069	9074	9079
81	9085	9090	9096	9101	9106	9112	9117	9122	9128	9133
82	9138	9143	9149	9154	9159	9165	9170	9175	9180	9186
83	9191	9196	9201	9206	9212	9217	9222	9227	9232	9238
84	9243	9248	9253	9258	9263	9269	9274	9279	9284	9289
85	9294	9299	9304	9309	9315	9320	9325	9330	9335	9340
86	9345	9350	9355	9360	9365	9370	9375	9380	9385	9390
87	9395	9400	9405	9410	9415	9420	9425	9430	9435	9440
88	9445	9450	9455	9460	9465	9469	9474	9479	9484	9489
89	9494	9499	9504	9509	9513	9518	9523	9528	9533	9538
90	9542	9547	9552	9557	9562	9566	9571	9576	9581	9586
91	9590	9595	9600	9605	9609	9614	9619	9624	9628	9633
92	9638	9643	9647	9652	9657	9661	9666	9671	9675	9680
93	9685	9689	9694	9699	9703	9708	9713	9717	9722	9727
94	9731	9736	9741	9745	9750	9754	9759	9763	9768	9773
95	9777	9782	9786	9791	9795	9800	9805	9809	9814	9818
96	9823	9827	9832	9836	9841	9845	9850	9854	9859	9863
97	9868	9872	9877	9881	9886	9890	9894	9899	9903	9908
98	9912	9917	9921	9926	9930	9934	9939	9943	9948	9952
99	9956	9961	9965	9969	9974	9978	9983	9987	9991	9996
100	0000	0004	0009	0013	0017	0022	0026	0030	0035	0039
N	0	1	2	3	4	5	6	7	8	9

Table III Temperature Conversion Table Centigrade to Fahrenheit

°C.	0	1	2	3	4	5	6	7	8	9
–50	**–58.0**	**–59.8**	**–61.6**	**–63.4**	**–65.2**	**–67.0**	**–68.8**	**–70.6**	**–72.4**	**–74.2**
–40	–40.0	–41.8	–43.6	–45.4	–47.2	–49.0	–50.8	–52.6	–54.4	–56.2
–30	–22.0	–23.8	–25.6	–27.4	–29.2	–31.0	–32.8	–34.6	–36.4	–38.2
–20	– 4.0	– 5.8	– 7.6	– 9.4	–11.2.	–13.0	–14.8	–16.6	–18.4	–20.2
–10	+14.0	+12.2	+10.4	+ 8.6	+ 6.8	+ 5.0	+ 3.2	+ 1.4	– 0.4	– 2.2
– 0	+32.0	+30.2	+28.4	+26.6	+24.8	+23.0	+21.2	+19.4	+17.6	+15.8
0	**32.0**	**33.8**	**35.6**	**37.4**	**39.2**	**41.0**	**42.8**	**44.6**	**46.4**	**48.2**
10	50.0	51.8	53.6	55.4	57.2	59.0	60.8	62.6	64.4	66.2
20	68.0	69.8	71.6	73.4	75.2	77.0	78.8	80.6	82.4	84.2
30	86.0	87.8	89.6	91.4	93.2	95.0	96.8	98.6	100.4	102.2
40	104.0	105.8	107.6	109.4	111.2	113.0	114.8	116.6	118.4	120.2
50	122.0	123.8	125.6	127.4	129.2	131.0	132.8	134.6	136.4	138.2
60	**140.0**	**141.8**	**143.6**	**145.4**	**147.2**	**149.0**	**150.8**	**152.6**	**154.4**	**156.2**
70	158.0	159.8	161.6	163.4	165.2	167.0	168.8	170.6	172.4	174.2
80	176.0	177.8	179.6	181.4	183.2	185.0	186.8	188.6	190.4	192.2
90	194.0	195.8	197.6	199.4	201.2	203.0	204.8	206.6	208.4	210.2
100	212.0	213.8	215.6	217.4	219.2	221.0	222.8	224.6	226.4	228.2
110	**230.0**	**231.8**	**233.6**	**235.4**	**237.2**	**239.0**	**240.8**	**242.6**	**244.4**	**246.2**
120	248.0	249.8	251.6	253.4	255.2	257.0	258.8	260.6	262.4	264.2
130	266.0	267.8	269.6	271.4	273.2	275.0	276.8	278.6	280.4	282.2
140	284.0	285.8	287.6	289.4	291.2	293.0	294.8	296.6	298.4	300.2
150	302.0	303.8	305.6	307.4	309.2	311.0	312.8	314.6	316.4	318.2
160	**320.0**	**321.8**	**323.6**	**325.4**	**327.2**	**329.0**	**330.8**	**332.6**	**334.4**	**336.2**
170	338.0	339.8	341.6	343.4	345.2	347.0	348.8	350.6	352.4	354.2
180	356.0	357.8	359.6	361.4	363.2	365.0	366.8	368.6	370.4	372.2
190	374.0	375.8	377.6	379.4	381.2	383.0	384.8	386.6	388.4	390.2
200	392.0	393.8	395.6	397.4	399.2	401.0	402.8	404.6	406.4	408.2
210	**410.0**	**411.8**	**413.6**	**415.4**	**417.2**	**419.0**	**420.8**	**422.6**	**424.4**	**426.2**
220	428.0	429.8	431.6	433.4	435.2	437.0	438.8	440.6	442.4	444.2
230	446.0	447.8	449.6	451.4	453.2	455.0	456.8	458.6	460.4	462.2
240	464.0	465.8	467.6	469.4	471.2	473.0	474.8	476.6	478.4	480.2
250	482.0	483.8	485.6	487.4	489.2	491.0	492.8	494.6	496.4	498.2

°F. = °C. × 9/5 + 32   °C. = °F. – 32 × 5/9

Table IV Autoclave Steam Pressures and Corresponding Temperatures

Steam Pressure lbs./sq. in.	Temperature °C.	Temperature °F.	Steam Pressure lbs./sq. in.	Temperature °C.	Temperature °F.	Steam Pressure lbs./sq. in.	Temperature °C.	Temperature °F.
0	100.0	212.0						
1	101.9	215.4	11	116.4	241.5	21	126.9	260.4
2	103.6	218.5	12	117.6	243.7	22	127.8	262.0
3	105.3	221.5	13	118.8	245.8	23	128.7	263.7
4	106.9	224.4	14	119.9	247.8	24	129.6	265.3
5	108.4	227.1	15	121.0	249.8	25	130.4	266.7
6	109.8	229.6	16	122.0	251.6	26	131.3	268.3
7	111.3	232.3	17	123.0	253.4	27	132.1	269.8
8	112.6	234.7	18	124.1	255.4	28	132.9	271.2
9	113.9	237.0	19	125.0	257.0	29	133.7	272.7
10	115.2	239.4	20	126.0	258.8	30	134.5	274.1

Figures are for steam pressure only and the presence of any air in the autoclave invalidates temperature readings from the above table.

Table V Autoclave Temperatures as Related to the Presence of Air

Gauge Pressure, lb.	Pure steam, complete air discharge		Two-thirds air discharge, 20-in. vacuum		One-half air discharge, 15-in. vacuum		One-third air discharge, 10-in. vacuum		No air discharge	
	°C.	°F.	°C.	°F.	°C.	°F.	°C.	°F.	°C.	°F.
5	109	228	100	212	94	202	90	193	72	162
10	115	240	109	228	105	220	100	212	90	193
15	121	250	115	240	112	234	109	228	100	212
20	126	259	121	250	118	245	115	240	109	228
25	130	267	126	259	124	254	121	250	115	240
30	135	275	130	267	128	263	126	259	121	250

Table VI MPN Determination from Multiple Tube Test

NUMBER OF TUBES GIVING POSITIVE REACTION OUT OF			MPN Index per 100 ml.	95 PERCENT CONFIDENCE LIMITS	
3 of 10 ml. each	3 of 1 ml. each	3 of 0.1 ml. each		Lower	Upper
0	0	1	3	<0.5	9
0	1	0	3	<0.5	13
1	0	0	4	<0.5	20
1	0	1	7	1	21
1	1	0	7	1	23
1	1	1	11	3	36
1	2	0	11	3	36
2	0	0	9	1	36
2	0	1	14	3	37
2	1	0	15	3	44
2	1	1	20	7	89
2	2	0	21	4	47
2	2	1	28	10	150
3	0	0	23	4	120
3	0	1	39	7	130
3	0	2	64	15	380
3	1	0	43	7	210
3	1	1	75	14	230
3	1	2	120	30	380
3	2	0	93	15	380
3	2	1	150	30	440
3	2	2	210	35	470
3	3	0	240	36	1,300
3	3	1	460	71	2,400
3	3	2	1,100	150	4,800

From: **Standard Methods for the Examination of Water and Wastewater,** Twelfth edition. (New York: The American Public Health Association, Inc., p. 608.)

Table VII Antimicrobic Zone of Inhibition Evaluation (Kirby-Bauer Method) Significance of Zone Diameters When Using High-Potency Antimicrobial Sensitivity Disks

Antimicrobial Agent	Disk Potency	R Resistant *mm.*	I Intermediate *mm.*	S Sensitive *mm.*
Amikacin	10 mcg	<12	12-13	>13
Ampicillin				
Gram-negative organisms and enterococci	10 mcg	<12	12-13	>13
Staphylococci and penicillin G susceptibles	10 mcg	<21	21-28	>28
Bacitracin	10 units	< 9	9-12	>12
Carbenicillin				
For *Proteus spp.* and *E. coli*	50 mcg	<18	18-22	>22
For *Pseudomonas aeruginosa*	50 mcg	<13	13-14	>14
Cephalothin				
For cephaloglycin only	30 mcg	<15		>14
For other cephalosporins	30 mcg	<15	15-17	>17
Chloramphenicol	30 mcg	<13	13-17	>17
Clindamycin	2 mcg	<15	15-16	>16
Colistin	10 mcg	< 9	9-10	>10
Erythromycin	15 mcg	<14	14-17	>17
Gentamicin				
For *Ps. aeruginosa*	10 mcg	<13		>12
Kanamycin	30 mcg	<14	14-17	>17
Lincomycin (Clindamycin)	2 mcg	<17	17-20	>20
Methicillin				
(Penicillinase-resistant penicillin class)	5 mcg	<10	10-13	>13
Nafcillin	1 mcg	<11	11-12	>12
Nalidixic Acid	30 mcg	<14	14-18	>18
Neomycin	30 mcg	<13	13-16	>16
Nitrofurantoin	300 mcg	<15	15-16	>16
Novobiocin	30 mcg	<18	18-21	>21
Oleandomycin	15 mcg	<21	12-16	>16
Oxolinic Acid	2 mcg	<11		>10
Penicillin G				
For staphylococci	10 units	<21	21-28	>28
For other organisms	10 units	<12	12-21+	>21
Polymyxin B	300 units	< 9	9-11	>11
Rifampin (for *Neisseria meningitidis* only)	5 mcg	<25		>24
Streptomycin	10 mcg	<12	12-14	>14
Tetracycline	30 mcg	<15	15-18	>18
Tobramycin	10 mcg	<12	12-13	>13
Triple Sulfa	250 mcg	<13	13-16	>16
Vancomycin	30 mcg	<10	10-11	>11

Table VIII Indicators of Hydrogen Ion Concentration

Many of the following indicators are used in the media of certain exercises in this manual. This table indicates the pH range of each indicator and the color changes which occur. To determine the exact pH within a particular range one should use a set of standard colorimetric tubes that are available from the prep room. Consult your lab instructor.

Indicators	Full Acid Color	Full Alkaline Color	pH Range
Cresol Red	red	yellow	0.2 – 1.8
Meta Cresol Purple (acid range)	red	yellow	1.2 – 2.8
Thymol Blue	red	yellow	1.2 – 2.8
Brom Phenol Blue	yellow	blue	3.0 – 4.6
Brom Cresol Green	yellow	blue	3.8 – 5.4
Chlor Cresol Green	yellow	blue	4.0 – 5.6
Methyl Red	red	yellow	4.4 – 6.4
Chlor Phenol Red	yellow	red	4.8 – 6.4
Brom Cresol Purple	yellow	purple	5.2 – 6.8
Brom Thymol Blue	yellow	blue	6.0 – 7.6
Neutral Red	red	amber	6.8 – 8.0
Phenol Red	yellow	red	6.8 – 8.4
Cresol Red	yellow	red	7.2 – 8.8
Meta Cresol Purple (alk. range)	yellow	purple	7.4 – 9.0
Thymol Blue (alkaline range)	yellow	blue	8.0 – 9.6
Cresolphthalein	colorless	red	8.2 – 9.8
Phenolphthalein	colorless	red	8.3 –10.0

Appendix Indicators, Stains, Reagents

INDICATORS

All the indicators used in this manual can be made by (1) dissolving a measured amount of the indicator in 95% ethanol, (2) adding a measured amount of water, and (3) filtering with filter paper. The following chart provides the correct amounts of indicator, alcohol, and water for various indicator solutions.

Indicator Solution	Indicator (gm)	95% Ethanol (ml)	Distilled H_2O (ml)
Bromcresol green	0.4	500	500
Bromcresol purple	0.4	500	500
Bromthymol blue	0.4	500	500
Cresol red	0.2	500	500
Methyl red	0.2	500	500
Phenolphthalein	1.0	50	50
Phenol red	0.2	500	500
Thymol blue	0.4	500	500

STAINS AND REAGENTS

Acid-Alcohol (for Ziehl-Neelsen stain)

3 ml concentrated hydrochloric acid in 100 ml of 95% ethyl alcohol.

Alcohol, 70% (from 95%)

Alcohol, 95% .... 368.0 ml
Distilled water .... 132.0 ml

Barritt's Reagent (Voges-Proskauer test)

Solution A: 6 gm alpha-naphthol in 100 ml 95% ethyl alcohol.
Solution B: 16 gm potassium hydroxide in 100 ml water.

Carbolfuchsin Stain (Ziehl's)

Solution A: Dissolve 0.3 gm of basic fuchsin (90% dye content) in 10 ml 95% ethyl alcohol.
Solution B: Dissolve 5 gm of phenol in 95 ml of water.
Mix solutions A and B.

Crystal Violet Stain (Hucker modification)

Solution A: Dissolve 2.0 gm of crystal violet (85% dye content) in 20 ml of 95% ethyl alcohol.
Solution B: Dissolve 0.8 gm ammonium oxalate in 80.0 ml distilled water.
Mix solutions A and B.

Diphenylamine Reagent (nitrate test)

Dissolve 0.7 gm diphenylamine in a mixture of 60 ml of concentrated sulfuric acid and 28.8 ml of distilled water.

Cool and add slowly 11.3 ml of concentrated hydrochloric acid. After the solution has stood for 12 hours some of the base separates, showing that the reagent is saturated.

Ferric Chloride Reagent (Ex. 73)

$FeCl_3 \cdot 6H_2O$ .... 12 gm
2% Aqueous HCl .... 100 ml

Make up the 2% aq. HCl by adding 5.4 ml of concentrated HCl (37%) to 94.6 ml H_2O. Inoculate with two or three colonies of beta hemolytic streptococci, incubate at 35° C for 20 or more hours. Centrifuge the medium to pack the cells, and pipette 0.8 ml of the clear supernate into a Kahn tube. Add 0.2 ml of the ferric chloride reagent to the Kahn tube and mix well. If a heavy precipitate remains longer than 10 minutes, the test is positive.

Gram's Iodine (Lugol's)

Dissolve 2.0 gm of potassium iodide in 300 ml of distilled water and then add 1.0 gm iodine crystals.

Kovacs' Reagent (indole test)

n-amyl alcohol .... 75.0 ml
Hydrochloric acid (conc.) .... 25.0 ml
ρ-dimethylamine-benzaldehyde ..... 5.0 gm

Lactophenol Cotton Blue Stain

Phenol crystals .... 20 gm
Lactic acid .... 20 ml
Glycerol .... 40 ml
Cotton blue .... 0.05 gm
Dissolve the phenol crystals in the other ingredients by heating the mixture gently under a hot water tap.

Malachite Green Solution (spore stain)

Dissolve 5.0 gm malachite green oxalate in 100 ml distilled water.

McFarland Nephelometer Barium Sulfate Standards (Ex. 45)

Prepare 1% aqueous barium chloride and 1% aqueous sulfuric acid solutions.

Add the amounts indicated in table 1 to clean, dry ampoules. Ampoules should have the same diameter as the test tube to be used in subsequent density determinations.

Seal the ampoules and label them.

Table 1 Amounts for Standards

Tube	Barium Chloride 1% (ml)	Sulfuric Acid 1% (ml)	Corresponding Approx. Density of Bacteria (million/ml)
1	0.1	9.9	300
2	0.2	9.8	600
3	0.3	9.7	900
4	0.4	9.6	1200
5	0.5	9.5	1500
6	0.6	9.4	1800
7	0.7	9.3	2100
8	0.8	9.2	2400
9	0.9	9.1	2700
10	1.0	9.0	3000

Methylene Blue (Loeffler's)

Solution A: Dissolve 0.3 gm of methylene blue (90% dye content) in 30.0 ml ethyl alcohol (95%).

Solution B: Dissolve 0.01 gm potassium hydroxide in 100.0 ml distilled water. Mix solutions A and B.

Naphthol, alpha

5% alpha-naphthol in 95% ethyl alcohol

Caution: Avoid all contact with human tissues. Alpha-naphthol is considered to be carcinogenic.

Nessler's Reagent (ammonia test)

Dissolve about 50 gm of potassium iodide in 35 ml of cold ammonia-free distilled water. Add a saturated solution of mercuric chloride until a slight precipitate persists. Add 400 ml of a 50% solution of potassium hydroxide. Dilute to 1 liter, allow to settle, and decant the supernatant for use.

Nigrosine Solution (Dorner's)

Nigrosine, water soluble .... 10 gm
Distilled water .... 100 ml
Boil for 30 minutes. Add as a preservative 0.5 ml formaldehyde (40%). Filter twice through double filter paper and store under aseptic conditions.

Nitrate Test Reagent (see Diphenylamine)

Nitrite Test Reagents

Solution A: Dissolve 8 gm sulfanilic acid in 1000 ml 5N acetic acid (1 part glacial acetic acid to 2.5 parts water).

Solution B: Dissolve 5 gm dimethyl-alpha-naphthylamine in 1000 ml 5N acetic acid. Do not mix solutions.

Caution: Although at this time it is not known for sure, there is a possibility that dimethyl-α-naphthylamine in solution B may be carcinogenic. For reasons of safety, avoid all contact with tissues.

Oxidase Test Reagent

Mix 1.0 gm of dimethyl-ρ-phenylenediamine hydrochloride in 100 ml of distilled water.

Preferably, the reagent should be made up fresh, daily. It should not be stored longer than one week in the refrigerator. Tetramethyl-ρ-phenylenediamine dihydrochloride (1%) is even more sensitive, but is considerably more expensive and more difficult to obtain.

Phenolized Saline

Dissolve 8.5 gm sodium chloride and 5.0 gm phenol in 1 liter distilled water.

Physiological Saline

Dissolve 8.5 gm sodium chloride in 1 liter distilled water.

Safranin (for Gram staining)

Safranin O (2.5% sol'n in 95% ethyl alcohol)	10.0 ml
Distilled water	100.0 ml

Trommsdorf's Reagent (nitrite test)

Add slowly, with constant stirring, 100 ml of a 20% aqueous zinc chloride solution to a mixture of 4.0 gm of starch in water. Continue heating until the starch is dissolved as much as possible, and the solution is nearly clear. Dilute with water and add 2 gm of potassium iodide. Dilute to 1 liter, filter, and store in amber bottle.

Vaspar

Melt together 1 pound of Vaseline and 1 pound of paraffin. Store in small bottles for student use.

Voges-Proskauer Test Reagent (see Barritt's)

C Appendix Media

Conventional Media The following media are used in the experiments of this manual. All of these media are available in dehydrated form from either Difco Laboratories, Detroit, Michigan, or Baltimore Biological Laboratory (BBL), a division of Becton, Dickson & Co., Cockeysville, Maryland. Compositions, methods of preparation, and usage will be found in their manuals which are supplied upon request at no cost. The source of each medium is designated as (B) for BBL and (D) for Difco.

- Bile esculin (D)
- Brewer's anaerobic agar (D)
- Desoxycholate citrate agar (B,D)
- Desoxycholate lactose agar (B,D)
- Endo agar (B,D)
- Eugonagar (B,D)
- Fluid thioglycollate medium (B,D)
- Heart infusion agar (D)
- Kligler iron agar (B,D)
- Lead acetate agar (D)
- Levine's EMB agar (B,D)
- Lipase reagent (D)
- Litmus milk (B,D)
- Lowenstein-Jensen medium (B,D)
- Mannitol salt agar (B,D)
- MR-VP medium (D)
- Mueller-Hinton medium (B,D)
- Nitrate broth (D)
- Nutrient agar (B,D)
- Nutrient broth (B,D)
- Nutrient gelatin (B,D)
- Phenol red sucrose broth (B,D)
- Phenylalanine agar (D)
- Phenylethyl alcohol medium (B)
- Russell double sugar agar (B,D)
- Sabouraud's glucose (dextrose) agar (D)
- Semisolid medium (B)
- Simmons citrate agar (B,D)
- Snyder test agar (D)
- Sodium hippurate (D)
- Spirit blue agar (D)
- SS agar (B,D)
- *m*-Staphylococcus broth (D)
- Staphylococcus medium 110 (D)
- Starch agar (D)
- Trypticase soy agar (B)
- Trypticase soy broth (B)
- Tryptone glucose extract agar (B,D)
- Urea (urease test) broth (B,D)
- Veal infusion agar (B,D)

Special Media The following media are not included in the manuals that are supplied by Difco and BBL; therefore, methods of preparation are presented here.

Bile Esculin Slants (Ex. 52)

Heart infusion agar	40.0 gm
Esculin	1.0 gm
Ferric chloride	0.5 gm
Distilled water	1000.0 ml

Dispense into sterile 15 × 125 mm screw-capped tubes, sterilize in autoclave at 121° C for 15 minutes, and slant during cooling.

Blood Agar

Trypticase soy agar powder	40 gm
Distilled water	1000 ml
Final pH of 7.3	
Defibrinated sheep or rabbit blood	50 ml

Liquefy and sterilize 1000 ml of trypticase soy agar in a large Erlenmeyer flask. While the TSA is being sterilized, warm up 50 ml of defibrinated blood to 50° C. After cooling the TSA to 50° C, aseptically transfer the blood to the flask and mix by gently rotating the flask (cold blood many cause lumpiness).

Pour 10–12 ml of the mixture into sterile petri plates. If bubbles form on the surface of the medium, flame the surface gently with a Bunsen burner before the medium solidifies. It is best to have an assistant to lift off the petri plate lids while pouring the medium into the plates. A full flask of blood agar is somewhat cumbersome to handle with one hand.

Deca-Strength Phage Broth (Ex. 22)

Peptone	100 gm
Yeast extract	50 gm
NaCl	25 gm
K_2HPO_4	80 gm
Distilled water	1000 ml

Final pH 7.6

Emmons' Culture Medium for Fungi

C. W. Emmons developed the following recipe as an improvement over Sabouraud's glucose agar for the cultivation of fungi. Its principal advantage is that a neutral pH does not inhibit certain molds that have difficulty growing on Sabouraud's agar (pH 5.6). Instead of relying on a low pH to inhibit bacteria, it contains chloramphenicol, which does not adversely affect the fungi.

Glucose	20 gm
Neopeptone	10 gm
Agar	20 gm
Chloramphenicol	40 mg
Distilled water	1000 ml

After the glucose, peptone, and agar are dissolved, heat to boiling, add the chloramphenicol which has been suspended in 10 ml of 95% alcohol and remove quickly from the heat. Autoclave for only 10 minutes.

Glucose–Minimal Salts Agar (Ex. 32, Ames test)

This medium is made from glucose, agar, and Vogel-Bonner medium E (50×).

Vogel-Bonner Medium E (50×)

Distilled water (45° C)	670 ml
$MgSO_4 \cdot 7H_2O$	10 gm
Citric acid monohydrate	100 gm
K_2HPO_4 (anhydrous)	500 gm
Sodium ammonium phosphate ($NaHNH_4PO_4 \cdot 4H_2O$)	175 gm

Add salts in the order indicated to warm water (45° C) in a 2-liter beaker or flask placed on a magnetic stirring hot plate. Allow each salt to dissolve completely before adding the next. Adjust the volume to 1 liter. Distribute into two 1-liter glass bottles. Autoclave, loosely capped, for 20 minutes at 121° C.

Plates of Glucose–Minimal Salts Agar

Agar	15 gm
Distilled water	930 ml
50× V-B salts	20 ml
40% glucose	50 ml

Add 15 gm of agar to 930 ml of distilled water in a 2-liter flask. Autoclave for 20 minutes using slow exhaust. When the solution has cooled slightly, add 20 ml of sterile 50× V-B salts and 50 ml of sterile 40% glucose. For mixing, a large magnetic stir bar can be added to the flask before autoclaving. After all the ingredients have been added, the solution should be stirred thoroughly. Pour 30 ml into each petri plate.

Important: The 50× V-B salts and 40% glucose should be autoclaved separately.

***m* Endo MF Broth** (Ex. 47)

This medium is extremely hygroscopic in the dehydrated form and oxidizes quickly to cause deterioration of the medium after the bottle has been opened. Once a bottle has been opened it should be dated and discarded after one year. If the medium becomes hardened within that time it should be discarded. Storage of the bottle inside a larger bottle which contains silica gel will extend shelf life.

Failure of Exercise 47 can often be attributed to faulty preparation of the medium. It is best to make up the medium the day it is to be used. It should not be stored over 96 hours prior to use. The Millipore Corporation recommends the following method for preparing this medium. (These steps are not exactly as stated in the Millipore Application Manual AM302.)

1. Into a 250-ml screw-cap Erlenmeyer flask place the following:

Distilled water	50 ml
95% ethyl alcohol	2 ml
Dehydrated medium (*m* Endo MF broth)	4.8 gms

   Shake the above mixture by swirling the flask until the medium is dissolved and then add another 50 ml of distilled water.
2. Cap the flask loosely and immerse it into a pan of boiling water. As soon as the medium begins to simmer, remove the flask from the water bath. Do not boil the medium any further.
3. Cool the medium to 45° C, and adjust the pH to between 7.1 and 7.3.
4. If the medium must be stored for a few days, place it in the refrigerator at 2–10° C, with screw-cap tightened securely.

Milk Salt Agar (15% NaCl)

Prepare three separate beakers of the following ingredients:

1. Beaker containing 200 grams of sodium chloride.
2. Large beaker (2000 ml size) containing 50 grams of skim milk powder in 500 ml of distilled water.
3. Glycerol-peptone agar medium:

$MgSO_4 \cdot 7H_2O$	5.0	gm
$MgNO_3 \cdot 6H_2O$	1.0	gm
$FeCl_3 \cdot 7H_2O$	0.025	gm
Difco proteose-peptone #3	5.0	gm
Glycerol	10.0	gm
Agar	30.0	gm
Distilled water	500.0	ml

Sterilize the above three beakers separately. The milk solution should be sterilized at 113–115° C (8 lbs. pressure) in autoclave for 20 minutes. The salt and glycerol-peptone agar can be sterilized at conventional pressure and temperature. After the milk solution has cooled to 55° C, add the sterile salt which should also be cooled down to a moderate temperature. If the salt is too hot, coagulation may occur. Combine the milk-salt and glycerol-peptone agar solutions by gently swirling with a glass rod. Dispense aseptically into petri plates.

Skim Milk Agar

Skim milk powder	100	gm
Agar	15	gm
Distilled water	1000	ml

Dissolve the 15 gm of agar into 700 ml of distilled water by boiling. Pour into a large flask and sterilize at 121° C, 15 lbs. pressure.

In a separate container, dissolve the 100 gm of skim milk powder into 300 ml of water heated to 50° C. Sterilize this milk solution at 113–115° C (8 lbs. pressure) for 20 minutes.

After the two solutions have been sterilized, cool to 55° C and combine in one flask, swirling gently to avoid bubbles. Dispense into sterile petri plates.

Sodium Chloride (6.5%) Tolerance Broth (Ex. 52)

Heart infusion broth	25	gm
NaCl	60	gm
Indicator (1.6 gm bromcresol purple in 100 ml 95% ethanol)	1	ml
Dextrose	1	gm
Distilled water	1000	ml

Add all reagents together up to 1000 ml (final volume). Dispense in 15×125 mm screw-capped tubes and sterilize in an autoclave 15 minutes at 121° C.

A positive reaction is recorded when the indicator changes from purple to yellow or when growth is obvious even though the indicator does not change.

Sodium Hippurate Broth (Ex. 52)

Heart infusion broth	25	gm
Sodium hippurate	10	gm
Distilled water	1000	ml

Sterilize in autoclave at 121° C for 15 minutes after dispensing in 15 × 125 mm screw-capped tubes. Tighten caps to prevent evaporation.

Soft Nutrient Agar (for bacteriophage)

Dehydrated nutrient broth	8	gm
Agar	7	gm
Distilled water	1000	ml

Sterilize in autoclave at 121° C for 20 minutes.

Spirit Blue Agar (Ex. 39)

This medium is used to detect lipase production by bacteria. Lipolytic bacteria cause the medium to change from pale lavender to deep blue.

Spirit blue agar (Difco)	35	gm
Lipase reagent (Difco)	35	ml
Distilled water	1000	ml

Dissolve the spirit blue agar in 1000 ml of water by boiling. Sterilize in autoclave for 15 minutes at 15 psi (121° C). Cool to 55° C and slowly add the 35 ml of lipase reagent, agitating to obtain even distribution. Dispense into sterile petri plates.

Top Agar (Ex. 32, Ames Test)

Tubes containing 2 ml of top agar are made up just prior to using from bottles of top agar base and his/bio stock solution.

His/Bio Stock Solution

D-Biotin (F.W. 247.3)	30.9	mg
L-Histidine·HCl (F.W. 191.7)	24.0	mg
Distilled water	250	ml

Dissolve by heating the water to the boiling point. This can be done in a microwave oven. Sterilize by filtration through 0.22 μm membrane filter, or autoclave for 20 minutes at 121° C. Store in a glass bottle at 4° C.

Top Agar Base

Agar	6	gm
Sodium chloride (NaCl)	5	gm
Distilled water	1000	ml

The agar may be dissolved in a steam bath or microwave oven, or by autoclaving briefly. Mix thoroughly and transfer 100-ml aliquots to 250-ml glass bottles with screw caps. Autoclave for 20 minutes with loosened caps. Slow exhaust. Cool the agar and tighten caps.

Just before using, add 10 ml of the his/bio stock solution to bottle of 100 ml of liquefied top agar base (45° C). After thoroughly mixing, distribute, aseptically, 2 ml of this mixture to sterile tubes (13 mm × 100 mm). Hold tubes at 45° C until used.

Tryptone Agar

Tryptone	10	gm
Agar	15	gm
Distilled water	1000	ml

Tryptone Broth

Tryptone	10	gm
Distilled water	1000	ml

Tryptone Yeast Extract Agar

Tryptone	10	gm
Yeast extract	5	gm
Dipotassium phosphate	3	gm
Sucrose	50	gm
Agar	15	gm
Water	1000	ml

pH 7.4

Appendix D
Identification Charts

Chart I Interpretation of Test Results on API 20E System

	Interpretation of Reactions			
Tube		**Positive**	**Negative**	**Comments**
ONPG		Yellow	Colorless	(1) Any shade of yellow is a positive reaction. (2) VP tube, before the addition of reagents, can be used as a negative control.
ADH	Incubation 18–24 h 36–48 h	Red or Orange Red	Yellow Yellow or Orange	Orange reactions occurring at 36–48 hours should be interpreted as negative.
LDC	18–24 h 36–48 h	Red or Orange Red	Yellow Yellow or Orange	Any shade of orange within 18–24 hours is a positive reaction. At 36–48 hours, orange decarboxylase reactions should be interpreted as negative.
ODC	18–24 h 36–48 h	Red or Orange Red	Yellow Yellow or Orange	Orange reactions occurring at 36–48 hours should be interpreted as negative.
CIT		Turquoise or Dark Blue	Light Green or Yellow	(1) Both the tube and cupule should be filled. (2) Reaction is read in the aerobic (cupule) area.
H_2S		Black Deposit	No Black Deposit	(1) H_2S production may range from a heavy black deposit to a very thin black line around the tube bottom. Carefully examine the bottom of the tube before considering the reaction negative. (2) A "browning" of the medium is a negative reaction unless a black deposit is present. "Browning" occurs with TDA-positive organisms.
URE	18–24 h 36–48 h	Red or Orange Red	Yellow Yellow or Orange	A method of lower sensitivity has been chosen. *Klebsiella, Proteus,* and *Yersinia* routinely give positive reactions.
TDA	Add 1 drop 10% ferric chloride	Brown-Red	Yellow	(1) Immediate reaction. (2) Indole positive organisms may produce a golden orange color due to indole production. This is a negative reaction.
IND	Add 1 drop Kovacs' reagent	Red Ring	Yellow	(1) The reaction should be read within 2 minutes after the addition of the Kovacs' reagent and the results recorded. (2) After several minutes, the HCl present in Kovacs' reagent may react with the plastic of the cupule resulting in a change from a negative (yellow) color to a brownish-red. This is a negative reaction.
VP	Add 1 drop of 40% potassium hydroxide, then 1 drop of 6% alpha-naphthol.	Red	Colorless	(1) Wait 10 minutes before considering the reaction negative. (2) A pale pink color (after 10 min.) should be interpreted as negative. A pale pink color which appears immediately after the addition of reagents but which turns dark pink or red after 10 min. should be interpreted as positive.
				Motility may be observed by hanging drop or wet mount preparation.
GEL		Diffusion of the pigment	No Diffusion	(1) The solid gelatin particles may spread throughout the tube after inoculation. Unless diffusion occurs, the reaction is negative. (2) Any degree of diffusion is a positive reaction.
GLU		Yellow or Gray	Blue or Blue-Green	Comments for all Carbohydrates: **Fermentation** (Enterobacteriaceae, *Aeromonas, Vibrio*) (1) Fermentation of the carbohydrates begins in the most anaerobic portion (bottom) of the tube. Therefore, these reactions should be read from the bottom of the tube to the top. (2) A yellow color at the bottom of the tube only indicates a weak or delayed positive reaction. **Oxidation** (Other Gram-negatives) (1) Oxidative utilization of the carbohydrates begins in the most aerobic portion (top) of the tube. Therefore, these reactions should be read from the top to the bottom of the tube. (2) A yellow color in the upper portion of the tube and a blue in the bottom of the tube indicates oxidative utilization of the sugar. This reaction should be considered positive **only** for non-Enterobacteriaceae gram-negative rods. This is a negative reaction for fermentative organisms such as Enterobacteriaceae.
MAN INO SOR RHA SAC MEL AMY ARA		Yellow	Blue or Blue-Green	
GLU Nitrate Reduction	After reading GLU reaction, add 2 drops 0.8% sulfanilic acid and 2 drops 0.5% N, N dimethylalpha-naphthylamine NO_2 N_2 gas	Red Bubbles; Yellow after reagents and zinc	Yellow Orange after reagents and zinc	(1) Before addition of reagents, observe GLU tube (positive or negative) for bubbles. Bubbles are indicative of reduction of nitrate to the nitrogenous (N_2) state. (2) A positive reaction may take 2–3 minutes for the red color to appear. (3) Confirm a negative test by adding zinc dust or 20-mesh granular zinc. A pink-orange color after 10 minutes confirms a negative reaction. A yellow color indicates reduction of nitrates to nitrogenous (N_2) state.
MAN INO SOR Catalase	After reading carbohydrate reaction, add 1 drop 1.5% H_2O_2	Bubbles	No bubbles	(1) Bubbles may take 1–2 minutes to appear. (2) Best results will be obtained if the test is run in tubes which have no gas from fermentation.

Courtesy of Analytab Products, Plainview, N.Y.

Chart II Symbol Interpretation of API 20E System

Tube	Chemical/Physical Principles	Components		Ref.
		Reactive Ingredients	Quantity	
ONPG	Hydrolysis of ONPG by beta-galactosidase releases yellow orthonitrophenol from the colorless ONPG; ITPG (isopropylthiogalactopyranoside) is used as inducer.	ONPG ITPG	0.2 mg 8.0 μg	12 13 14
ADH	Arginine dihydrolase transforms arginine into ornithine, ammonia, and carbon dioxide. This causes a pH rise in the acid-buffered system and a change in the indicator from yellow to red.	Arginine	2.0 mg	15
LDC	Lysine decarboxylase transforms lysine into a basic primary amine, cadaverine. This amine causes a pH rise in the acid-buffered system and a change in the indicator from yellow to red.	Lysine	2.0 mg	15
ODC	Ornithine decarboxylase transforms ornithine into a basic primary amine, putrescine. This amine causes a pH rise in the acid-buffered system and a change in the indicator from yellow to red.	Ornithine	2.0 mg	15
CIT	Citrate is the sole carbon source. Citrate utilization results in a pH rise and a change in the indicator from green to blue.	Sodium Citrate	0.8 mg	21
H_2S	Hydrogen sulfide is produced from thiosulfate. The hydrogen sulfide reacts with iron salts to produce a black precipitate.	Sodium Thiosulfate	80.0 μg	6
URE	Urease releases ammonia from urea; ammonia causes the pH to rise and changes the indicator from yellow to red.	Urea	0.8 mg	7
TDA	Tryptophane deaminase forms indolepyruvic acid from tryptophane. Indolepyruvic acid produces a brownish-red color in the presence of ferric chloride.	Tryptophane	0.4 mg	22
IND	Metabolism of tryptophane results in the formation of indole. Kovacs' reagent forms a colored complex (pink to red) with indole.	Tryptophane	0.2 mg	10
VP	Acetoin, an intermediary glucose metabolite, is produced from sodium pyruvate and indicated by the formation of a colored complex. Conventional VP tests may take up to 4 days, but by using sodium pyruvate, API has shortened the required test time. Creatine intensifies the color when tests are positive.	Sodium Pyruvate Creatine	2.0 mg 0.9 mg	3
GEL	Liquefaction of gelatin by proteolytic enzymes releases a black pigment which diffuses throughout the tube.	Kohn Charcoal Gelatin	0.6 mg	9
GLU MAN INO SOR RHA SAC MEL AMY ARA	Utilization of the carbohydrate results in acid formation and a consequent pH drop. The indicator changes from blue to yellow.	Glucose Mannitol Inositol Sorbitol Rhamnose Sucrose Melibiose Amygdalin (l +) Arabinose	2.0 mg 2.0 mg 2.0 mg 2.0 mg 2.0 mg 2.0 mg 2.0 mg 2.0 mg 2.0 mg	5 6 12
GLU Nitrate Reduction	Nitrites form a red complex with sulfanilic acid and N, N-dimethylalpha-naphthylamine. In case of negative reaction, addition of zinc confirms the presence of unreduced nitrates by reducing them to nitrites (pink-orange color). If there is no color change after the addition of zinc, this is indicative of the complete reduction of nitrates through nitrites to nitrogen gas or to an anaerogenic amine.	Potassium Nitrate	80.0 μg	6
MAN INO SOR Catalase	Catalase releases oxygen gas from hydrogen peroxide.			24

Courtesy of Analytab Products, Plainview, N.Y.

Chart III Characterization of Gram-Negative Rods—The API 20E System

	ORGANISM	ONPG	ADH	LDC	ODC	CIT	H_2S	URE	TDA	IND	VP	GEL	GLU	MAN	INO	SOR	RHA	SAC	MEL	AMY	ARA	OXI
Escherichieae	*E. coli*	98.2	1.0	90.2	67.3	0	1.0	0	0	85.0	0	0	100	98.4	0.1	95.5	84.5	41.1	88.4	0.1	95.0	0
	Shigella dysenteriae	27.8	0	0	0	0	0	0	0	33.0	0	0	100	0.1	0	0	22.2	0	61.1	0	16.7	0
	Sh. flexneri	5.3	0	0	0	0	0	0	0	15.0	0	0	100	94.7	0	78.9	0	0	21.1	0	36.8	0
	Sh. boydii	5.0	0	0	0	0	0	0	0	20.0	0	0	100	60.0	0	53.3	1.0	0	33.3	0	66.7	0
	Sh. sonnei	96.7	0	0	80.0	0	0	0	0	0	0	0	100	99.0	0	39.9	80.0	0	50.0	0	96.7	0
	Edwardsiella tarda	0	0	99.0	99.0	0	55.0	0	0	100	0	0	100	0	0	0	0	0	50.0	0	1.1	0
Salmonelleae	*Salmonella enteritidis*	1.9	1.0	89.2	95.4	15.4	76.9	0	0	3.1	0	0	100	98.7	4.6	95.2	95.4	4.6	96.9	0	94.5	0
	Sal. typhi	0	0	90.0	0	0	0.1	0	0	0	0	0	100	99.0	0	99.0	1.8	0	100	0	27.0	0
	Sal. paratyphi A	0	0	0	100	0	0.2	0	0	0	0	0	100	99.0	0	99.0	99.0	0	40.0	0	80.0	0
	Arizona-S. arizonae	94.7	1.0	95.0	98.5	15.0	85.0	0	0	0	0	0	100	99.0	0	87.0	96.1	0	89.5	0	95.0	0
	Citrobacter freundii	97.0	10.0	0	60.0	10.0	81.0	0	0	6.0	0	0	100	98.0	1.0	96.0	87.0	59.0	77.0	30.0	98.0	0
	C. diversus-Levinea	97.0	10.0	0	90.0	10.0	0	0	0	91.0	0	0	100	97.0	14.5	88.0	99.0	51.0	47.0	34.0	99.0	0
	C. amalonaticus	97.0	10.0	0	95.0	10.0	0	0	0	99.0	0	0	100	97.0	0.1	93.0	99.0	29.4	53.0	80.0	93.8	0
Klebsielleae	*Klebsiella pneumoniae*	99.0	0	80.0	0	13.9	0	10.0	0	0	72.0	0	100	98.0	30.0	95.0	91.0	99.0	99.0	98.0	99.0	0
	K. oxytoca	98.0	0	83.0	0	13.0	0	10.0	0	100	60.0	1.0	100	99.0	29.0	92.0	98.0	99.0	99.0	98.0	99.0	0
	K. ozaenae	85.0	0	38.0	0	1.0	0	0	0	0	0	0	100	69.0	1.0	76.0	69.0	15.0	92.0	99.0	84.0	0
	K. rhinoscleromatis	0	0	0	0	0	0	0	0	0	0	0	100	99.0	1.0	86.0	53.0	33.0	66.0	99.0	95.0	0
	Enterobacter aerogenes	99.0	0	98.0	98.0	8.9	0	0	0	0	56.0	0	100	99.0	28.0	90.0	90.0	85.0	97.0	96.0	98.0	0
	Ent. cloacae	97.0	51.9	0	65.0	9.0	0	0	0	0	80.0	0	100	99.0	1.0	92.0	90.0	98.0	92.0	65.0	95.0	0
	Ent. agglomerans	90.0	0	0	0	5.4	0	0	0	50.0	20.0	0	100	99.0	1.0	80.0	60.0	60.0	70.0	70.0	95.0	0
	Ent. gergoviae	99.0	0	61.0	99.0	8.2	0	75.0	0	0	75.0	0	100	99.0	1.0	8.3	99.0	99.0	99.0	99.0	99.0	0
	Ent. sakazakii	97.0	51.6	0	59.0	8.6	0	0	0	4.0	85.0	0	100	99.0	4.0	8.5	90.0	95.0	90.0	76.0	95.0	0
	Serratia liquefaciens	85.0	0	85.0	95.0	8.9	0	1.0	0	0	50.0	60.0	100	99.0	1.0	99.0	30.0	85.0	80.7	80.0	92.9	0
	Ser. marcescens	83.0	0	88.0	94.0	8.0	0	1.0	0	0	58.0	72.0	100	96.0	1.0	97.0	2.0	98.0	37.0	72.0	18.0	0
	Ser rubidaea	96.0	0	60.5	0.1	8.2	0	0	0	0	70.0	75.6	100	99.0	10.0	75.0	13.4	99.0	82.6	96.0	85.8	0
	Ser odorifera 1	99.0	0	95.0	100	9.5	0	0	0	90.0	63.0	62.0	100	99.0	10.0	99.0	85.0	100	99.0	90.0	99.0	0
	Ser odorifera 2	99.0	0	92.0	0	9.1	0	0	0	90.0	80.0	78.0	100	99.0	10.0	99.0	95.0	0	99.0	85.4	99.0	0
	Hafnia alvei	60.0	0	99.0	99.0	1.0	0	0	0	0	25.0	0	99.0	99.0	0	35.0	75.0	0	50.0	30.0	95.0	0

Chart III continued

	ORGANISM	ONPG	ADH	LDC	ODC	CIT	H_2S	URE	TDA	IND	VP	GEL	GLU	MAN	INO	SOR	RHA	SAC	MEL	AMY	ARA	OXI
Proteeae	*Proteus vulgaris*	0.5	0	0	0	4.1	75.3	91.0	95.0	75.3	0	75.3	100	0	0.1	0	0	83.0	1.0	20.0	4.0	0
	Prot. mirabilis	1.0	0	0	90.0	5.8	66.0	97.0	90.0	1.0	0	93.0	100	0	0	0	1.0	9.6	10.0	1.0	27.0	0
	Providencia alcalifaciens	0	0	0	0	9.8	0	0	95.0	94.0	0	0	100	0	0	0	0	0	0	0	25.0	0
	Prov. stuartii	1.0	0	0	0	8.5	0	0	95.0	86.0	0	0	100	0	8.0	0	0.8	3.7	34.0	0	30.0	0
	Prov. stuartii URE +	1.0	0	0	0	6.9	0	99.0	99.0	95.0	0	0	100	15.0	5.0	0	0.5	65.0	20.0	0	20.0	0
	Prov. rettgeri	1.0	0	0	0	7.1	0	80.0	95.0	90.0	0	0	100	85.0	1.0	30.0	40.0	5.0	0	40.0	10.0	0
	Morganella morganii	1.0	0	0	87.0	0.2	0	78.0	92.0	92.0	0	0	98.0	0	0	0	0	0	0	0	1.0	0
Yersiniae	*Yersinia enterocolitica*	81.0	0	0	36.0	0	0	59.0	0	54.0	0.4	0	100	99.0	1.0	95.0	9.0	78.0	40.4	31.0	76.6	0
	Y. pseudotuberculosis	80.0	0	0	0	0	0	88.0	0	0	0	0	100	94.0	0	76.0	58.0	0	5.0	0	52.0	0
	Y. pestis	93.0	0	0	0	0	0	0	0	0	1.0	0	93.0	87.0	0	56.0	0	0	0.6	25.0	87.0	0
	API Group 1	99.0	0	58.8	99.0	9.2	0	0	0	99.0	0	0	100	99.0	0	75.4	82.4	82.4	94.1	97.0	94.1	0
	API Group 2	99.0	2.0	7.3	0	0	0	0	0	0	0	0	100	99.0	0	2.3	30.8	5.6	90.0	38.5	92.3	0
Other Gram-negatives	*Pseudomonas maltophilia*	62.0	0	5.0	0	7.6	0	0	0	0	0	50.0	0.5	0	0	0	0	0	0	0	22.0	4.8
	Ps. cepacia	61.0	0	5.0	5.0	7.5	0	0	0	0	1.0	46.0	33.0	1.0	0	1.0	0	7.0	0	1.0	1.0	90.7
	Ps. paucimobilis	40.0	0	0	0	1.0	0	0	0	0	0	0	0.5	0	0	0	0	0.5	0	0	0.5	50.0
	A. calco. var. anitratus	0	0	0	0	2.8	0	0	0	0	1.0	0.1	85.0	0	0	0	0	0.1	77.0	0	60.0	0
	A. calco. var. lwoffii	0	0	0	0	0	0	0	0	0	0.1	0	0	0	0	0	0	0	0	0	0	0
	CDC Group VE-1	90.0	1.0	0	0	7.7	0	0	0	0	1.0	1.3	33.0	0	1.0	0	1.0	0.1	1.0	1.0	16.0	0
	CDC Group VE-2	0	0	0	0	7.9	0	0	0	0	1.0	0.1	4.5	0	1.0	0	0	0	1.0	0	5.0	0

Courtesy of Analytab Products, Plainview, N.Y.

Chart IV Characterization of Enterobacteriaceae—The Enterotube II System

Groups			Glucose	Gas Production	Lysine	Ornithine	H_2S	Indole	Adonitol	Lactose	Arabinose	Sorbitol	Voges-Proskauer	Dulcitol	Phenylalanine Deaminase	Urea	Citrate
ESCHERICHIEAE		*Escherichia*	+ 100.0	+J 92.0	d 80.6	d 57.8	−K 4.0	+ 96.3	− 5.2	+J 91.6	+ 91.3	± 80.3	− 0.0	d 49.3	− 0.1	− 0.1	− 0.2
ESCHERICHIEAE		*Shigella*	+ 100.0	−A 2.1	− 0.0	∓B 20.0	− 0.0	∓ 37.8	− 0.0	−B 0.3	± 67.8	∓ 29.1	− 0.0	d 5.4	− 0.0	− 0.0	− 0.0
EDWARDSIELLEAE		*Edwardsiella*	+ 100.0	+ 99.4	+ 100.0	+ 99.0	+ 99.6	+ 99.0	− 0.0	− 0.0	∓ 10.7	− 0.2	− 0.0	− 0.0	− 0.0	− 0.0	− 0.0
SALMONELLEAE		*Salmonella*	+ 100.0	+C 91.9	+H 94.6	+I 92.7	+E 91.6	− 1.1	− 0.0	− 0.8	± 89.2	+ 94.1	− 0.0	d D 86.5	− 0.0	− 0.0	d F 80.1
SALMONELLEAE		*Arizona*	+ 100.0	+ 99.7	+ 99.4	+ 100.0	+ 98.7	− 2.0	− 0.0	d 69.8	+ 99.1	+ 97.1	− 0.0	− 0.0	− 0.0	− 0.0	+ 96.8
SALMONELLEAE	*CITROBACTER*	*freundii*	+ 100.0	+ 91.4	− 0.0	d 17.2	± 81.6	− 6.7	− 0.0	d 39.3	+ 100.0	+ 98.2	− 0.0	d 59.8	− 0.0	dw 89.4	+ 90.4
SALMONELLEAE	*CITROBACTER*	*amalonaticus*	+ 100.0	+ 97.0	− 0.0	+ 97.0	− 0.0	+ 99.0	− 0.0	± 70.0	+ 99.0	+ 97.0	− 0.0	∓ 11.0	− 0.0	± 81.0	+ 94.0
SALMONELLEAE	*CITROBACTER*	*diversus*	+ 100.0	+ 97.3	− 0.0	+ 99.8	− 0.0	+ 100.0	+ 100.0	d 40.3	+ 98.0	+ 98.2	− 0.0	± 52.2	− 0.0	dw 85.8	+ 99.7
PROTEEAE	*PROTEUS*	*vulgaris*	+ 100.0	±G 86.0	− 0.0	− 0.0	+ 95.0	+ 91.4	− 0.0	− 0.0	− 0.0	− 0.0	− 0.0	− 0.0	+ 100.0	+ 95.0	d 10.5
PROTEEAE	*PROTEUS*	*mirabilis*	+ 100.0	+G 96.0	− 0.0	+ 99.0	+ 94.5	− 3.2	− 0.0	− 2.0	− 0.0	− 0.0	∓ 16.0	− 0.0	+ 99.6	± 89.3	± 58.7
PROTEEAE	*MORGANELLA*	*morganii*	+ 100.0	±G 86.0	− 0.0	+ 97.0	− 0.0	+ 99.5	− 0.0	− 0.0	− 0.0	− 0.0	− 0.0	− 0.0	+ 95.0	+ 97.1	−L 0.0
PROTEEAE	*PROVIDENCIA*	*alcalifaciens*	+ 100.0	d G 85.2	− 0.0	− 1.2	− 0.0	+ 99.4	+ 94.3	− 0.3	− 0.7	− 0.6	− 0.0	− 0.0	+ 97.4	− 0.0	+ 97.9
PROTEEAE	*PROVIDENCIA*	*stuartii*	+ 100.0	− 0.0	− 0.0	− 0.0	− 0.0	+ 98.6	∓ 12.4	− 3.6	− 4.0	− 3.4	− 0.0	− 0.0	+ 94.5	∓ 20.0	+ 93.7
PROTEEAE	*PROVIDENCIA*	*rettgeri*	+ 100.0	∓G 12.2	− 0.0	− 0.0	− 0.0	+ 95.9	+ 99.0	d 10.0	− 0.0	− 1.0	− 0.0	− 0.0	+ 98.0	+ 100.0	+ 96.0
KLEBSIELLEAE	*ENTEROBACTER*	*cloacae*	+ 100.0	+ 99.3	− 0.0	+ 93.7	− 0.0	− 0.0	∓ 28.0	± 94.0	+ 99.4	+ 100.0	+ 100.0	d 15.2	− 0.0	± 74.6	98.[illegible]
KLEBSIELLEAE	*ENTEROBACTER*	*sakazakii*	+ 100.0	+ 97.0	− 0.0	+ 97.0	− 0.0	∓ 16.0	− 0.0	+ 100.0	+ 100.0	− 0.0	+ 97.0	− 6.0	− 0.0	− 0.0	+ 94.0
KLEBSIELLEAE	*ENTEROBACTER*	*gergoviae*	+ 100.0	+ 93.0	± 64.0	+ 100.0	− 0.0	− 0.0	− 0.0	∓ 42.0	+ 100.0	− 0.0	+ 100.0	− 0.0	− 0.0	+ 100.0	+ 96.0
KLEBSIELLEAE	*ENTEROBACTER*	*aerogenes*	+ 100.0	+ 95.9	+ 97.5	+ 95.9	− 0.0	− 0.8	+ 97.5	+ 92.5	+ 100.0	+ 98.3	+ 100.0	− 4.1	− 0.0	− 0.0	+ 92.6
KLEBSIELLEAE	*ENTEROBACTER*	*agglomerans*	+ 100.0	∓ 24.1	− 0.0	− 0.0	− 0.0	∓ 19.7	− 7.5	d 52.9	+ 97.5	d 26.3	± 64.8	d 12.9	∓ 27.6	d 34.1	d 84.2
KLEBSIELLEAE	*HAFNIA*	*alvei*	+ 100.0	+ 98.9	+ 99.6	+ 98.6	− 0.0	− 0.0	− 0.0	d 2.8	+ 99.3	− 0.0	± 65.0	− 2.4	− 0.0	− 3.0	d 5.6
KLEBSIELLEAE	*SERRATIA*	*marcescens*	+ 100.0	±G 52.6	+ 99.6	+ 99.6	− 0.0	−w 0.1	∓ 56.0	− 1.3	− 0.0	+ 99.1	+ 98.7	− 0.0	− 0.0	d w 39.7	+ 97.6
KLEBSIELLEAE	*SERRATIA*	*liquefaciens*	+ 100.0	d 72.5	± 64.2	+ 100.0	− 0.0	−w 1.8	− 8.3	d 15.6	+ 97.3	+ 97.3	∓ 49.5	− 0.0	− 0.9	d w 3.7	+ 93.6
KLEBSIELLEAE	*SERRATIA*	*rubidaea*	+ 100.0	d G 35.0	± 61.0	− 0.0	− 0.0	−w 2.0	± 88.0	+ 100.0	+ 100.0	− 8.0	+ 92.0	− 0.0	− 0.0	d w 4.0	± 88.0
KLEBSIELLEAE	*KLEBSIELLA*	*pneumoniae*	+ 100.0	+ 96.0	+ 97.2	− 0.0	− 0.0	− 0.0	± 89.0	+ 98.7	+ 99.9	+ 99.4	+ 93.7	∓ 33.0	− 0.0	+ 95.4	+ 96.8
KLEBSIELLEAE	*KLEBSIELLA*	*oxytoca*	+ 100.0	+ 96.0	+ 97.2	− 0.0	− 0.0	+ 100.0	± 89.0	∓ 98.7	+ 100.0	+ 98.0	+ 93.7	∓ 33.0	− 0.0	∓ 95.4	∓ 96.8
KLEBSIELLEAE	*KLEBSIELLA*	*ozaenae*	+ 100.0	d 55.0	∓ 35.8	− 1.0	− 0.0	− 0.0	+ 91.8	d 26.2	+ 100.0	± 78.0	− 0.0	− 0.0	− 0.0	d 14.8	d 28.1
KLEBSIELLEAE	*KLEBSIELLA*	*rhinoschleromatis*	+ 100.0	− 0.0	− 0.0	− 0.0	− 0.0	− 0.0	+ 98.0	d 6.0	+ 100.0	+ 98.0	− 0.0	− 0.0	− 0.0	− 0.0	− 0.0
YERSINIAE	*YERSINIA*	*enterocolitica*	+ 100.0	− 0.0	− 0.0	+ 90.7	− 0.0	∓ 26.7	− 0.0	− 0.0	+ 98.7	+ 98.7	− 0.1	− 0.0	− 0.0	+ 90.7	− 0.0
YERSINIAE	*YERSINIA*	*pseudotuberculosis*	+ 100.0	− 0.0	− 0.0	− 0.0	− 0.0	− 0.0	− 0.0	− 0.0	± 55.0	− 0.0	− 0.0	− 0.0	− 0.0	+ 100.0	− 0.0

Courtesy of Roche Diagnostics, Nutley, N.J.

E. *S. enteritidis* bioserotype Paratyphi A and some rare biotypes may be H_2S negative.

F. *S. typhi*, *S. enteritidis* bioserotype Paratyphi A and some rare biotypes are citrate-negative and *S. cholerae-suis* is usually delayed positive.

G. The amount of gas produced by *Serratia*, *Proteus* and *Providencia alcalifaciens* is slight; therefore, gas production may not be evident in the ENTEROTUBE II.

H. *S. enteritidis* bioserotype Paratyphi A is negative for lysine decarboxylase.

I. *S. typhi* and *S. gallinarum* are ornithine decarboxylase-negative.

J. The Alkalescens-Dispar (A-D) group is included as a biotype of *E. coli*. Members of the A-D group are generally anaerogenic, non-motile and do not ferment lactose.

K. An occasional strain may produce hydrogen sulfide.

L. An occasional strain may appear to utilize citrate.

Chart V Characterization of Oxidative-Fermentative Gram-Negative Rods

Legend:
+ = Positive
− = Negative
V = Variable (11%−89% positive)
☐ Data based on literature only.

		OF ANAEROBIC DEXTROSE	ARGININE DIHYDROLASE	NITROGEN GAS	H_2S	INDOLE	OF XYLOSE	UREA	CITRATE	OXIDASE	OF DEXTROSE
ACHROMOBACTER	*SPECIES Bio. 1*	−	−	+	−	−	V	−	V	V	+
	Bio. 2	−	V	V	−	−	V	V	+	V	+
	XYLOSOXIDANS	−	−	V	−	−	V	−	−	+	+
ACINETOBACTER	*ANITRATUS*	V	−	−	−	−	+	+	V	V	−
	LWOFFII	−	−	−	−	−	−	−	V	V	−
AEROMONAS HYDROPHILA		+	V	−	−	V	−	+	−	V	+
ALCALIGENES FAECALIS		−	−	V	−	−	−	−	V	V	+
BORDETELLA BRONCHISEPTICA		−	−	−	−	−	−	−	+	+	+
FLAVOBACTERIUM SPECIES		V	−	−	−	V	−	V	−	−	+
GROUP	*2F* FLAVOBACTER-	−	−	−	−	+	−	−	−	−	+
	2J IUM-LIKE	−	−	−	−	+	−	−	+	−	+
	2K-1 PSEUDOMONAS-	−	−	−	−	−	−	−	−	−	V
	2K-2 LIKE	+	−	−	−	−	+	V	+	+	+
	4E ALCALIGENES-LIKE	−	−	−	−	−	−	−	V	V	+
	5A-1	−	−	+	−	−	+	+	V	+	+
	5A-2 PSEUDOMONAS	−	−	V	−	−	V	+	+	V	+
	5E-1 LIKE	+	+	−	−	−	+	+	V	+	−
	5E-2	V	−	−	−	−	+	+	V	V	−
	M-4 MORAXELLA-	−	−	−	−	−	−	−	−	+	+
	M-4f LIKE	−	−	−	−	−	−	−	+	+	+
MORAXELLA	*SPECIES*	−	−	−	−	−	−	−	−	−	+
	PHENYLPYRUVICA	−	−	−	−	−	−	−	+	−	+
PASTEURELLA	*HAEMOLYTICA*	V	−	−	−	−	−	−	−	−	+
	MULTOCIDA	−	−	−	−	+	−	−	−	−	+
	UREAE	V	−	−	−	−	−	−	V	−	+
PLESIOMONAS SHIGELLOIDES		+	+	−	−	+	−	+	−	−	+
PSEUDOMONAS	*AERUGINOSA*	−	+	V	−	−	V	V	V	V	+
	ACIDOVORANS	−	−	−	−	−	−	−	−	V	+
	ALCALIGENES	−	−	−	−	−	−	−	V	V	+
	CEPACIA	V	−	−	−	−	V	V	V	+	V
	DIMINUTA	−	−	−	−	−	−	−	−	−	+
	FLUORESCENS	−	V	−	−	−	V	V	V	+	+
	MALLEI	−	+	−	−	−	−	+	−	−	−
	MALTOPHILIA	−	−	−	−	−	−	V	V	−	V
	PSEUDOALCALIGENES	−	−	−	−	−	−	−	−	V	+
	PSEUDOMALLEI	−	+	V	−	−	−	+	V	+	+
	PUTIDA	V	+	−	−	−	+	+	V	+	+
	PUTREFACIENS	−	−	−	+	−	−	−	V	−	+
	STUTZERI	−	V	V	−	−	V	V	−	+	+
	TESTOSTERONI	−	−	−	−	−	−	−	−	V	+
	VESICULARIS	V	−	−	−	−	V	V	−	−	+
VIBRIO	*ALGINOLYTICUS*	+	−	−	−	V	−	+	−	−	+
	CHOLERAE	+	−	−	−	+	V	+	−	V	+
	PARAHAEMOLYTICUS	+	−	−	−	+	−	V	−	V	+
CHROMOBACTERIUM VIOLACEUM		+	+	−	−	−	−	+	−	V	V

Courtesy of Roche Diagnostics, Nutley, N.J.

Chart VI Reaction Interpretations for API Staph-Ident

MICROCUPULE		INTERPRETATION OF REACTIONS		
#	SUBSTRATE	POSITIVE	NEGATIVE	COMMENTS AND REFERENCES
1	PHS	Yellow	Clear or straw-colored	A positive result should be recorded only if significant color development has occurred.(3)
2	URE	Purple to Red-Orange	Yellow or Yellow-Orange	Phenol red has been added to the urea formulation to allow detection of alkaline end products resulting from urea utilization.(1)
3	GLS	Yellow	Clear or straw-colored	A positive result should be recorded only if significant color development has occurred.
4 5 6 7	MNE MAN TRE SAL	Yellow or Yellow-Orange	Red or Orange	Cresol red has been added to each carbohydrate to allow detection of acid production if the respective carbohydrates are utilized. (1,7)
8	GLC	Yellow	Clear or straw-colored	A positive result should be recorded only if significant color development has occurred.
9	ARG	Purple to Red-Orange	Yellow or Yellow-Orange	Phenol red has been added to the arginine formulation to allow detection of alkaline end products resulting from arginine utilization.(1)
10	NGP	Add 1–2 drops of STAPH-IDENT REAGENT Plum-Purple (Mauve)	 Yellow or colorless	Color development will begin within 30 seconds of reagent addition. (1,5)

Courtesy of Analytab Products, Plainview, N.Y.

Abbreviation	Test
PHS	Phosphatase
URE	Urea utilization
GLS	β-Glucosidase
MNE	Mannose utilization
MAN	Mannitol utilization
TRE	Trehalose utilization
SAL	Salicin utilization
GLC	β-Glucuronidase
ARG	Arginine utilization
NGP	β-Galactosidase

Chart VII Biochemistry of API Staph-Ident Tests

MICROCUPULE				
#	SUBSTRATE	CHEMICAL/PHYSICAL PRINCIPLES	REACTIVE INGREDIENTS	QUANTITY
1	PHS	Hydrolysis of p-nitrophenyl-phosphate, disodium salt, by alkaline phosphatase releases yellow paranitrophenol from the colorless substrate.	p-nitrophenyl-phosphate, disodium salt	0.2%
2	URE	Urease releases ammonia from urea; ammonia causes the pH to rise and changes the indicator from yellow to red.	Urea	1.6%
3	GLS	Hydrolysis of p-nitrophenyl-β-D-glucopyranoside by β-glucosidase releases yellow para-nitrophenol from the colorless substrate.	p-nitrophenyl-β-D-glucopyranoside	0.2%
4 5 6 7	MNE MAN TRE SAL	Utilization of carbohydrate results in acid formation and a consequent pH drop. The indicator changes from red to yellow.	Mannose Mannitol Trehalose Salicin	1.0% 1.0% 1.0% 1.0%
8	GLC	Hydrolysis of p-nitrophenyl-β-D-glucuronide by β-glucuronidase releases yellow para-nitrophenol from the colorless substrate.	p-nitrophenyl-β-D-glucuronide	0.2%
9	ARG	Utilization of arginine produces alkaline end products which change the indicator from yellow to red.	Arginine	1.6%
10	NGP	Hydrolysis of 2-naphthol-β-D-galactopyranoside by β-galactosidase releases free β-naphthol which complexes with STAPH-IDENT REAGENT to produce a plum-purple (mauve) color.	2-naphthol-β-D-galactopyranoside	0.3%

Courtesy of Analytab Products, Plainview, N.Y.

CHART VIII API Staph-Ident Profile Register*

Profile	Identification	
0 040	STAPH CAPITIS	
0 060	STAPH HAEMOLYTICUS	
0 100	STAPH CAPITIS	
0 140	STAPH CAPITIS	
0 200	STAPH COHNII	
0 240	STAPH CAPITIS	
0 300	STAPH CAPITIS	
0 340	STAPH CAPITIS	
0 440	STAPH HAEMOLYTICUS	
0 460	STAPH HAEMOLYTICUS	
0 600	STAPH COHNII	
0 620	STAPH HAEMOLYTICUS	
0 640	STAPH HAEMOLYTICUS	
0 660	STAPH HAEMOLYTICUS	
1 000	STAPH EPIDERMIDIS	
1 040	STAPH EPIDERMIDIS	
1 300	STAPH AUREUS	
1 540	STAPH HYICUS (An)	
1 560	STAPH HYICUS (An)	
2 000	STAPH SAPROPHYTICUS	NOVO R
	STAPH HOMINIS	NOVO S
2 001	STAPH SAPROPHYTICUS	
2 040	STAPH SAPROPHYTICUS	NOVO R
	STAPH HOMINIS	NOVO S
2 041	STAPH SIMULANS	
2 061	STAPH SIMULANS	
2 141	STAPH SIMULANS	
2 161	STAPH SIMULANS	
2 201	STAPH SAPROPHYTICUS	
2 241	STAPH SIMULANS	
2 261	STAPH SIMULANS	
2 341	STAPH SIMULANS	
2 361	STAPH SIMULANS	
2 400	STAPH HOMINIS	NOVO S
	STAPH SAPROPHYTICUS	NOVO R
2 401	STAPH SAPROPHYTICUS	
2 421	STAPH SIMULANS	
2 441	STAPH SIMULANS	
2 461	STAPH SIMULANS	
2 541	STAPH SIMULANS	
2 561	STAPH SIMULANS	
2 601	STAPH SAPROPHYTICUS	
2 611	STAPH SAPROPHYTICUS	
2 661	STAPH SIMULANS	
2 721	STAPH COHNII (SSP1)	
2 741	STAPH SIMULANS	
2 761	STAPH SIMULANS	
3 000	STAPH EPIDERMIDIS	
3 040	STAPH EPIDERMIDIS	
3 140	STAPH EPIDERMIDIS	
3 540	STAPH HYICUS (An)	
3 541	STAPH INTERMEDIUS (An)	
3 560	STAPH HYICUS (An)	
3 601	STAPH SIMULANS	NOVO S
	STAPH SAPROPHYTICUS	NOVO R
4 060	STAPH HAEMOLYTICUS	
4 210	STAPH SCIURI	
4 310	STAPH SCIURI	
4 420	STAPH HAEMOLYTICUS	
4 440	STAPH HAEMOLYTICUS	
4 460	STAPH HAEMOLYTICUS	
4 610	STAPH SCIURI	
4 620	STAPH HAEMOLYTICUS	
4 660	STAPH HAEMOLYTICUS	

Profile	Identification	
4 700	STAPH AUREUS	COAG +
	STAPH SCIURI	COAG −
4 710	STAPH SCIURI	
5 040	STAPH EPIDERMIDIS	
5 200	STAPH SCIURI	
5 210	STAPH SCIURI	
5 300	STAPH AUREUS	COAG +
	STAPH SCIURI	COAG −
5 310	STAPH SCIURI	
5 600	STAPH SCIURI	
5 610	STAPH SCIURI	
5 700	STAPH AUREUS	COAG +
	STAPH SCIURI	COAG −
5 710	STAPH SCIURI	
5 740	STAPH AUREUS	
6 001	STAPH XYLOSUS	XYL + ARA +
	STAPH SAPROPHYTICUS	XYL − ARA −
6 011	STAPH XYLOSUS	
6 021	STAPH XYLOSUS	
6 101	STAPH XYLOSUS	
6 121	STAPH XYLOSUS	
6 221	STAPH XYLOSUS	
6 300	STAPH AUREUS	
6 301	STAPH XYLOSUS	
6 311	STAPH XYLOSUS	
6 321	STAPH XYLOSUS	
6 340	STAPH AUREUS	COAG +
	STAPH WARNERI	COAG −
6 400	STAPH WARNERI	
6 401	STAPH XYLOSUS	XYL + ARA +
	STAPH SAPROPHYTICUS	XYL − ARA −
6 421	STAPH XYLOSUS	
6 460	STAPH WARNERI	
6 501	STAPH XYLOSUS	
6 521	STAPH XYLOSUS	
6 600	STAPH WARNERI	
6 601	STAPH SAPROPHYTICUS	XYL − ARA −
	STAPH XYLOSUS	XYL + ARA +
6 611	STAPH XYLOSUS	
6 621	STAPH XYLOSUS	
6 700	STAPH AUREUS	
6 701	STAPH XYLOSUS	
6 721	STAPH XYLOSUS	
6 731	STAPH XYLOSUS	
7 000	STAPH EPIDERMIDIS	
7 021	STAPH XYLOSUS	
7 040	STAPH EPIDERMIDIS	
7 141	STAPH INTERMEDIUS (An)	
7 300	STAPH AUREUS	
7 321	STAPH XYLOSUS	
7 340	STAPH AUREUS	
7 401	STAPH XYLOSUS	
7 421	STAPH XYLOSUS	
7 501	STAPH INTERMEDIUS (An)	COAG +
	STAPH XYLOSUS	COAG −
7 521	STAPH XYLOSUS	
7 541	STAPH INTERMEDIUS (An)	
7 560	STAPH HYICUS (An)	
7 601	STAPH XYLOSUS	
7 621	STAPH XYLOSUS	
7 631	STAPH XYLOSUS	
7 700	STAPH AUREUS	
7 701	STAPH XYLOSUS	
7 721	STAPH XYLOSUS	
7 740	STAPH AUREUS	

*Date of Publication: March, 1984

Courtesy of Analytab Products, Plainview, N.Y.

Appendix The Streptococci: Classification, Habitat, Pathology, and Biochemical Characteristics

To fully understand the characteristics of the various species of medically important streptococci, this appendix has been included as an adjunct to Exercise 52. The table of streptococcal characteristics on this page is the same one that is shown on page 199 of Exercise 52. It is also the basis for much of the discussion that follows.

The first system that was used for grouping the streptococci was based on the type of hemolysis and was proposed by J. H. Brown in 1919. In 1933, R. C. Lancefield proposed that these bacteria be separated into groups A, B, C, etc., on the basis of precipitation-type serological testing. Both hemolysis and serological typing still play predominant roles today in our classification system. Note below that the Landsteiner groups are categorized with respect to the type of hemolysis that is produced on blood agar.

Beta Hemolytic Groups

Using a streak-stab technique, a blood agar plate is incubated aerobically at 37° C for 24 hours. Isolates that have colonies surrounded by clear zones completely free of red blood cells are characterized as being *beta hemolytic*. Three serological groups of streptococci fall in this category: groups A, B, and C; a few species in group D are also beta hemolytic.

Table I Physiological Tests for Streptococcal Differentiation

GROUP	Type of Hemolysis	Bacitracin Susceptibility	CAMP Reaction or Hippurate Hydrolysis	SXT Sensitivity	Bile Esculin Hydrolysis	Tolerance to 6.5% NaCl	Optochin Susceptibility	Bile Solubility
Group A *S. pyogenes*	beta	+	−	R	−	−	−	−
Group B *S. agalactiae*	beta	−*	+	R	−	±	−	−
Group C *S. equi* *S. equisimilis* *S. zooepidemicus*	beta	−*	−	S	−	−	−	−
Group D (enterococci) *S. faecalis* *S. faecium* etc.	alpha beta none	−	−	R	+	+	−	−
Group D (nonenterococci) *S. bovis* etc.	alpha none	−	−	R/S	+	−	−	−
Viridans *S. mitis* *S. salivarius* *S. mutans* etc.	alpha none	−*	−*	S	−	−	−	−
Pneumococci *S. pneumoniae*	alpha	±	−		−	−	+	+

*Exceptions occur occasionally

Group A Streptococci

This group is represented by only one species: *Streptococcus pyogenes.* Approximately 25% of all upper respiratory infections (URI) are caused by this species; another 10% of URI are caused by other streptococci; most of the remainder (65%) are caused by viruses. Since no unique clinical symptoms can be used to differentiate viral from streptococcal URI, and since successful treatment relies on proper identification, it becomes mandatory that throat cultures be taken in an attempt to prove the presence or absence of streptococci. It should be added that if streptococcal URI are improperly treated, serious sequelae such as pneumonia, acute endocarditis, rheumatic fever, and glomerularnephritis can result.

S. pyogenes is the only beta hemolytic streptococcus that is primarily of *human origin.* Although the pharynx is the most likely place to find this species, it may be isolated from the skin and rectum. Asymptomatic pharyngeal and anal carriers are not uncommon. Outbreaks of postoperative streptococcal infections have been traced to both pharyngeal and anal carriers among hospital personnel.

These coccoidal bacteria (0.6–1.0 μm diameter) occur as pairs and as short to moderate-length chains in clinical specimens; in broth cultures, the chains are often longer.

When grown on blood agar, the colonies are small (0.5 mm dia.), transparent to opaque, and domed; they have a smooth or semimatt surface and an entire edge; complete hemolysis (beta type) occurs around each colony, usually two to four times the diameter of the colony.

S. pyogenes produces two hemolysins: streptolysin S and streptolysin O. The beta-type hemolysis on blood agar is due to the complete destruction of red blood cells by the *streptolysin S.*

There is no group of physiological tests that can be used with *absolute* certainty to differentiate *S. pyogenes* from other streptococci; however, if an isolate is beta hemolytic and sensitive to bacitracin, one can be 95% certain that the isolate is *S. pyogenes.* The characteristics of this organism are the first ones tabulated in table I on the previous page.

Group B Streptococci

The only recognized species of this group is ***S. agalactiae.*** Although this organism is frequently found in milk and associated with *mastitis in cattle,* the list of human infections caused by it is as long as the one for *S. pyogenes:* abscesses, acute endocarditis, impetigo, meningitis, neonatal sepsis, and pneumonia are just a few. Like *S. pyogenes,* this pathogen may also be found in the pharynx, skin, and rectum; however, it is more likely to be found in the genital and intestinal tracts of healthy adults and infants. It is not unusual to find the organism in vaginal cultures of third-trimester pregnant women.

Cells are spherical to ovoid (0.6–1.2 μm diameter) and occur in chains of seldom less than four cells; long chains are frequently present. Characteristically, the chains appear to be composed of paired cocci.

Colonies of *S. agalactiae* on blood agar often produce double zone hemolysis. After 24 hours incubation colonies exhibit zones of beta hemolysis. After cooling, a second ring of hemolysis forms which is separated from the first by a ring of red blood cells.

Reference to table I emphasizes the significant characteristics of *S. agalactiae.* Note that this organism gives a positive CAMP reaction, hydrolyzes hippurate, and is not (usually) sensitive to bacitracin. It is also resistant to SXT. Presumptive identification of this species relies heavily on a positive CAMP test or hippurate hydrolysis, even if beta hemolysis is not clearly demonstrated.

Group C Streptococci

Three species fall in this group: *S. equisimilis, S. equi,* and *S. zooepidemicus.* Although all of these species may cause human infections, the diseases are not usually as grave as those caused by groups A and B. Some group C species have been isolated from impetiginous lesions, abscesses, sputum, and the pharynx. There is no evidence that they are associated with acute glomerularnephritis, rheumatic fever, or even pharyngitis.

Presumptive differentiation of this group from *S. pyogenes* and *S. agalactiae* is based primarily on (1) resistance to bacitracin, (2) inability to hydrolyze hippurate or bile esculin, and (3) a negative CAMP test. There are other groups that have some of these same characteristics, but they will not be studied here. Tables 12.16 and 12.17 on page 1049 of *Bergey's Manual,* vol. 2, provide information about these other groups.

Alpha Hemolytic Groups

Streptococcal isolates that have colonies with zones of incomplete lysis around them are said to be **alpha hemolytic.** These zones are often greenish; sometimes they are confused with beta hemolysis. *The* *only way to be certain that such zones are not beta hemolytic is to examine the zones under 60× microscopic magnification.* Figure 52.4, page 197, illustrates the differences between alpha and beta hemolysis. If some red blood cells are seen in the zone, the isolate is classified as being alpha hemolytic.

The grouping of streptococci on the basis of alpha hemolysis is not as clear-cut as it is for beta hemolytic groups. Note in table I that the bottom four groups that have alpha hemolytic types may also have beta hemolytic or nonhemolytic strains. Thus, we see that hemolysis in these four groups can be a misleading characteristic in identification.

Alpha hemolytic isolates from the pharynx are usually *S. pneumoniae,* viridans streptococci, or group D. Our primary concern here in this experiment is to identify isolates of *S. pneumoniae.* To accomplish this goal, it will be necessary to differentiate any alpha hemolytic isolate from group D and viridans streptococci.

Streptococcus pneumoniae (Pneumococcus)

This organism is the most frequent cause of bacterial pneumonia, a disease which has a high mortality rate among the aged and debilitated. It is also frequently implicated in conjunctivitis, otitis media, pericarditis, subacute endocarditis, meningitis, septicemia, empyema, and peritonitis. Thirty to 70% of normal individuals carry this organism in the pharynx.

Spherical or ovoid, these cells (0.5–1.25 μm diameter) occur typically as pairs, sometimes singly, often in short chains. Distal ends of the cells are pointed or lancet-shaped and are heavily encapsulated with polysaccharide on primary isolation.

Colonies on blood agar are small, mucoidal, opalescent, and flattened with entire edges surrounded by a zone of greenish discoloration (alpha hemolysis). In contrast, the viridans streptococcal colonies are smaller, gray to whitish gray, and opaque with entire edges.

Presumptive identification of *S. pneumoniae* can be made with the optochin and bile solubility tests. On the optochin test, the pneumococci exhibit sensitivity to ethylhydrocupreine (optochin). With the bile solubility test, pneumococci are dissolved in bile (2% sodium desoxycholate). Table I reveals that except for bacitracin susceptibility (±), *S. pneumoniae* is negative on all other tests used for differentiation of streptococci.

Viridans Group

Streptococci that fall in this group are primarily alpha hemolytic; some are nonhemolytic. Approximately ten species are included in this group. All of them are highly adapted parasites of the upper respiratory tract. Although usually regarded as having low pathogenicity, they are opportunistic and sometimes cause serious infections. Two species (*S. mutans* and *S. sanguis*) are thought to be the primary cause of dental caries, since they have the ability to form dental plaque. Viridans streptococci are implicated more often than any other bacteria in subacute bacterial endocarditis.

When it comes to differentiation of bacteria of this group from the pneumococci and enterococci, we will use the optochin, bile solubility, and salt tolerance tests. See table I.

Group D Streptococci (Enterococci)

The enterococci of serological group D may be alpha hemolytic, beta hemolytic, or nonhemolytic. The four species of group D that are classified as enterococci are *S. faecalis, S. faecium, S. durans,* and *S. avium.*

Subacute endocarditis, pyelonephritis, urinary tract infections, meningitis, and biliary infections are caused by these organisms. All five of these species have been isolated from the intestinal tract. Approximately 20% of subacute bacterial endocarditis and 10% of urinary tract infections are caused by members of this group. Differentiation of this group from other streptococci in systemic infections is mandatory because *S. faecalis, S. faecium,* and *S. durans* are resistant to penicillin and require combined antibiotic therapy.

Since *S. faecalis* can be isolated from many food products (not connected with fecal contamination), it can be a transient in the pharynx and show up as an isolate in throat cultures. Morphologically, the cells are ovoid (0.5–1.0 μm diameter), occurring as pairs in short chains. Hemolytic reactions of *S. faecalis* on blood agar will vary with the type of blood used in the medium. Some strains produce beta hemolysis on agar with horse, human, and rabbit blood; on sheep blood agar the colonies will always exhibit alpha hemolysis. Other streptococci are consistently either beta, alpha, or nonhemolytic.

Cells of *S. faecium* are morphologically similar to *S. faecalis* except that motile strains are often encountered. A strong alpha type hemolysis is usually seen around colonies of *S. faecium* on blood agar.

Although presumptive differentiation of group D enterococcal streptococci from groups A, B, and C is not too difficult with physiological tests, it is more laborious to differentiate the individual species within group D. As indicated in table I, the enterococci (1) hydrolyze bile esculin, (2) are CAMP negative, and (3) grow well in 6.5% NaCl broth.

Differentiation of the five species within this group involves nine or ten physiological tests.

Group D Streptococci (Nonenterococci)

The only medically significant nonenterococcal species of group D is *S. bovis*. This organism is found in the intestinal tract of humans as well as in cows, sheep, and other ruminants. It can cause meningitis, subacute endocarditis, and urinary tract infections. On blood agar, the organism is usually alpha hemolytic; occasionally, it is nonhemolytic. The best way to differentiate it from the group D enterococci is to test its tolerance to 6.5% NaCl. Note in table I that *S. bovis* will not grow in this medium, but all enterococci will.

Reading References

General Information

Alcamo, I. Edward. *Fundamentals of Microbiology.* Menlo Park, Calif: Benjamin Cummings, 1987.

Atlas, R. M. *Microbiology: Fundamentals and Applications.* New York: Macmillan Publishing, 1986.

Baron, S. *Medical Microbiology.* 2nd ed. Reading, Mass.: Addison-Wesley Publishing, 1985.

Brock, T. D.; Smith, D. W.; and Madigan, M. T. *Biology of Microorganisms.* 4th ed. Englewood Cliffs, N.J.: Prentice-Hall, 1984.

Fuerst, R. *Frobisher and Fuerst's Microbiology.* Philadelphia: W. B. Saunders, 1983.

Jakoby, W., and Pastan, I. *Methods in Enzymology.* New York: Academic Press, 1987.

Ketchum, P. A. *Microbiology: Introduction for Health Professionals.* New York: John Wiley and Sons, 1984.

———. *Microbiology: Concepts and Applications.* New York: John Wiley and Sons, 1988.

Myrvik, Quentin N., and Weiser, Russell S. *Fundamentals of Medical Mycology.* Philadelphia: Lea and Febiger, 1988.

Norton, Cynthia F. *Microbiology.* 2nd ed. Reading, Mass.: Addison-Wesley Publishing, 1985.

Pelczar, M. J., and Chan, E. C. *Microbiology.* 5th ed. New York: McGraw-Hill, 1985.

Rippon, J. W. *Medical Mycology.* 3rd ed. Philadelphia: W. B. Saunders, 1988.

Ross, Frederick C. *Introductory Microbiology.* 2nd ed. Glenview, Ill.: Scott, Foresman, & Co., 1986.

Starr, M. P., et al. *The Prokaryotes: A Handbook on Habitats, Isolation and Identification of Bacteria.* Vols. 1 and 2. New York: Springer-Verlag, 1981.

Tortora, Gerard J.; Funke, B. R.; and Case, C. L. *Microbiology: An Introduction.* 3rd ed. Menlo Park, Calif.: Benjamin/Cummings Publishing, 1988.

Volk, W. A., and Wheeler, M. F. *Basic Microbiology,* 6th ed. New York: Harper and Row, 1987.

Laboratory Procedures

American Type Culture Collections. *Catalog of Cultures,* 8th ed. Rockville, Md. n.d.

Koneman, Elmer W., and Roberts, Glenn D. *Practical Laboratory Mycology.* 3rd ed. Baltimore, Md.: Williams & Wilkins, 1985.

Lennette, E. H., et al. *Manual of Clinical Microbiology,* 4th ed. Bethesda: American Society for Microbiology, 1985.

MacFaddin, J. F. *Biochemical Tests for Identification of Medical Bacteria.* Baltimore: Williams and Wilkins, 1989.

McGinnis, M. R. *Laboratory Handbook of Medical Mycology.* New York: Academic Press, 1980.

Miller, Brinton M., et al. *Laboratory Safety: Principles and Practices.* Washington, D.C.: American Society for Microbiology, 1986.

Norris, J. R., and Ribbons, D. W. *Methods in Microbiology.* Vols. 1–3. New York: Academic Press, 1969.

Shapton, D. A., and Board, R. G. *Isolation Methods for Anaerobes.* New York: Academic Press, 1971.

Shapton, D. A., and Gould, G. W. *Isolation Methods for Microbiologists.* New York: Academic Press, 1969.

Rose, Noel R.; Friedman, Herman; and Fahey, John L. *Manual of Clinical Laboratory Immunology.* 3rd ed. Washington, D.C.: American Society for Microbiology, 1986

Stains and Staining Techniques

BBL Manual of Products and Laboratory Procedures, 5th ed. Cockeysville, Md.: Becton Dickson, and Co., 1968.

Bullock, G. R., and Petrusz, P. *Techniques in Immunocytochemistry,* vol. 3. New York: Academic Press, 1985.

Difco Manual of Dehydrated Culture Media and Reagents, 10th ed. Detroit, Mich.: Difco Laboratories, 1984.

Lennette, E. H., et al. *Manual of Clinical Microbiology,* 4th ed. Bethesda: American Society for Microbiology, 1985.

Identification of Microorganisms

Coonrod, J. Donald. *The Direct Detection of Microorganisms in Clinical Samples.* New York: Academic Press, 1983.

Gibbons, N. E.; Pattee, K. B.; and Holt, J. B. *Supplement to Index Bergeyana.* Baltimore: Williams and Wilkins, 1981.

Goodfellow, M., and Board, R. G. *Microbiological Classification and Identification.* New York: Academic Press, 1980.

———, et al. *Computer Assisted Bacterial Systematics.* New York: Academic Press, 1985.
Holt, J. G. *The Shorter Bergey's Manual of Determinative Bacteriology.* Baltimore: Williams and Wilkins, 1977.
Jahn, T. L., et al. *How to Know the Protozoa.* Dubuque, Iowa: Wm. C. Brown Publishers, 1978.
Krieg, Noel R., et al. *Bergey's Manual of Systematic Bacteriology,* vol. 1. Baltimore, Md.: Williams & Wilkins, 1984.
Lennette, E. H., et al. *Manual of Clinical Microbiology.* 4th ed. Bethesda: American Society for Microbiology, 1985.
Skinner, F. A., and Lovelock, D. W. *Identification Methods for Microbiologists.* 2nd ed. New York: Academic Press, 1980.
Sneath, Peter H. A., et al. *Bergey's Manual of Systematic Bacteriology,* vol. 2. Baltimore, Md.: Williams & Wilkins, 1986.
Staley, James T., et al. *Bergey's Manual of Systematic Bacteriology,* vol. 3. Baltimore, Md.: Williams & Wilkins, 1989.
Williams, Stanley T. *Bergey's Manual of Systematic Bacteriology,* vol. 4. Baltimore, Md.: Williams & Wilkins, 1989.

Sanitary and Medical Microbiology

Greenberg, Arnold E., et al. *Standard Methods for the Examination of Water and Wastewater,* 16th ed. Washington, D.C.: American Public Health Association, 1985.
Jay, James M. *Modern Food Microbiology,* 3rd ed. New York: Van Nostrand, 1986.
Lennette, E. H., et al. *Manual of Clinical Microbiology,* 4th ed. Bethesda: American Society for Microbiology, 1985.
Richardson, Gary H. *Standard Methods for the Examination of Dairy Products.* 15th ed. Washington, D.C.: American Public Health Association, 1985.
Russell, A. D. *The Destruction of Bacterial Spores.* New York: Academic Press, 1982.

Index